ELECTRONIC LIGHT MICROSCOPY

The Principles and Practice of Video-Enhanced Contrast, Digital Intensified Fluorescence, and Confocal Scanning Light Microscopy

TECHNIQUES IN MODERN BIOMEDICAL
MICROSCOPY

SERIES EDITOR
DAVID SHOTTON
DEPARTMENT OF ZOOLOGY UNIVERSITY OF OXFORD OXFORD, ENGLAND

ELECTRONIC LIGHT MICROSCOPY
THE PRINCIPLES AND PRACTICE OF VIDEO-ENHANCED CONTRAST, DIGITAL INTENSIFIED FLUORESCENCE, AND CONFOCAL SCANNING LIGHT MICROSCOPY

ELECTRONIC LIGHT MICROSCOPY

TECHNIQUES IN MODERN BIOMEDICAL MICROSCOPY

The Principles and Practice of Video-Enhanced Contrast, Digital Intensified Fluorescence, and Confocal Scanning Light Microscopy

EDITED BY

DAVID SHOTTON

DEPARTMENT OF ZOOLOGY UNIVERSITY OF OXFORD OXFORD, ENGLAND

WILEY-LISS

A JOHN WILEY & SONS, INC., PUBLICATION

New York • Chichester • Brisbane • Toronto • Singapore

Address All Inquiries to the Publisher
Wiley-Liss, Inc., 605 Third Avenue, New York, NY 10158-0012

Copyright © 1993 Wiley-Liss, Inc.

Printed in the United States of America

Library of Congress Cataloging-in-Publication Data

Electronic light microscopy : the principles and practice of video-
    enhanced contrast, digital intensified fluorescence, and confocal
    scanning light microscopy / edited by David M. Shotton.
        p. cm. — (Techniques in modern biomedical microscopy)
    Includes bibliographical references and index.
    ISBN 0-471-56077-4
    1. Microscopy. 2. Imaging systems—Image quality. 3. Confocal
microscopy. I. Shotton, David II. Series
    [DNLM: 1. Image Enhancement. 2. Microscopy, Electronic Light.
QH212.E4 E38879]
QH205.2.E54 1992
578'.4—dc20
DNLM/DLC
for Library of Congress   92–24459
                                                                          CIP

The text of this book is printed on acid-free paper

Contents

Color Figure Section

Contributors

Stephen R. Bolsover, Department of Physiology, University College London, London WC1E 6BT, England [181]

A. Boyde, Department of Anatomy and Developmental Biology, Hard Tissue Research Unit, University College London, London WC1E 6BT, England [289]

Kenneth R. Castleman, Perceptive Systems, Inc., 2525 South Shore Boulevard, Suite 100, League City, TX 77573 [71]

F. Cornelissen, Department of Life Sciences, Division of Cellular Biology, Janssen Pharmaceutical Research Labs, Turnhoutseweg 30, B-2340 Beerse, Belgium [141]

M. De Brabander, Department of Life Sciences, Division of Cellular Biology, Janssen Pharmaceutical Research Labs, Turnhoutseweg 30, B-2340 Beerse, Belgium [141]

A. Draaijer, Department of Electro-Optical Systems, TNO Division of Technology for Society, Postbus 217, 2600 AE Delft, The Netherlands [273]

H. Geerts, Department of Life Sciences, Division of Cellular Biology, Janssen Research Foundation, Turnhoutseweg 30, B-2340 Beerse, Belgium [141]

P.M. Houpt, TNO Division of Technology for Society, Postbus 217, 2600 AE Delft, The Netherlands [273]

A. Hung Leo, European Molecular Biology Laboratory, Postfach 1022.40, D-6900 Heidelberg, Germany [157]

Theodore Inoué, Universal Imaging Corporation, 502 Brandywine Parkway, West Chester, PA 19380 [95]

Gordon S. Kino, Department of Applied Physics, Edward L. Ginton Laboratory, W.W. Hansen Laboratories, Stanford University, Stanford, CA 94305 [315]

T.E. Kreis, European Molecular Biology Laboratory, Postfach 1022.40 D-6900 Heidelberg, Germany [157]

K. Lahteenmaki, European Molecular Biology Laboratory, D-6900 Heidelberg, Germany [157]

Willi Maile, Institute for Zoology, Technical University of Munich, Lichtenbergstrasse 4, D-8046 Garching, Germany [105]

Barry R. Masters, Department of Anatomy and Cell Biology, Uniformed Services University of the Health Sciences, F.E. Herbert School of Medicine, 4301 Jones Bridge Road, Bethesda, MD 20814-4799 [315]

R. Nuydens, Department of Life Sciences, Division of Cellular Biology, Janssen Pharmaceutical Research Labs, Turnhoutseweg 30, B-2340 Beerse, Belgium **[141]**

R. Nuyens, Department of Life Sciences, Division of Cellular Biology, Janssen Pharmaceutical Research Labs, Turnhoutseweg 30, B-2340 Beerse, Belgium **[141]**

Peter J. Shaw, Department of Cell Biology, John Innes Institute, John Innes Centre for Plant Science Research, Colney Lane, Norwich NR4 7UH, England **[211]**

David Shotton, Department of Zoology, University of Oxford, South Parks Road, Oxford OX1 3PS, England **[ix,1,39]**

R. Angus Silver, Department of Physiology, University College London, Gower Street, London WC1E 6BT, England **[181]**

A.W.M. Smeulders, Department of Molecular Cell Biology, Section of Molecular Cytology, University of Amsterdam, Plantage Muidergracht 14, 1018 TV Amsterdam, The Netherlands **[247]**

H.T.M. van der Voort, Department of Molecular Cell Biology, Section of Molecular Cytology, University of Amsterdam, Plantage Muidergracht 14, 1018 TV Amsterdam, The Netherlands **[247]**

Dieter G. Weiss, Institute for Zoology, Technical University of Munich, Lichtenbergerstrasse 4, D-8046 Garching, Germany **[105]**

Michael Whitaker, Department of Physiology, University College London, Gower Street, London WC1E 6BT, England **[181]**

T. Wilson, Department of Engineering Science, University of Oxford, Parks Road, Oxford OX1 3PJ, England **[231]**

Preface

The Current Renaissance in Biological Light Microscopy

The modern research light microscope is an excellent scientific instrument. Yielding high-quality flat-field images with a spatial resolution close to the theoretical optimum, utilizing a wide range of optical contrast generation techniques, and with built-in facilities for photomicrography, it is both long-lasting and has low maintenance costs. It is an essential tool in any biological laboratory interested in the structure of biological tissue. Nevertheless, the major inherent advantage of the light microscope (LM) over the electron microscope (EM), namely its utility for the study of dynamic processes in living cells, has until recently been severely compromised by its inability to image very weak fluorescent signals, to make visible minute cellular structures with the contrast generation methods available for unstained cytoplasm, and to obtain accurate images of three-dimensional specimens. One consequence of this has been the dominance of electron microscopic investigations of fixed cells and tissues during the postwar development of cell biology.

Within the last ten years, however, combinations of novel optical microscopic techniques with advanced electronic imaging and digital image processing technology have resulted in dramatic improvements in the quality of light microscopic images, leading to a true renaissance in the use of light microscopy for the imaging of both fixed and living biological specimens. Now things hitherto invisible can be clearly seen and measured, three-dimensional image can be obtained accurately, internal physiological parameters can be measured using a new generation of ratio imaging reporter dyes, and the recording of dynamic changes within living specimens has become significantly easier than in the past.

Until recently, these advances almost exclusively employed video cameras as the means of electronic acquisition of microscope images, and went under the general description of *video microscopy*. While video cameras remain very important, particularly in studies of dynamic processes, the recent development of other electronic imaging devices, such as high-resolution solid state charge-coupled device (CCD) array cameras and confocal scanning optical systems, and the near-universal inclusion of some form of image digitization and digital image processing into the experimental system, requires that this field now be redefined more generally as *electronic light microscopy*.

Many of these recent developments have been technologically driven. For years, desirable experiments have been clear to research workers, but only recently have the tools been available to permit these experiments to be conducted, in the form of affordable electronic imaging equipment, digital image processing hardware, and personal computers of sufficient power and memory capacity. However, this is only part of the story. As in all major scientific developments, the insight, intelligence, imagination, and energy of particular pioneering individuals have been crucial in this renaissance in the applications of light microscopy to the biological sciences. Curiously, most of those individuals involved in the development of video microscopy have been from the United States, while the pioneers of confocal microscopy are almost without exception Europeans.

As is to be expected in such a new science, almost all the principal protagonists in the field of electronic light microscopy are still actively engaged in research. However, one of its greatest pioneers, Robert Day Allen, tragically died of cancer in 1986, during the most exciting and productive period of his research career. A brief account of his discovery of video-enhanced contrast microscopy will serve to illustrate the importance of individual astuteness and discernment in the birth of this new science.

In 1981, while teaching his short course *Optical Microscopy* at the Marine Biological Laboratory in Woods Hole, Massachusetts, Bob Allen made an accidental discovery which subsequently revolutionized both our use of the light microscope for the study of living cytoplasm and our understanding of cell motility. To enable all his students to see his demonstration of differential interference contrast optics simultaneously, he had rigged a video camera onto his demonstration microscope. Then, looking down the eyepieces, he adjusted the analyzer away from its normal position of near-extinction with the polarizer, and reported that the optical image contrast was fading. However, thanks to a fortuitously faulty black-level adjustment on the video camera, his students saw the image sharpen remarkably on the video monitor. He recalled his first reaction: "I went round the table and looked at the monitor. The image was fantastic, something I'd never seen before. All kinds of details not visible to the naked eye leapt out at us."[1]

This happy accident, and Bob's vivid appreciation of its cause and significance, led to the development and subsequent widespread use of video-enhanced contrast microscopy, for which he received worldwide recognition. While many others have also played major roles, it is not my purpose to review their contributions here. However, I consider that the publication of Shinya Inoué's book *Video Microscopy*,[2] encapsulating years of his own extensive experience, marked a significant turning point from the early development of this area of electronic light microscopy, using prototype experimental setups in a few select laboratories, to its maturation into an established discipline open for use by anyone. The publication in 1989 of the original edition of The Handbook of Confocal Microscopy, edited by James Pawley,[3] played a similar role for the field of confocal scanning light microscopy.

This present volume seeks to describe these important new developments in the whole field of electronic light microscopy. It is in no way intended to mimic or duplicate specialist volumes on individual techniques nor, alternatively, to be merely a topical collection of research reports that will rapidly become outdated. Rather, its purpose is two-fold: first to introduce the principles of electronic light microscopy and the digital image processing that almost invariably accompanies it, and second to exemplify applications of the methods outlined, for the benefit of those wishing to understand these new approaches and contemplating whether they might benefit their own studies.

---

[1] Allen RD (1985): Quoted in an article by Richard Higgins entitled "A New Window into Living Cells," *Boston Globe*, August 12, 1985.

[2] Inoué S (1986): "Video Microscopy."New York, London: Plenum Press.

[3] Pawley J (ed) (1989): "The Handbook of Biological Confocal Microscopy." University of Wisconsin, Madison: IMR Press. A revised edition of this handbook, also edited by James Pawley, has been subsequently published in 1990 by Plenum Press (New York, London).

The book is divided conceptually into four sections. In the first introductory section, Chapter 1, "An Introduction to the Electronic Acquisition of Light Microscope Images," and Chapter 2, "An Introduction to Digital Image Processing and Image Display in Electronic Light Microscopy," provide general introductions to the subject matter of the following chapters for those totally new to the field, giving background material that is assumed in the remainder of the book. Chapter 3 sets out the fundamentals of digital sampling theory as applied to biological light microscopic image acquisition and image display, and Chapter 4 addresses the very practical requirement for an effective interface between the user and the electronic light microscope. The three subsequent sections deal in detail with the techniques of video-enhanced contrast microscopy, intensified fluorescence microscopy, and confocal scanning light microscopy, respectively. Together they form an accurate description of the present state of the art of electronic light microscopy as applied to the life sciences.

It is my hope that this volume, by bringing together for the first time fundamental information concerning the principles and practice of these three major forms of electronic light microscopy, will prove of value both to established research workers and to graduate students entering this field throughout the decade ahead, and will serve as a text for advanced microscopy courses. It is intended to be didactic and to provide a conceptual education, rather than to lay down detailed technical specifications. The pace of current developments in both electronic imaging technology and digital image processing is tremendous, with new types of cameras and more powerful image processors being launched onto the market at more competitive prices all the time, and it is thus inevitable that the technological frontiers will move somewhat between the submission of these chapters for publication and the appearance of the printed volume. Nevertheless, the principles of electronic light microscopy and digital image processing set forth in this book will remain unchanged.

I am deeply grateful to my co-authors in this volume, all eminent authorities in their respective fields of endeavor, for their willing collaboration and hard work in writing and assisting me in editing their excellent contributions. Without their involvement this book would not exist. My own induction into and ongoing involvement in the field of electronic light microscopy was catalyzed by helpful discussions and collaborations with many friends and colleagues, particularly David Agard, the late Robert Day Allen, Brad Amos, Andrew Dixon, Shinya Inoué, Bachara Kachar, Bas Ploem, Phil Presley, Bruce Schapp, John Sedat, Peter Shaw, Jon Singer, Hans Tanke, Lans Taylor, and Nick White. Their helpfulness, wisdom, and scientific excellence have contributed indirectly to the contents of my own chapters, and it is a pleasure to acknowledge my indebtedness to them. I am most grateful to Amir Attaran for his critical reading of my chapters and his helpful suggestions for their improvement and clarification. The responsibility for remaining shortcomings, errors, and omissions remains, of course, entirely my own.

David Shotton
Oxford
February, 1992

CHAPTER 1

An Introduction to the Electronic Acquisition of Light Microscope Images

David Shotton

Electronic Light Microscopy, pages 1–38 © *1993 Wiley-Liss, Inc.*

I. WIDE-FIELD LIGHT MICROSCOPY

A. Principles and Types

1. Definitions. All forms of electronic light microscopy involve electronic encoding of an optical image, but they may be divided into two fundamentally distinct types, *wide-field microscopy* and *scanning microscopy* (Fig. 1).

Wide-field microscopy encompasses all those forms of electronic light microscopy in which a conventional wide-field two-dimensional optical image of the specimen is formed and focused onto the image plane of an electronic two-dimensional imaging device, such as the faceplate of a conventional video camera tube or the semiconductor array of a charge-coupled device (CCD) array camera, from which it is read out sequentially, by the raster scanning of the faceplate or the sequential charge transfer of the CCD array, as a series of electronically encoded image lines.

By contrast, scanning light microscopy involves the physical scanning of the object plane (i.e., of the specimen itself) with a diffraction-limited point of light, or in some cases with a one- or two-dimensional array of such points. All forms of confocal scanning light microscopy (CSLM) described below are of this type. In such microscopy, a continuous two-dimensional optical image of the specimen is not formed within the microscope, but rather the result of the interaction of the scanned light beam(s) with successive regions of the specimen is measured and recorded. How this is achieved is outlined in Section II below.

2. Video-enhanced contrast microscopy. The recent advances in wide-field electronic light microscopy fall into two related categories. The first is video-enhanced contrast microscopy (VECM), which involves the electronic amplification of the minute contrast variations created by subresolution phase objects until these become visible, usually followed by real-time digital subtraction of a pre-recorded background image in order to enhance the signal-to-noise ratio (SNR) by removing various unwanted imperfections in the optical image, collectively termed "mottle," which otherwise degrades the image to a significant extent at the high

A   Wide field

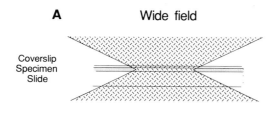

Coverslip
Specimen
Slide

B   Point scanning

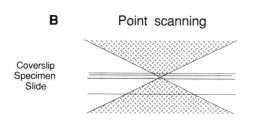

Coverslip
Specimen
Slide

Fig. 1. A comparison of the illumination experienced by the specimen from a high numerical aperture oil-immersion objective during **(A)** wide-field illumination in a conventional microscope and **(B)** point-scanning illumination at the focal plane by a diffraction limited point of light in a single beam scanning light microscope. The volume occupied by the specimen between the coverslip and the microscope slide is stippled, and the cone of illuminating light is shown cross-hatched. For simplicity, refraction effects are not shown. (Reproduced from Shotton [1988b] with permission from the Company of Biologists.)

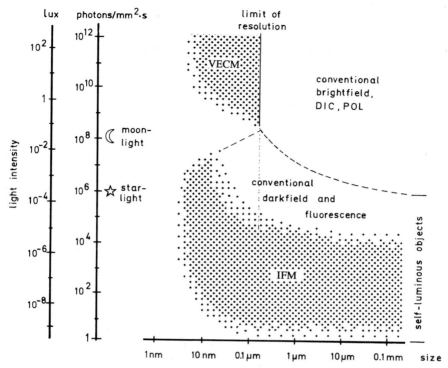

Fig. 2. A plot of detectable specimen size against sensitivity for various modes of light microscopy, showing how video-enhanced contrast microscopy (VECM) at high light levels can be used to visualize objects an order of magnitude smaller than the resolution limit (about 0.2 μm) of the conventional light microscope, while intensified fluorescence microscopy (IFM) can detect fluorescent or phosphorescent objects of similar sizes which have a light emission intensity several orders of magnitude fainter than that detectable by conventional fluorescence microscopy. (Reproduced with permission of Oxford University Press from Weiss et al. [1989].)

contrast levels achieved after amplification. For this to be successfully achieved, careful attention must be paid to the capabilities and performance of the microscope and camera employed, and to the magnification of the optical interface between them, as discussed in Sections ID and IE below.

The original development by Allen, Inoué, and others of VECM produced one of the most powerful techniques now available to the biologist for the study of living cells. Major gains of image quality permit objects as small as individual microtubules and rapid processes such as fast axonal transport to be clearly seen in real time with high contrast, despite the fact that the objects observed may have dimensions an order of

magnitude smaller than the resolution limit of the visible light microscope (Fig. 2). VECM is thus far more than the simple use of a video camera to observe and to permit the video recording of a conventional light microscope image as an economical alternative to cinematography. Rather, it represents a radically new way of using the compound light microscope for the high-resolution imaging of dynamic events and microscopic structures within living biological cells or cytoplasmic extracts.

The development of VECM is a very recent phenomenon [reviewed and detailed by Allen, 1985; Inoué, 1986; Schnapp, 1987; Shotton, 1988a; and Weiss et al., 1989], and the details of its principles, implementation, and applications are described in Chapter 5 of this volume by Dieter Weiss and Willi Maile.

3. Intensified fluorescence microscopy. The second major technique in wide-field electronic light microscopy is that of intensified fluorescence microscopy (IFM), pioneered by Reynolds [1968, 1972], Willingham and Pastan [1978], and Reynolds and Taylor [1980]. Formerly referred to as video-intensified fluorescence microscopy (VIFM), this now increasingly involves the use of high-resolution slow-scan cooled CCD array cameras in place of intensified video cameras, so the term video has been dropped. One or other type of highly sensitive low light level imaging camera is attached to a conventional fluorescence microscope in order to amplify the signal obtained from a fluorescently labeled specimen, permitting the visualization of faint fluorescence which could not be clearly seen by eye nor recorded by direct photomicrography (Fig. 2).

Alternatively, such is the exquisite sensitivity of IFM systems that the technique may be used to observe well-labeled cells over extended periods of time by intentionally reducing the exposure to the fluorescence excitation illumination to a level at which photobleaching ceases to be a major problem. This allows one, for instance, to make a time-lapse study of the redistribution of labeled surface ligands or intracellular molecules over extended periods of time. For many applications, direct image intensification is all that is required, making this the simplest form of electronic light microscopy. The output from a high-sensitivity video camera may simply be recorded and used directly, without further enhancement, as originally detailed by Willingham and Pastan [1983]. However, particularly for weakly fluorescing specimens, temporal image averaging using a digital image processor is employed to enhance the signal-to-noise ratio, in which case IFM is sometimes alternatively described as digital imaging fluorescence microscopy.

Good introductions to fluorescence microscopy are to be found in Ploem and Tanke [1987], Wang and Taylor [1989], and Taylor and Wang [1989]. Descriptions of research applications of VECM, IFM, and confocal microscopy reported at recent major international meetings are to be found in *Cell Motility and the Cytoskeleton* (vol. 10, nos. 1, 2 [1988]) and in Herman and Jacobson [1990].

B. Specimen Preparation Techniques for Intensified Fluorescence Microscopy

1. Fluorescence immunocytochemical labeling and in situ hybridization. While, with the exception of colloidal gold labeling detailed in Section I F2 below, specimens for VECM are normally unstained, relying upon the microscope's ability to generate contrast from refractive index or birefringence differences within the specimen, for instance by differential interference contrast (DIC) and polarization microscopy, the development of new fluorescence labeling methods has been of the utmost significance in intensified fluorescence microscopy. Certain specimens possess sufficient autofluorescence that staining with exogenous fluorescent chromophores is not required, but for most applications some form of fluorescent dye is associated with the specimen prior to observation. Conventional immunofluorescence labeling is an ideal way to localize specific cellular protein antigens, but on living cells care must be taken to avoid artifacts affecting the normal physiological behavior of the antigen in question brought about by the crosslinking effect of divalent immunoglobulins. Monovalent F_{ab} fragments of the primary antibody directly conjugated to a fluorochrome may avoid this problem, at the expense of a lower intensity of fluorescence labeling. Alternatively, one may observe the disruption and recovery of normal cellular function following the intracellular microinjection of a crosslinking fluorescently labeled primary antibody. Obviously, surface antigens on isolated cells are straightforward to label, but labeling of internal antigens demands the introduction of the label into the cell cytoplasm with a minimum of trauma, by direct microinjection, by electroporation, by fusion with liposomes or erythrocyte ghosts, or by one of a variety of more esoteric methods reviewed by McNeil [1989].

In situ hybridization using a specific oligonucleotide probe may be used to localize particular sequences of nucleic acid in fixed and permeabilized cells, the probe being either directly fluorescently labeled or, more usually, tagged with biotin or a hapten that can subsequently be stained using a fluorescently labeled avidin or antibody conjugate. An innovative method of simultaneously recognizing many probes by employing probes labeled with different ratios of blue, green, and red fluorescing chromophores has recently been developed by Nederlof et al. [1989, 1990].

2. Fluorescence analogue cytochemistry. As an alternative to studying the fates of exogenous molecules, the distributions of selected endogenous molecules or organelles within living cells may be observed by the technique of fluorescence analogue cytochemistry, pioneered by Taylor and Wang [1978, 1980], Wehland and Weber [1980], and Wang et al. [1982], and reviewed most recently in Wang and Taylor [1989]. In this, a normal intracellular protein or lipid component is first biochemically purified and labeled with a fluorochrome. It is checked for native properties and is then reincorporated into living cells. Once the cells have been thus loaded and checked for viability, IFM may be used to study the incorporation of the fluorescent analogue into normal cellular structures, and its subsequent redistribution or behavior in response to physiological and pharmacological stimuli.

3. Vital dyes and observation of dynamic processes. The study of dynamic processes in living cells is also possible using vital fluorescent stains that do not compromise the normal functioning of the cell. For instance, one may study the intracellular distributions of fluorescent dyes which label specific intra-cellular components, such as NBD ceramide which labels the Golgi complex, or rhodamine 123 and other dyes which selectively stain mitochondria. This type of experimental approach is elegantly illustrated by Thomas Kreis and his colleagues in Chapter 7 of this volume. The intensified fluorescence microscopic techniques thus permit extended observation of dynamic cellular processes, which previously could be seen only as isolated "snapshots" of different cells taken by conventional photomicrography at what the experimentalist hoped were appropriate time intervals, each requiring such intense illumination and prolonged exposure as to invalidate subsequent study of that particular cell.

4. Fluorescence ratio imaging. Fluorescence ratio imaging [Tsien and Poenie, 1986; Bright et al., 1989] is an elegant technique that permits the measurement of an increasing variety of intracellular physiological parameters such as intracellular free calcium concentrations or intracellular pH within different regions of individual cells. By calculating the ion concentration of interest from the *ratio* between the fluorescence excitation or fluorescence emission of the reporter dye at two different wavelengths, for instance at an ion-sensitive wavelength and at an ion-insensitive wavelength, one automatically makes correction for two important experimental variables, variation in thickness of the cell from which the signal is being recorded in different regions of the image, and inhomogeneity in reporter dye concentration in different regions of the cell, which would otherwise invalidate the measurements made. This important major development in intensified fluorescence microscopy is discussed in Chapter 8 by Michael Whitaker and his colleagues, who point out the need for proper experimental controls and warn against a variety of pitfalls for the unwary novice.

5. New fluorochromes and multiparameter imaging. The full promise of fluorescence labeling of living cells has been demonstrated by DeBiasio et al. [1987] and Waggoner et al. [1989], who have used the capability of a cooled CCD array camera to image from the blue into the near-infrared (i.e., well beyond the range of the human eye) to undertake multiparameter fluorescence imaging on living cells. Five chromophores with nonoverlapping fluorescence excitation and emission spectra, including two newly synthesized dyes which emit in the far red and the infrared, staining different components or compartments within the cell, were sequentially imaged at 450, 530, 620, 670, and 780 nm within the same cell by changing wavelength-selection filters. One practical advantage of multiparameter imaging, in contrast to multidimensional imaging discussed below, is that the data volume scales linearly with the number of parameters, rather than as the power of the number of dimensions.

One way of avoiding troublesome autofluorescence is to employ delayed fluorescence imaging, using chromophores which continue to emit for several hundred microseconds after being briefly excited. By gating the visualization or detector system to occlude the prompt fluorescence emission characteristic of autofluorescence (and most conventional fluorescent dyes), the delayed fluorescence of the specific label is seen against a black background [Song et al., 1990].

Further details of labeling and imaging methods for fluorescence microscopy, and of applications of these methods to a variety of cellular systems, may be found in a wide-ranging review by Arndt-Jovin et al. [1985] and in the two recent excellent comprehensive volumes of *Methods in Cell Biology* edited by Lans Taylor and Yu-Li Wang [Wang and Taylor, 1989; Taylor and Wang, 1989].

C. Information Flow and Hardware Requirements

While differing in their optical and electronic imaging principles, the two major forms of wide-field electronic light microscopy share a common approach to informa-

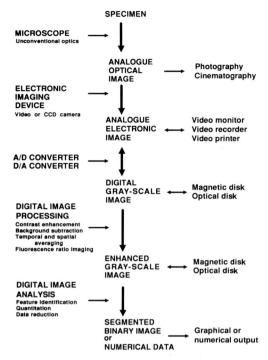

Fig. 3. Stages in information flow (thick vertical arrows) in wide-field electronic light microscopy. The left column shows the device or process employed, the center column the form of the data, and the right column the various methods of recording and retrieving the data (see text for details). (Reproduced with permission from Shotton [1991].)

tion processing, illustrated in Figure 3, and to this extent have much in common, both in the principles of image processing employed and in the hardware and software required to achieve them.

A typical setup suitable for both VECM and IFM microscopy is shown diagrammatically in Figure 4, and its components are described in detail elsewhere [Shotton, 1988a; Weiss et al., 1989] and by Weiss and Maile in Chapter 5 of this volume. A large, stable, high-quality optical microscope is required, ideally mounted upon a vibration isolation table in a dust-free environment. Microscopes having a direct optical path free from prisms and accessory lenses are generally to be preferred for fluorescence studies, since such additional optical com-

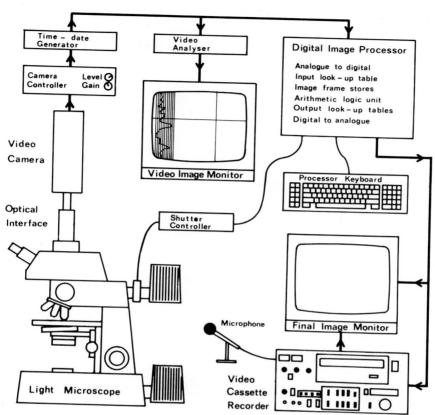

Fig. 4. A diagrammatic representation of the principal equipment components of a system capable of video-enhanced contrast microscopy (VECM) and intensified fluorescence microscopy (IFM). For simplicity, a noninverted microscope and only a single video camera are shown. An additional SIT video or cooled CCD array camera would in practice by required for low light level fluorescence imaging, either substituting for the video camera used for VECM or mounted on a separate optical port. The flow of analogue video information is indicated by the heavy lines. Individual components are described in the text, except for the digital image processor, which is described in Chapter 2. (Reproduced with permission from Shotton [1988a].)

ponents introduce unwanted surface reflections and consequent loss of fluorescence intensity, while inverted microscopes are often used for VECM studies because of their mechanical stability and ease of access to the specimen. In the former case, an electronic shutter on the epi-illumination pathway is particularly advantageous to prevent unnecessary illumination between observations, since this will reduce fluorochrome bleaching. The objectives employed usually combine high magnification with the highest possible numerical aperture (NA), in order to maximize both the in-plane spatial resolution and image brightness. When used for VECM with transmitted light, they are matched with a condenser of at least equal numerical aperture to avoid reducing the effective NA of the objective and hence impairing the resolution. In contrast to conventional practice, a mercury arc lamp with a DC power supply is generally employed to provide a stable light source of very high illuminating intensity, since the technique requires both a high image magnification to ensure faithful electronic sampling of the optical image (discussed in Section IE, below) and a high image brightness to operate the video camera at near-saturation. The great advantages of a fiber-optic illuminator for VECM [Ellis, 1985] are described in Chapter 5. Conversely, a tungsten or quartz halogen lamp of much lower brightness may be used for IFM instead of a mercury lamp, because of the sensitivity of the detection system and the desire to avoid photobleaching.

To the microscope is attached one or more electronic imaging cameras (video or CCD), discussed in Section ID below, ideally arranged with fully reflecting mirrors or prisms in the microscope so that 100% of the light may be directed to the selected camera, using a light-tight optical interface. The importance of the correct magnification within this optical interface is stressed in Section IE. The output signal from the camera is usually fed via a time–date generator to one or more monochrome video monitors, to allow observation of the initial "raw" video image and to permit accurate analogue contrast and brightness adjustment, the importance of which for VECM is described in Section ID.

The optimized video signal is then passed to a digital image processor, where the analogue video signal is digitally sampled and converted into a series of 8-bit integer numbers, each representing one of 256 possible gray levels from 0 (black) to 255 (white) for that particular picture element (*pixel*). Digital image processors normally come with a wide variety of image processing software algorithms for data acquisition and digital image processing. However, the appropriateness of this software for electronic light microscopy, and the convenience and "user-friendliness" of the user interface, vary considerably between different models, and one should consider and test carefully before purchasing. Chapter 4 by Theodore Inoué is dedicated to a consideration of this matter. A brief introduction to the functioning of a digital image processor and the basics of digital image processing is given in Chapter 2 (Shotton).

To record the processed images, a high-resolution video cassette recorder (VCR) is generally employed. Although better than that of domestic ½-inch VCRs, the dynamic bandwidth of the ¾-inch "low-band" U-matic VCRs commonly used for such recording is limited, so that the horizontal resolution of the video image becomes restricted upon recording to about 340 TV lines. (Horizontal video resolution is rather curiously defined as the number of equivalent vertical TV lines that could be resolved over a horizontal distance equal to the picture height, i.e., over ¾ of a horizontal line.) This then becomes the limiting resolution when the recorded images are played back. Fortunately, the recent introduction of the ½-inch super-VHS recording format has improved this situation somewhat. For studies of slow dynamic processes such as cell division, a time-lapse VCR is usually required. Most presently available time-lapse VCRs are designed for security surveillance, have poor resolution, and tend to give a very noisy dynamic replay. However, a recently introduced super-VHS format time-lapse VCR (Panasonic model AG6720) is a good machine for both monochrome and color time-lapse recording.

Digital optical memory disk recorders (OMDRs) with storage capacity of up to 2 Gigabytes (2,000 Mbytes) are also available, and look set to revolutionize the time-lapse recording and archiving of high-resolution digitized video images. A 2-Gbyte disk can store 8,000 images each of 512^2 pixels by 8 bits, or substantially more if digital image-compression algorithms are employed. If each image represented a digitized video frame, the storage of 8,000 such images would represent over 2 hr of time-lapse recording made at ¹⁄₂₅th normal speed, i.e., one frame per second. OMDRs have a further substantial advantage of giving high-quality full-frame still replay, compared to the reduced vertical resolution of the single-field "freeze" replay mode of a conventional time-lapse VCR. Unfortunately, they are still significantly more expensive than VCRs.

The ultimate revolution, digital video, is at present only just beginning to emerge from the development labs.

D. Video and CCD Cameras

1. Information encoding in a conventional monochrome video signal. Few types of visual data transmission are more familiar to us in the present age than television, in which the impression of motion is conveyed to the brain by presenting the eyes with a rapid succession of stationary images, each of which constitutes a single video frame. These images are built up point by point on the television or video monitor screen by an electron beam which is scanned in a succession of parallel, almost horizontal lines across the face of the cathode ray tube, filling the screen from top to bottom.

Each video frame lasts 40 msec and is divided into 625 lines, each of 64 μsec duration.[1] During the first 12 μsec of each line, termed the horizontal blanking interval, no visual signal is displayed and the horizontal scan coils of the television or video monitor are reset from projecting the right-hand end of the previous line in preparation for projecting the left-hand beginning of the new line (Fig. 5). During this interval, a 4.7 μsec line synchronization ("sync") pulse

and (in the case of PAL-encoded color images) a 2.25 μsec reference frequency "color burst" are transmitted. The image-brightness information for that particular video line is then transmitted as analogue voltage fluctuations with time over the next 52 μsec, varying between +0.3 (black) and +1.0 volt (white). Upon projection of the image within the television or video monitor, these voltage fluctuations become translated into variations in the intensity of the electron beam, and hence into variations in image brightness along that particular video line on the tube's phosphor screen, as the electron beam is scanned across it at uniform velocity from left to right (as seen by the viewer standing in front of the screen). Since the downward deflection on the electron beam is being steadily incremented during this horizontal scanning period, each video line is imperceptibly inclined (at approximately 0.15° to the horizontal), with the right end lower.

Of the 625 lines comprising one video frame, only 575 "active" lines are actually displayed. These are presented as two interlaced fields, each of 287.5 lines, with a 50 Hz field repetition frequency, in order to eliminate the flicker which would otherwise be present if full frames were displayed

---

[1]For simplicity, these numerical values refer only to the 625 lines, 25 frames per second CCIR monochrome (PAL color) video standard employed in Britain. Those for the 525 lines, 30 Hz RS-170 monochrome (NTSC color) video standard employed in Japan and the United States differ slightly, but the principles employed in analogue and digital image processing of video signals are common to both systems. Table I gives a comparison of the two standards. The European SECAM standard employed in France and elsewhere also operates at 625 lines per frame and 25 frames per second, but differs in its color encoding. Variations of these standards are used in different regions of the world.

Not to be confused with video standards, which refer to the electronic encoding of the video information as a temporally varying analogue voltage, are the tape formats used for recording the video signal onto magnetic tape. There are two formats widely used for electronic light microscopy, the VHS format, commonly employed in domestic video machines and using a ½-inch tape, and the "semiprofessional" ¾-inch U-matic

format. The tape format employed by a particular VCR is quite independent of the video standard which that VCR is electronically designed to handle, so that, for instance, a VHS tape recorded on an NTSC standard VCR cannot be played on a U-matic VCR of any type, nor on a VHS machine designed to play only PAL standard tapes. The super VHS (sVHS) format is compatible with VHS, but permits higher-resolution color and monochrome recording.

Triple standard (NTSC/PAL/SECAM) VCRs enable one conveniently to play, although not to re-record, material taped on the same format using a different video standard, and it is possible to have material transcribed between different video standards commercially, but otherwise the three principal video standards are, for all practical low-budget purposes, incompatible. High-definition digital television (HDTV) and digital recording will soon replace these present medium-resolution standards and formats, but it appears likely that the golden opportunity presented by this technological development, of uniting the world under a single HDTV standard, will not be taken.

TABLE I. Principal Characteristics of the CCIR and RS-170 Monochrome Video Standards

Characteristics	CCIR	RS-170
Frame		
Aspect ratio (horizontal:vertical)	4:3	4:3
Interlace	2:1	2:1
Frequency		
Per frame	25 Hz	30 Hz
Per field	50 Hz	60 Hz
Period		
Per frame	40.00 ms	33:33 ms
Per field	20.00 ms	16.67 ms
Video lines per frame	625	525
Active lines per frame	575	485[1]
Video lines per field	312.5	262.5
Active lines per field	287.5	242.5[1]
Inactive lines per field	25	20[1]
Vertical blanking interval	1.61 ms	1.28ms[1]
Vertical field active scan period	18.39 ms	15.39 ms[1]
Line		
Line frequency	15,625 Hz	15,750 Hz
Line period	64.00 μs	63.49 μs
Horizontal blanking interval	12.05 μs	11.11 μs[1]
Horizontal active scan period	51.95 μs	52.38 μs[1]
Luminance		
Peak white level	+1.0 v	+0.714 v
Blanking (black) level	+0.3 v	0.0 v[2]
Sync level	0.0 v	−0.286 v
Broadcast bandwidth[3]	5.5 MHz	4.2 MHz

[1]Nominal values. Differences exist between broadcast and closed-circuit TV (CCTV) signals.
[2]In the RS-170 standard, the optical black level is set slightly above the blanking level, by an amount (≤ 0.1 v) known as the *setup voltage*.
[3]CCTV systems are not limited by the broadcast bandwidth, and can thus operate at higher bandwidths than those shown here, thereby offering enhanced horizontal resolution.

without interlace at a repetition frequency of 25 Hz. The first field is composed of the odd-numbered lines 1, 3, 5, . . . , 573 and the first half of line 575, which appears at the bottom of the left half of the screen, while the second field comprises the second half of line zero, at the top right of the screen, together with the even-numbered lines 2, 4, 6, . . . , 574 (Fig. 6). Each field thus fills the screen, but with half the vertical line resolution of the whole frame. The two "missing" 25-line periods of each video frame comprise the field vertical blanking intervals, during which the deflection coils return the electron beam from the bottom to the top of the screen in preparation for projection of the next field.

In broadcast television, these "raw" video images are encoded as frequency modulations of an ultrahigh frequency (UHF) carrier radio wave, with a bandwidth of 5.5 MHz, equivalent to a horizontal resolution of about 440 TV lines. This carrier wave conveys the information from the broadcast television transmitter to the aerial of the domestic television receiver, where the radio frequency video image is demodulated before display. In contrast, closed-circuit television (CCTV) systems, including those employed in electronic light microscopy,

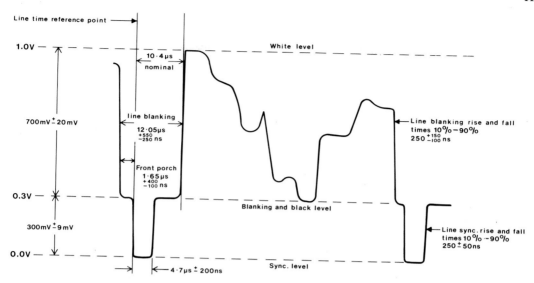

Fig. 5. A time/voltage representation of a single monochrome video line from a CCIR standard video signal (horizontal axis: time; vertical axis: voltage), showing the initial line blanking synchronization signals occupying 12 μsec, followed by the picture signal, a continuously varying analogue voltage be- tween +0.3 volt (black) and +1.0 volt (white) during the subsequent 52 μsec. (Reproduced from the British Department of Trade and Industry publication Specification of Television Standards for 625-Line System I Transmissions in the United Kingdom [1984], with permission of the controller of HMSO.)

convey video images directly within coaxial cables from the camera to other system components as unmodified analogue voltage fluctuations, with the magnitude and time constants shown in Figure 5, and are not limited in horizontal resolution to that of the broadcast video bandwidth [see Inoué, 1986; Chapter 6, De Brabander et al., this volume, for a full discussion of video equipment bandwidth and resolution].

2. Analogue video contrast enhancement. The contrast of a video image is determined by the efficiency with which the voltage fluctuations encoding it fill the available dynamic range of the video signal: A high-contrast image spans the entire range from black (+0.3 volt) to white (+1.0 volt). Since voltage signals may easily be modified electronically, analogue contrast enhance-ment of a low-contrast video image is easily achieved. Indeed such basic image-process-ing procedures are familiar to us all from ad-justment of the "brightness" and "contrast" controls of domestic television receivers.

Of particular importance for VECM is the understanding that the input light intensi-ties to which the output voltage values of "black" (+0.3 volt) and "white" (+1.0 volt) correspond are not fixed, but rather are set electronically within the TV camera, by adjustment of the black level (or offset, pedestal, or bias) control and of the gain (or amplification) control. In broadcast tele-vision and most CCTV video cameras, this is done rapidly and automatically with an efficiency equal to that of the adaptive responses of the human eye, enabling shots moving from bright sunlight to deep shade to be recorded without significant loss of contrast. In VECM, however, where the initial optical image contrast is extremely low, it is important that the operator should have control of these parameters, and thus video cameras used for this application should be fitted with manual black level and gain controls situated in a control box sepa-rate from the camera head on the microscope. An analogue video analyzer, which shows as a graphical overlay on the video monitor how fully the intensity fluctuations of the image fill the available dynamic range of the video signal from black to white, is very

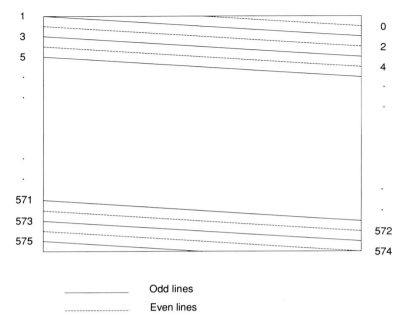

Fig. 6. A diagram to show the nature of the 2:1 interlace of video lines from the odd and even fields that comprise a single video frame. The line-numbering system adopted here shows the 575 displayed lines of the full video frame in sequential order from top to bottom of the screen. For clarity, only the first few and last few lines are shown, at ×25 the normal spacing and slope. Lines 1, 3, 5, . . . , 573, 575 comprising the initial odd field are displayed first, while lines 0, 2, 4, . . . , 572, 574 comprising the second even field are displayed 20 msec later, positioned in the spaces between the odd lines. The 50 nondisplayed lines that occupy the two vertical blanking intervals are not shown. (Note that television engineers adopt a different line-numbering system, in which lines are numbered in temporal rather than spatial sequence, including the nondisplayed lines. In that scheme, field 1 comprises lines 1, 2, 3, . . . , 311, 312 and the first half of line 313, while field 2 comprises the second half of line 313 and lines 314, 315, . . . , 624, 625.)

useful for such optimization (see Fig. 4, and Chapter 5, Fig. 6).

Similar contrast manipulations may be achieved digitally, and may be of particular value for VECM of images acquired from sources other than conventional video cameras. However, inaccuracies will be introduced by digital contrast enhancement of a low-contrast analogue image, since this will fill only a few of the 256 available gray levels upon digitization, resulting in step discontinuities when this limited digital data set is expanded to fill the entire dynamic range. It is thus important that analogue contrast be optimized prior to digitization.

3. Video cameras. Conventionally, video images are initially formed by an electron beam moving, in an identical manner to that described above for a video monitor, across the target faceplate of the television camera tube, resulting in the generation of voltage fluctuations with time in proportion to the integrated intensity of the optical image falling upon each point of the camera target during the preceding 40 msec (one frame time). Indeed, it is the line and field synchronization (sync) signals generated by the camera and transmitted with the video image which entrain the display beam of the receiver or monitor.

The principal hardware difference between VECM and IFM is in the nature of the image detection system employed. Most high light level VECM applications, where the intent is to optimize the image quality of individual video frames for real-time video recording of rapid cellular events, employ 1-inch Newvicon or Chalnicon monochrome video cameras, which have relatively high light sensitivity (300–500 nA/lux) and reasonably good horizontal resolution of about 700 TV lines in the center of the image, falling to about 500 lines at the edges. These cameras produce signals conforming to one or the other of the conventional video standards described above, enabling their use with standard low-cost accessories. However, there is now increasing use of video cameras of higher resolution and line frequency, and of high frame rate solid-state CCD array cameras with high spatial resolution, which require nonstandard monitors and digitizers to handle their signals.

For low light level IFM applications, one of several types of image-intensifying video tube have traditionally been used to detect low-intensity fluorescence emissions. Silicon intensifier target (SIT) image tubes of very high sensitivity (typically 30,000 nA/lux, some two orders of magnitude greater than that of a conventional video camera) and moderate horizontal resolution (approx. 500 lines at 10^{-2} lux faceplate illumination, falling to 200 lines at 2×10^{-5} lux) have been most commonly employed. Intensified SIT (ISIT) tubes increase this sensitivity by another order of magnitude, although their resolution is somewhat lower. Individual video frames from such tubes, particularly the latter, are inherently noisy, both because of the statistical fluctuations in photon emission from very dim fluorescent specimens and, more importantly, because of the presence of substantial electronic "white" noise generated by the SIT and ISIT intensifying devices themselves. In addition, SIT and ISIT cameras show even worse geometric distortion and shading than most other video cameras.

The possibilities of low light level imaging are extended by the availability of ultra-high-sensitivity video cameras and photon-counting cameras, incorporating two- or three-stage multichannel plate (MCP) image intensifiers, which are less noisy and less sensitive to "blooming" (oversaturation) on image bright spots, have good gain uniformity and photometric linearity characteristics, low lag, and low geometric distortion.

In these, the optical microscopic image is focused onto a photocathode, causing the emission of electrons which in turn are focused onto the MCP image intensifier, in which they are amplified some millionfold before striking a phosphor screen. The intensified image so produced is then optically projected onto the target of a conventional high-resolution video camera or CCD camera, usually by direct fiber-optics coupling, to produce a standard video signal with an overall enhancement in sensitivity which is maximally more than two orders of magnitude greater than that of a SIT camera, albeit at present with a maximum horizontal resolution of only some 400 TV lines, limited by the structure of the MCP itself. This structure often results in a fixed noise pattern of a hexagonal grid appearing superimposed upon the image, which can, however, subsequently be removed by digital image filtering (see Chapter 2, Shotton, this volume).

4. CCD array cameras. Even more important for low light level imaging has been the recent advent at affordable prices of cooled slow-scan solid-state charge-coupled device array cameras with high spatial resolution (often exceeding $1{,}024^2$ pixels), in which the primary optical image falls onto a two-dimensional array of silicon photodetector "electron wells," from which, at the end of an exposure, the accumulated electrons are read out and measured sequentially, pixel by pixel and row by row. Cooling the CCD array chip greatly reduces the electronic dark-current noise in such images. This enables prolonged integrating exposures to be made to detect low photon fluxes, with an extremely high signal-to-noise ratio. Because they also exhibit low geometric distortion, shading, and lag, and have high photometric linearity and quantum efficiency, they are almost ideal imaging devices where video frame rates are not required. Table II lists the principal advantages and disadvantages of cooled CCD array cameras in comparison with SIT video cameras. Current improvements in the image-scan frequency of such cooled CCD array cameras, from about 0.1 Hz to better than 4 Hz, now make them increasingly

TABLE II. Comparison of SIT and Cooled CCD Array Cameras for Low Light Level Imaging

Characteristics	SIT Video	Cooled CCD
Sensitivity	Good	Excellent
Electronic noise	Bad	Very little
Averaging	Subsequently, in video frame store	During exposure, on CCD array chip
Photometric linearity	Moderate	Excellent
Dynamic range	Limited (6–8 bits)	Large (12–14 bits)
Spatial resolution	Moderate (200–500 lines)	Good to excellent (up to 2,048 × 2,048 pixels)
Geometric linearity	Poor	Excellent
Temporal resolution and exposure time	Video rate (0.04 sec)	Slow scan (usually 0.1–10 seconds; may be up to several minutes)
Lag	Poor	Absent
Accidental damage by excess light	Prone	Immune
Approximate cost	$5,000	$20,000

attractive for the study of slower dynamic processes.

Although the initial CCD readout is an analogue electrical current from each photon well on the silicon surface of the CCD array, these cameras are designed as digital instruments, and these analogue currents are immediately converted into digital form. Since the readout from cooled CCD array cameras is not at video rates, it is possible for the image to be digitized to a much greater precision than can be achieved for a video signal using a conventional image processor (i.e., to 12 or 14 bits, rather than to 8 bits). Coupled with their low dark current and readout noise, this gives such CCD images a tremendous dynamic range compared with video images. While conventional video monitor display of such a CCD image (or of at least part of it!) is possible, cooled CCD array cameras usually have their own digital image memories, or are able to feed the digital information directly into the memory of a computer for subsequent processing. The use of such cameras is detailed in this volume by Peter Shaw (Chapter 9) and by Barry Masters and Gordon Kino (Chapter 14).

Most uncooled CCD television cameras, in which the CCD array is scanned out at video rates, generate a conventional analogue video signal. While cheap and compact, these have limited spatial resolution and dynamic range, and are not generally used for high-quality microscopy. However, one uncooled CCD array camera with high spatial resolution (1,320 × 1,035 pixels), moderate frame rates (2 to 10 frames per second), and analogue or 8-bit digital video output is now on the market (from Kodak/Videk), and it is likely that such cameras will find increasing use in VECM.

Detailed descriptions of low light level video cameras and cooled CCD array cameras are given by Spring and Lowy [1989] and by Aikens et al. [1989], respectively, and a recent quantitative comparison of these various low light level imaging devices is given by Tsay et al. [1990].

E. Spatial Resolution and Electronic Sampling of the Optical Image

The fidelity with which the optical image is recorded as an analogue electronic image, and later digitized, is a function of the relative spatial resolutions of the optical, analogue, and digital systems employed. So important is this topic that it is specially addressed in Chapter 3 of this volume by Kenneth Castleman, one of the pioneers of digital image processing and its application to biological systems [Castleman, 1979, 1987]. It is often naively claimed that video cameras lack sufficient resolution to record microscopic images accurately. While it is self-evident that present-day video images are of more limited absolute resolution than photographic prints of comparable size, this claim of insufficient resolution is not necessarily true, but depends upon the magnification at which the optical image of the microscopic specimen is projected onto the target of the video camera. For accurate high-resolution wide-field electronic image capture, the limiting spatial resolution of the electronic imaging system must be made twice that of the optical image falling upon it. This can be achieved by magnifying the optical image falling on the video camera faceplate to such an extent that even the low absolute resolution of the video system oversamples the highest resolution information in the optical image. Thus if the magnification of the optical interface between the microscope's primary image plane and the camera target, termed the optical transfer factor, is made sufficiently large, the optical image from even a high numerical aperture objective will be sufficiently oversampled by the video camera, and the resulting video image will remain diffraction limited.

A single example will serve to introduce the subject. Rayleigh's criterion for the resolution limit of the optical microscope, d_{min}, the distance between two point objects at which the peak intensity of the Airy disk of the first object overlaps with the first minimum of the Airy disk of the second, is defined as $d_{min} = 1.22\lambda/2NA$, where λ is the wavelength of the light employed and NA is the numerical aperture of the objective lens and matched condenser [Spencer, 1982]. For light of $\lambda = 546$ nm wavelength, d_{min} is 238 nm measured at the specimen, when using a 1.4 NA oil-immersion objective.

The video cameras commonly employed for VECM have reasonably good horizontal spatial resolution of about 700 TV lines, determined by the size of the scanning spot

and their ability to modulate the signal within a single analogue video line at a reasonably high frequency. Their spatial resolution is limited vertically by the number of horizontal video lines present in the full video image, this being nominally 525 or 625 lines according to the video standard used, of which 40 or 50 lines per frame are not actually displayed, as described above.

Let us assume a worst-case situation of a video camera giving an NTSC video signal with 485 displayed TV lines. It is these lines that sample the continuous optical image in the vertical direction. The limiting vertical resolution of this system is obtained when these lines are alternately dark and light, thus being defined as 242.5 line pairs. Since the physical vertical dimension of the image-forming faceplate of a standard 1-inch video camera is 9.6 mm, the limiting resolution of the video camera is thus 25.26 line pairs per millimeter at the camera faceplate, assuming the diameter of the scanning spot is equal to the line spacing. The camera would thus produce images in which the highest resolution is the reciprocal of this figure, namely 39.59 μm at the faceplate, the distance covered by two video lines. To avoid resolution impairment by undersampling using this camera, the optical image on the faceplate must be magnified so that the spatial resolution of the video camera is at least twice that of the optical image, as demanded by the Shannon sampling theory [Castleman, 1979, and Chapter 3, this volume]. This overall magnification factor must take into account the microscope objective, any intermediate magnifying relay lenses present in the microscope, and the transfer factor of the optics between the microscope and the camera. Using a $\times 63/1.4$ NA objective, no intermediate magnification, and transfer optics between the microscope and the camera faceplate giving a further magnification of $\times 5.5$, the total magnification of the imaging system is $\times 346.5$, making the spatial resolution of the optical image on the camera faceplate 82.5 μm (238 nm $\times$ 346.5), i.e., just over twice the limiting faceplate resolution of the video camera. (Alternatively, as recommended by Castleman [Chapter 3], the camera resolution may be related back to the specimen plane by dividing the camera faceplate resolution by the total magnification, giving a camera resolution of 114.25 nm at the specimen plane.) With this magnification, the optical image is thus perfectly sampled at almost exactly twice its spatial frequency by the video system, with no loss of information. A higher transfer factor would not increase the spatial resolution of the video image. However, use of a lower transfer factor would result in some, perhaps significant, undersampling of the optical image by the electronic imaging device, with loss of high-frequency spatial information. In practice this would be detected by the loss of ability to image subresolution objects such as individual microtubules.

While already sampled by the video lines in the vertical direction, the optical image is further sampled in the horizontal direction when the analogue video lines are digitized. Here, too, there is the requirement that the sampling be at the Nyquist sampling frequency of twice the spatial resolution of the optical system, in order to represent the continuous optical image of the real continuous world exactly by its digitized samples [Castleman, 1979, and Chapter 3, this volume]. This is a function of the bandwidth of the digitizing hardware, discussed in Chapter 2 (Shotton, this volume).

The trade-off for the necessarily high optical magnification required for high-resolution VECM is, of course, a small field of view, typically only 20 μm across. Consequently, some workers have found it useful to incorporate a zoom lens in the transfer optics, so that the high transfer factor required for detailed high-resolution study can easily be decreased to obtain a larger field of view for survey work, without the necessity of changing the objective or other optical components.

The treatment given above oversimplifies the situation by ignoring the manner in which the microscope point spread function (psf) and the video psf combine to give the overall imaging system psf. More complete explanations may be found in Hansen [1986] and Castleman (Chapter 3, this volume). Greater detail on almost all other aspects of video technology and video microscopy are to be found in Shinya Inoué's excellent book Video Microscopy [Inoué, 1986].

F. Image Brightness and Visualization of Subresolution Objects

1. Fluorescent objects. Image brightness is inversely proportional to the square of the image magnification and, when using epifluorescence optics, is directly proportional to the fourth power of the numerical aperture. Thus the light-gathering power needs to be carefully considered when choosing a lens for a particular low light level fluorescence imaging application. A comparison of the nominal epifluorescence brightness of a few objective lenses is given in Table III. The actual magnification and nominal numerical aperture of a particular objective may vary considerably (±10%) from the nominal values. The actual fluorescence brightness transmitted by an objective lens of a particular magnification and NA is also a function of its construction, a plan apochromat objective with many internal lens elements transmitting less light than more simple fluorite lenses. Since many fluorescence imaging applications involve only a single fluorochrome, with only a limited Stokes shift between the excitation and emission wavelengths, the need for precise correction for chromatic aberration over a wide range of wavelengths afforded by plan apochromat objectives may in these cases be sacrificed for greater image brightness.

Secondary magnification of the primary optical image serves only to decrease the image brightness by the square of the secondary magnification power. The guiding principles in epifluorescence imaging should thus be (1) to use the lowest overall magnification possible, commensurate with the

TABLE III. Effect of Objective Magnification and Numerical Aperture on Relative Brightness During Epi-Fluorescence Microscopy: Data for 15 Commonly Used Objectives From the Zeiss and Nikon Ranges

Objective Magnification[1]	Numerical Aperture[1]	Objective Brightness Index[2]	Relative Brightness ($\times 10/0.25$ NA = 1.0)	Rank Order	Percentage Brightness Relative to $\times 60/1.4$ NA
10	0.25	3.9	1.0	15	3.7
10	0.50	62.5	16.0	5 =	59
16	0.40	10.0	2.6	14	9.4
16	0.50	24.4	6.3	9 =	23
20	0.50	15.6	4.0	13	15
20	0.75	79.1	20.3	3	74
25	0.60	20.7	5.3	11	19
25	0.80	65.5	16.8	4	61
40	0.75	19.8	5.1	12	19
40	1.00	62.5	16.0	5 =	59
60	1.40	106.7	27.3	1	100
63	1.25	61.5	15.7	7	58
63	1.40	96.8	24.8	2	91
100	1.25	24.4	6.3	9 =	23
100	1.40	38.4	9.8	8	36

[1]Nominal values. Actual objective magnification and numerical aperture may vary considerably from the values engraved upon the objective barrel.
[2]Objective brightness index = $(NA^4/mag^2) \times 10^5$.

need to resolve objects within the specimen; and (2) within this constraint, to choose an objective with the highest light-gathering power to maximize image brightness (Table III), normally an immersion objective of the highest possible numerical aperture in combination with little or no secondary magnification.

For most intensified fluorescence microscopy, it is necessary to strike an empirical balance, based upon the degree of labeling within one's specimen and the sensitivity of the low light level camera employed, between a high transfer factor in the relay optics which maximizes image resolution and a lower one which increases image brightness. A transfer factor of $\times 1$, for instance, which may be conveniently achieved by using a trinocular head and a direct, non-magnifying C-mount adaptor on a conventional epifluorescence microscope, such that the primary image plane of the microscope falls directly upon the camera tube target, will give a 30-fold enhancement of image

brightness over a secondary magnification transfer factor of $\times 5.5$. Although the optical image is consequently undersampled by the camera system, for certain specimens the resulting loss in absolute resolving power may not be a serious problem. Fluorescently labeled features in the specimen are self-luminous against a dark background, and hence can be clearly seen at any magnification (although not necessarily resolved from one another), provided they are of sufficient brightness to be detected, just as can stars in the night sky which subtend a solid angle on the retina smaller than the angular resolution limit of the unaided eye. Some workers even use a demagnifying secondary lens to transmit the primary optical image to the electronic imaging camera, in order to employ a higher NA objective and thus gather more light from the specimen. However, note from Table III that the brightness of a lens, unlike image resolution, is not a function of the NA alone, and that certain objectives of intermediate magnification have good light-transmission characteristics.

2. Video-enhanced contrast microscopic visualization of colloidal gold particles. As mentioned above, VECM, by virtue of its ability to amplify tiny variations in contrast, is able to make visible subcellular organelles and particles an order of magnitude smaller than the resolution limit of the light microscopy. Any such subresolution object will generate an optical image inflated by diffraction effects to have a diameter approximately equal to the resolution limit of the microscope optics employed, itself a function of the wavelength of light and the NA of the objective, minimally about 200 nm. Two such particles thus cannot be separately resolved if they lie closer together than the resolution limit of the microscope, since their images will overlap, but the single image that they form will be of greater intensity than that of a single object.

Marc De Brabander and his colleagues [De Brabander et al., 1985, 1986, 1989] have pioneered the application of VECM to biological specimens labeled with subresolution colloidal gold particles, widely used for electron microscopic immunocytochemical labeling of specific antigens [De Mey, 1983]. Using conventional bright-field or epipolarization optics, and the electronic contrast enhancement techniques described above, they have been able to visualize individual stationary colloidal gold particles in vitro as small as 5 nm diameter and moving ones on or within living cells of 20 nm diameter, the images of these particles being approximately 200 nm in diameter because of diffraction effects, as explained above. Smaller immunogold particles may be used to stain microtubules or other cellular structures, rendering them a delicate pink color in conventional bright-field microscopy [De Mey et al., 1982], and the contrast of such immunogold staining may be dramatically enhanced by VECM [De Brabander et al., 1986], although here the individual gold particles are not resolved. Since immunogold labeling can be undertaken simultaneously with immunofluorescence or fluorescence analogue labeling, the use of a microscope equipped with both high and low light level video cameras for VECM and IFM now makes possible sophisticated double or multiple labeling experiments.

The ability to coat gold particles with any chosen antibody, cellular protein, or ligand, and then to follow changes in their individual locations after incubation with or loading into living cells, represents a revolutionary step toward a new cell biology, namely the use of the light microscope for the dynamic study of known macromolecular interactions within their natural environment, the living cytoplasm. Furthermore, such gold-labeled cells may be subsequently processed by a variety of electron microscopic (EM) preparative techniques, permitting the subsequent transmission EM localization of these very same gold particles, and fulfilling a previously unobtainable dream of making direct correlations between dynamic processes observable only in the living cell and specific macromolecular interactions revealed only by electron microscopy. This permits a truly integrated experimental approach to many of the current problems of cell biology, and illustrates more clearly than ever before the complementary natures of optical and electron microscopy. Marc De Brabander and his colleagues detail the application of video-enhanced contrast microscopy for the visualization of colloidal gold particles, which they have named nanovid microscopy, in Chapter 6 of this volume.

II. CONFOCAL SCANNING LIGHT MICROSCOPY

Wide-field epifluorescence microscopy suffers from one major disadvantage, namely out-of-focus blur. The illumination of the entire field of view of the specimen with intense light at the excitatory wavelength (Fig. 1A) excites fluorescence emissions throughout the whole depth of the specimen, not just at the focal plane. Much of the emitted light coming from regions of the specimen above and below the focal plane is collected by the objective lens and thus contributes as out-of-focus blur to the final image of the specimen at that focal plane, seriously degrading it by reducing the contrast and sharpness of the image.

While it is possible to remove most of this blur from wide-field images by computational means, as outlined in Chapter 2 (Shotton) and detailed by Peter Shaw in Chapter 9 later in this volume, confocal scanning light microscopy (CSLM) presents an alternative method in which almost all the light causing this blur is prevented by optical means from contributing to the initial image. This enables one to use the confocal microscope directly for noninvasive serial optical sectioning (opto-digital microtomy) of suitably labeled biological specimens, yielding high-resolution images essentially free from out-of-focus blur. Direct three-dimensional cellular tomography, the acquisition and subsequent study of complete in-focus 3D data sets, can thus be undertaken with great facility.

A. Types of Scanning Systems

In any form of scanning light microscopy (SLM), an objective is used to focus one or more beams of light to a single diffraction-limited point or points at the focal plane within a three-dimensional specimen (Fig. 1B). This is commonly achieved by arranging for a laser beam to illuminate a single aperture or pinhole situated at the primary image plane of the microscope objective lens, and allowing the objective to image and demagnify the light transmitted by this illuminating aperture into the microscope's focal plane within the specimen (Fig. 7). Such an instrument, a single beam scanning light microscope, is discussed first. In its epi-illumination configuration, the same objective is then used, in conjunction with an appropriate beam splitter or dichroic mirror, to image the reflected light or fluorescence emission from the specimen onto a

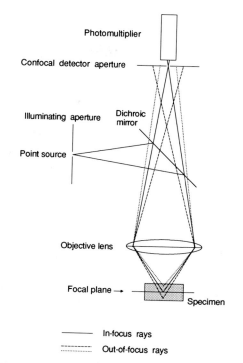

Fig. 7. The confocal principle in epifluorescence scanning optical microscopy. Excitatory laser light from the illuminating aperture passes through an excitation filter (not shown), is reflected by the dichroic mirror, and is focused by the microscope objective lens to a diffraction-limited spot at the focal plane within the three-dimensional specimen. Fluorescence emissions, excited both within the illuminated in-focus voxel and within the illuminated cones above and below it (shown in Fig. 1), are collected by the objective and pass through the dichroic mirror and the emission filter (not shown). However, only those emissions from the in-focus voxel (———) are able to pass unimpeded through the detector aperture to be detected by the photomultiplier. Fluorescence emissions from regions below the focal plane (– – – –) and from above it (·······) have different primary image plane foci and are thus severely attenuated by the detector aperture, contributing essentially nothing to the final confocal image.

nonimaging photomultiplier tube (PMT) light detector, which measures the instantaneous intensity of the specimen response to the scanned beam falling upon it. The SLM may also be used in transmission mode, in which case a second lens is used to image the light to the detector. In this simplest of situations, therefore, involving scanning the specimen with a single point of light, the SLM is the optical equivalent of a scanning electron microscope (SEM), although with particular advantages in terms of z-axis resolution when used confocally, as described below.

To produce an image using such a single beam SLM, as with any form of scanning microscopy, the illuminating point probe must be moved in a regular two-dimensional raster across or, more accurately for transparent three-dimensional biological preparations, through the specimen, and the instantaneous response of the light detector at all points in this scan must be displayed with equivalent spatial position and relative brightness on the synchronously scanned phosphor screen of a CRT monitor or on some other suitable imaging device. The image is thus built up point by point, as on the screen of an SEM monitor or of a domestic television, although the scanning rate for a single beam SLM is usually significantly slower than that of conventional video. This makes it easy to digitize the image "on the fly" and accumulate it into a digital image memory, where it may be subjected to digital image processing as discussed in Chapter 2 (Shotton).

In practice, scanning of a single laser beam may most conveniently be brought about either by the lateral movement of the specimen in the focal plane relative to a stationary optical path (scanned stage SLM), or by the angular movement of the illuminating beam filling the back focal plane of a stationary objective, which causes the focused light beam to move laterally in the focal plane relative to the stationary specimen (scanned beam SLM).

Scanned stage SLM has the important advantage of constant axial illumination, thus reducing optical aberrations and ensuring complete evenness of the optical response across the entire scanned field (space-invariant imaging), a feature desirable for optimal image quality and ease of subsequent image processing. In addition, it permits scanned fields of view much larger than the static field of view of the objective employed, and thus allows one to change between very low and very high magnification imaging while maintaining optimal resolution and light collection, by the simple expedient of altering the stage scan amplitude while using a single high-magnification objective of high numerical aperture. However, because accurate two-dimensional stage scanning cannot be done rapidly, such microscopes generally have a slow scan rate, usually taking at least 10 sec to collect an image of 512^2 pixels.

Scanned beam SLM, in which rotating or vibrating mirrors cause angular scanning of an expanded beam of light filling the aperture of the back focal plane of a plan (flat-field) objective, is more common, and allows higher scanning frame rates, typically of between 0.1 and 2 Hz for an image of 512^2 or 512×768 pixels. As in conventional wide-field light microscopy, the image field of view in such scanned beam SLMs is limited to that of the objective used. One practical advantage of this type of SLM is that it may be achieved by direct attachment of a scanning unit to a conventional compound microscope, enabling all the normal wide-field imaging modes to be retained and employed on the same specimen.

A third alternative method of scanning a single beam across the specimen is that of scanning the objective itself, relative to a stationary specimen and a stationary illuminating system. While suffering from a number of optical disadvantages, this method has been practically implemented in a number of cases. The final logical possibility for scanning is that of moving the illuminating aperture, or more accurately a very large array of apertures, relative to a stationary optical beam, a stationary objective, and a stationary specimen. This is the method employed in the other main type of SLM, the tandem scanning microscope (TSM), in which the specimen is simultaneously scanned by many focused light beams passing through a rotating Nipkow disk, which sweep out the entire field of view at a high frame rate, generating an apparently continuous optical image that may be viewed conventionally through oculars or captured by an electronic imaging camera. Such confocal microscopes are described in greater detail by Alan Boyde in Chapter 13 and by Barry Masters and Gordon Kino in Chapter 14 of this volume.

B. The Principle of Confocal Imaging

In the confocal epifluorescence mode of a single beam confocal scanning light microscope (Fig. 7), the fluorescent light emitted by excited fluorochrome molecules in the single diffraction-limited illuminated in-focus spot is imaged through the same objective that is used to focus the illuminating beam onto the specimen, via the dichroic mirror and suitable interference filters to an aligned detector aperture at the primary image plane of the objective, and hence to a photomultiplier tube. In this arrangement, the confocality of the detector aperture with the illuminated in-focus spot ensures that only light emanating from that region of the specimen is fully passed by the detector aperture and detected by the PMT. In contrast, fluorescence emissions excited by the illuminating beam within the conical illuminated regions of the specimen above and below the focal plane (Fig. 1B) come to focus elsewhere. Light from these regions of the specimen is defocused at the detector aperture plane (Fig. 7), and is thus almost totally prevented by the confocal detector aperture from reaching the PMT detector, contributing very little to the final image. As a consequence, only in-focus information from the illuminated spot is recorded. Reflection contrast confocal microscopy employs similar optics, except that a half-silvered beam splitter is substituted for the dichroic mirror. If the confocal detector aperture is removed, light from out-of-focus regions is also permitted to reach the photodetector, and depth discrimination is thereby abolished, the imaging properties of the non-confocal scanning light microscope being equivalent to those of a conventional wide-field light microscope. Figure 8 shows the actual arrangement of scanning and confocal imaging optics in one widely used single beam confocal laser scanning microscope, which employs mirrors rather than lenses to avoid chromatic aberrations and a long, folded optical path to permit the use of iris confocal detector apertures of adjustable diameter, allowing the user to optimize the trade-off between signal brightness and confocality (see Section IID below).

Although the optical arrangements differ, the same principle of out-of-focus blur exclusion by use of a limiting aperture (or apertures) applies to all other forms of confocal microscopy. Detailed theoretical discussions of confocal image formation are to be found in Wilson and Sheppard [1984] and Wilson [1990a], and in Chapter 10 by Tony Wilson in this volume.

C. Imaging Modes in Confocal Microscopy

The ability of the CSLM to record only in-focus information permits the instrument to be operated as though it had an infinite depth of field, rather than an extremely limited one. By slowly and progressively changing the focal plane through the entire three-dimensional specimen (an axial z scan) while performing successive x,y scans in the focal plane, and by automatically accumulating only in-focus data into a single photographic or digital image throughout this process, a complete in-focus projection image of the specimen down the z-axis may be recorded, as a result of the confocal rejection of out-of-focus information. Additionally, since the position along the z axis of a particular feature is by definition the in-focus position at which the intensity is maximal, axial scan data may be used to produce accurate height information of the specimen, which may be displayed either as a HEIGHT image (see Chapter 2, Shotton), in which

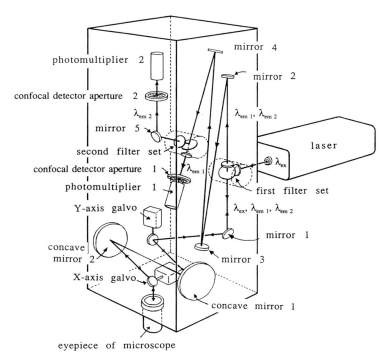

photomultiplier 2

confocal detector aperture 2

$\lambda_{em\,2}$

mirror 5

second filter set

confocal detector aperture 1

photomultiplier 1

Y-axis galvo

concave mirror 2

X-axis galvo

mirror 4

mirror 2

$\lambda_{em\,1}, \lambda_{em\,2}$

laser

$\lambda_{em\,1}$

λ_{ex}

first filter set

$\lambda_{ex}, \lambda_{em\,1}, \lambda_{em\,2}$

mirror 1

mirror 3

concave mirror 1

eyepiece of microscope

Fig. 8. Diagram of the optical path within the scanner/detector unit of the MRC 500/600 single-beam confocal laser scanning microscopy system (BioRad Microscience Ltd.). Two fast galvanometer-driven scanning mirrors cause the path of the incoming beam of excitatory laser light of wavelength λ_{ex} to oscillate about the optical axis. This beam reaches the microscope objective lens (not shown) and hence the specimen via a microscope eyepiece, which is usually mounted in the camera port of a trinocular microscope head. In the normal epi-illumination mode, the returning beam, containing fluorescence emissions at wavelengths λ_{em1} and λ_{em2} in the case of double fluorochrome labeling, is descanned by the same mirrors. Now stationary, this mixed beam passes through the two fluorescent filter sets, each of which comprises two bandpass filters and a dichroic mirror of appropriate selectivities. The first isolated the fluorescence emissions from any reflected laser light of the illuminating wavelength, while the second splits the two fluorescence emissions, one to each of the two photomultiplier detectors via the confocal detector apertures 1 and 2. For single channel imaging, the second filter set is removed, all the light being permitted to fall directly onto photomultiplier 1. High-efficiency front-surfaced mirrors are used rather than lenses, making the system achromatic. The long, folded optical path gives the output beam a diameter of several millimeters at the confocal detector aperture plane, which permits the use of variable iris diaphragms as detector apertures. This makes it possible to trade off confocality against signal strength, as discussed in the text. The two chan-nels can be independently adjusted in this respect. The scanner/detector unit is normally mounted above a conventional upright microscope, as shown. Alternatively, for use with an inverted microscope, it is used on its side, the scanning laser beam entering through a microscope side port. Detection of the scanned transmitted beam is possible using a fiber-optic device (not shown) which collects the light emerging "backwards" through the microscope condenser and introduces it into the light path leading to the second photomultiplier for measurement. The scanned transmission image generated in this way is in spatial register with the confocal epi-illumination image detected by the first photomultiplier, but is itself not confocal. The operation of the scanning galvanometers is controlled by a scan controller card within the host microcomputer (PC 286/386/486). This also contains a frame-store card which digitizes the incoming photomultiplier signal(s), and both stores and displays the resultant digital confocal image(s). At full resolution, it takes about 1 sec to acquire a normal in-plane (x,y) image, but faster rates may be achieved by lowering the number of scan lines or by decreasing the size of the scan. Repetitive scanning (in x) of a single line may be undertaken at 500 Hz, i.e., 20 times faster than video rate. Vertical (x,z) sections may be collected by stepping the z-focus position in small defined increments between sequential x scans. (Figure courtesy of Brad Amos who, with John White, invented this confocal design at the Medical Research Council Laboratory of Molecular Biology, Cambridge.)

lighter areas correspond to points closer to the observer, or as three-dimensional height profile plot.

Vertical sectioning may also be undertaken when imaging confocally, by scanning in x and z at a particular y position, thus generating optical sections parallel to the optical axis of the microscope, something impossible to achieve by conventional wide-field microscopy except by resampling a complete deblurred 3D image, as described in Chapter 2. Scanning in z may be achieved by a motor drive on the microscope's focus knob, or by using a piezoelectric z translator between the microscope stage and the specimen.

As in flow cytometry, it is possible, by the appropriate use of dichroic beam splitters and emission filters, and the addition of one or more additional photomultiplier tubes, to convert a single-wavelength CSLM into a multiparameter instrument capable of simultaneously imaging two or more wavelengths. In biology, this has two major applications: imaging of specimens after multiple labeling, and fluorescence ratio imaging. The former may be used for the simultaneous visualization of at least two labels, for instance, colloidal gold particles imaged at 488 nm by confocal reflection contrast together with fluorescein emitting at a longer wavelength in response to excitation at 488 nm, or fluorescein and Texas Red excited simultaneously at 514 nm. For dual fluorochrome imaging, high-performance bandpass interference filters are required. Any remaining "bleedthrough" from one channel to another, unavoidably caused by the long tail on the fluorescein emission spectrum, may be measured and subsequently corrected for accuracy.

Time-resolved fluorescence imaging [see Section IB5 above] may easily be achieved in single beam confocal scanning light microscopy by the simple expedient of laterally offsetting the confocal detector aperture along the x (most rapidly scanned) axis relative to the illuminating spot, hence imaging from that region of the specimen a defined number of microseconds after it has been illuminated by the excitatory beam.

Ratio imaging may be achieved confocally either by sequential image acquisition at two excitation wavelengths, as for wide-field microscopy, or by simultaneous dual-wavelength fluorescence emission imaging using a dual-channel confocal microscope fitted with the appropriate dichroics and filters in front of the two PMTs. Most excitation ratio imaging dyes, for example Fura-2 used to quantify intracellular free calcium ion concentrations, require excitation by ultraviolet light, unavailable on most confocal microscopes, which are equipped only with visible-light lasers and optics. There are, however, several new physiological indicators for dual-emission ratio imaging, including the SNAFL and SNARF pH indicators from Molecular Probes, which can be excited at 488 nm by the argon ion laser common to the majority of laser scanning confocal microscopes. The optical sectioning capability of the CSLM, which dispenses with the requirement for ratio imaging to correct for specimen thickness variation (although not for dye concentration heterogeneity), makes easier the use of nonratio imaging probes such as Fura Red, Fluo-3, and Rhod-2 for calcium measurement with a single argon ion 488 nm excitation wavelength and a single emission wavelength. [Minta et al., 1989]

If the CSLM is equipped with a nonconfocal detector to measure the light transmitted by the specimen from the scanning beam, one may use its two-channel capability to simultaneously collect a conventional transmitted scanned image (e.g., using Nomarski optics) in one channel and a confocal epifluorescence image of the same field of view in the other. The distribution of the fluorescent label may then be displayed in a suitable pseudocolor superimposed upon the monochrome transmission image.

One benefit of simultaneous dual-channel imaging using a scanning light microscope is that the images obtained are in exact spatial register, since corresponding pixels in each image are measured simultaneously during the illumination of a single point in the specimen. The same advantage holds for serial optical sectioning, provided the axial movement of the specimen is accurate. This simplifies any subsequent image processing involving pairs or stacks of images, described in Chapter 2 (Shotton). However, where necessary, computational alignment of out-of-register digital images may be undertaken without great difficulty.

D. Resolution and Aberrations in Confocal Microscopy

In confocal microscopy, both axial and lateral confocal resolution are maximized by use of a high-NA objective, since this determines the size of the diffraction-limited scanning spot. While scanned stage on-axis space-invariant imaging does not require a flat-field objective, scanned beam imaging does. Individual objectives of the same type differ considerably in their optical characteristics and aberrations, which may conveniently be measured by determining the point spread function of the lens (see Chapters 2, Shotton, and 3, Castleman). This may be done by imaging a plane front-surfaced mirror [Cogswell et al., 1990b; Sheppard and Cogswell, 1990] or a subresolution fluorescent bead or reflective gold particle (see Shaw, Chapter 9), making it possible to evaluate objectives and choose the best for confocal work.

While in conventional microscopy the image formation is determined by the point spread function (psf) (defined in Chapter 2) of the imaging objective alone, in a confocal microscope the image formation is determined by the psfs of both the illuminating and the imaging lenses. This is because both lenses play an equal role in image formation, the former in defining the size of the diffraction-limited in-focus illuminated spot and the latter in imaging this to the detector. The confocal system in-focus psf is thus the product of the individual psfs of the two lenses, and the confocal image is defined by the convolution of the object with this product. For the confocal epi-illumination arrangement (Fig. 7), these two psfs are of course identical, since the same objective is employed for both illumination and imaging, so that the system psf equals the psf of the objective lens squared.

This improved confocal system psf results in a small but significant increase in lateral spatial resolution (super-resolution) during confocal imaging, by a factor of approximately 1.4 over that achievable by conventional wide-field optimal microscopy (see Chapter 10 by Tony Wilson for a full discussion). However, the detector must closely approximate an ideal point detector if such true confocal operation and the associated improvement in lateral resolution are to be obtained. The ideally small confocal detector aperture required to produce this super-resolution permits very little light to reach the photodetector, and thus may give a poor signal-to-noise level with weakly fluorescing specimens. In practice, therefore, apertures larger than that required for true confocal imaging are often employed in order to increase sensitivity when imaging faint fluorescent specimens, and the benefit of increased lateral resolution is rapidly lost. Out-of-focus blur rejection is less sensitive to modest increases in confocal detector aperture diameter, so that moderately large apertures may be used to obtain a high signal-to-noise ratio while retaining reasonable out-of-focus blur rejection, at the expense of lateral super-resolution.

Since fluorescence emission occurs at a longer wavelength than that of the excitation, by a factor determined by the Stokes shift of the particular fluorochrome employed, the spatial resolution of confocal fluorescence imaging will be determined both by the illuminating wavelength (λ_1) and by the emission wavelength (λ_2), since in the confocal system it is the shorter wavelength excitatory light, λ_1, which determines the size of the illuminating spot. Thus in confocal fluorescence microscopy the highest spatial frequency which can be recorded is proportional to ($1/\lambda_1 + 1/\lambda_2$). This gives the CSLM, when used in fluorescence mode, an additional resolution advantage over the conventional wide-field fluorescence microscopy, in which the spatial resolution is determined solely by the longer wavelength, λ_2, of the fluorescence emission. Since resolution increases with decreasing wavelength, choice of fluorochromes with shorter excitation wavelengths and smaller Stokes shifts will maximize the spatial resolution of the images obtained.

There are various ways to estimate the improvement in axial resolution brought about by confocal imaging. These, based on modeling the imaging of a variety of theoretical object forms and on experimental determinations using actual test specimens (points, lines, planes, and extended structures), are more fully discussed by Tony Wilson in Chapter 10. One of these estimates is obtained by determining the variation of the integrated intensity of the image of a point source as one changes focus, which indicates how well the microscope discriminates against parts of the object not in the focal plane. In conventional microscopy, the integrated intensity of the light emanating from any one point in the specimen is unchanged as one moves a little away from focus, the light merely being redistributed in the final image, eventually contributing to a uniform background intensity which reduces specimen contrast, explaining why the conventional optical microscope has no real resolving power in the direction of the optical axis. The use of a circular detector aperture during confocal imaging causes the integrated light intensity to fall off sharply as one moves out of focus, dropping, for instance, to half the original value at a distance of $0.7\,\lambda$ from the focal plane along the optical (z) axis for a confocal transmission microscope with an objective of NA = 1.0 and an ideally small confocal aperture [Wilson and Sheppard, 1984]. In practice, this condition is rarely met, particularly for fluorescence imaging, where the size of the detector aperture is often increased to allow more light to reach the detector, resulting in a degree of out-

of-focus blur rejection that is somewhat less than optimal.

It is important to realize that high-NA oil-immersion objectives are corrected for observation of specimens close beneath the coverslip. Thus when used for optical sectioning deep within aqueous specimens, their performance may become severely compromised by spherical aberration. It is thus best to undertake confocal imaging with the specimen as close as possible to the coverslip and, if extensive serial optical sectioning of aqueous specimens is required, to undertake it using a high-NA water-immersion objective. This has the additional advantage that the optical depth moved through the specimen will equal the physical change in focal position, rather than being significantly in error, by the ratio of the refractive index of the immersion fluid relative to that of the specimen.

Absence of chromatic aberration is also important for good confocal fluorescence imaging, particularly for off-axis imaging in scanned beam confocal systems, since its presence will cause a wavelength- and scan position-dependent axial shift in the optimal confocal position of the detector aperture with respect to the illuminating aperture. This point is particularly important when undertaking simultaneous dual-wavelength confocal epifluorescence imaging, where the three wavelengths involved, the illuminating wavelength and the two emission wavelengths, need to be brought to the same focal point by the objective. Presence of significant chromatic aberration under these circumstances will result in one or both of the fluorescence emissions being collected from focal depths other than that being illuminated, with disastrous results on the overall three-dimensional point spread function of the confocal imaging system, a loss of intensity and optical sectioning capability, and a consequence that the two fluorescence emission images are not in register along the z-axis, having been collected from different z-planes. This situation would obviously be particularly disastrous for confocal fluorescence ratio imaging. Much aberration may, of course, be eliminated by reducing the effective NA of the objective, for instance by closing a diaphragm at the back focal plane of the objective, or in a conjugate image plane, but at the expense of image brightness and lateral resolution.

It is thus clear that the optical sectioning performance of a confocal microscope depends crucially upon a variety of experimental variables: the nature of the specimen itself, the wavelength of the illuminating light, the Stokes shift of the fluorochrome employed (in the case of fluorescence imaging), the numerical aperture and degree of aberration correction of the objective, and the size and shape of the confocal detector aperture. One important conclusion is that significant rejection of out-of-focus information, flare, and scattered light will be obtained both with circular detector apertures significantly larger than required for true confocal imaging, and with slit apertures, discussed in Section IIE below. It is thus advantageous to have a confocal microscope equipped with a variable imaging confocal aperture, so that its size may be optimized for each particular application.

As explained by Wilson in Chapter 10, the confocal spatial resolution along the z-axis is intrinsically less than the lateral resolution. In practice, axial and lateral resolutions of approximately 800 nm and 200 nm, respectively, may be attained routinely during single laser beam scanning confocal fluorescence imaging, provided that the focal step size between sections is sufficiently small to avoid undersampling the z-axis. Thus sampling in x and y at a pixel spacing of 100 nm, with an intersection spacing of 400 nm, approximates well the Shannon sampling requirement for accurate digitization of a continuous image discussed in Section IE above, yielding elongated image volume elements (*voxels*) with a 4:1 axial:lateral ratio. Subsequent three-dimensional image processing, particularly rotation and stereoscopic display discussed in

Chapter 2, must take account of the noncubic nature of these voxels.

E. Specimen Preparation Techniques for Confocal Microscopy

1. Immunogold. Colloidal gold labels can be imaged well by confocal reflection contrast microscopy, 5 nm-diameter gold beads being just detectable under optimal conditions with signal averaging, while larger individual gold particles are clearly seen, as for VECM [Section IF2 above]. Since the reflection signal obtained from gold particles is strong and resistant to photobleaching, immunogold labeling may be preferable to immunofluorescence labeling for 3D image collection where multiple scans are required. The slightly higher spatial resolution observed in practice for confocal reflection imaging as compared with confocal fluorescence imaging is also advantageous. (Since silver has a higher reflectivity than gold, even better staining, although not visualization of individual gold particles, may be expected by silver enhancement after saturation immunolabeling with small [1 to 5 nm] gold conjugates, which themselves would not provide a significant reflection signal.)

In practice, while immunogold particles labeling cell surface antigens are easy to image by confocal reflection imaging after the cells have been fixed, cleared, and mounted in a mountant of suitably high refractive index such as glycerol or DPX [White et al., 1989], the situation is less straightforward when imaging gold conjugates on living cells immersed in a tissue culture medium of low refractive index. Here reflections from the cell itself, particularly from the plasma membrane, are significant and make it difficult to identify individual 40 nm gold particles. Confocal bright-field transmission imaging of colloidal gold particles would be expected to result in clearer images from living cells, similar to those obtained by nanovid video-enhanced contrast microscopy described in Section IF above and in Chapter 6, De Brabander et al., this volume.

2. Surface replicas. Platinum-carbon surface replication is a widely used technique for high-resolution transmission electron microscopic (TEM) imaging of biological tissues and macromolecules after freeze-fracture, deep-etching, or spray-drying preparative techniques. Such surface replicas are excellent specimens for confocal reflection imaging, enabling the three-dimensional topographical detail of the surface to be determined with greater ease than by parallax calculations from stereoscopic pairs of transmission electron micrographs obtained by specimen tilting. This opens the way for sophisticated correlative microscopy. A cell may first be imaged by confocal microscopy in the living state after colloidal gold or fluorescent immunolabeling of its surface antigens. It may then be rapidly frozen, freeze-fractured, and replicated. This replica, after removal of the biological material and mounting on an EM grid, may then be studied by high-resolution TEM to determine ultrastructural detail. If the label-fracture preparation technique is used, the positions of surface immunogold particles may also be determined. After oil immersion, the same replica may subsequently be imaged by confocal reflection microscopy to determine its three-dimensional topography. With suitable image processing, these different image data may finally be combined.

By the same token, scanning electron microscope specimens prepared by gold sputter coating provide excellent confocal reflection images, with similar possibilities for correlative microscopy.

3. Fluorescence. Since, to a first approximation, the confocal fluorescence image contains only information from the focal plane, that from out-of-focus planes having been excluded, it may appear less bright than a conventional wide-field fluorescence image of the same specimen, although with an enhanced signal-to-noise ratio. It is thus often appropriate to label specimens more intensely than would be suitable for conventional fluorescence microscopy by using higher concentrations of reagents or employing a triple-layer labeling protocol to achieve greater amplification. To achieve this while avoiding spurious artifactual cross-reactivities, it is important that the antisera employed be affinity-purified, that the number of fluorochrome conjugates per immunoglobulin molecule be optimized, and that appropriate blocking steps and negative controls be used.

Since fluorochrome photobleaching will be a significant problem if the specimen is to be scanned many times in order to collect optical sections, confocal observation of fixed specimens is best undertaken after mounting in the presence of a bleach retardant such as DABCO (1,4 diazobicyclo 2,2,2 octane), N-propyl gallate or p-phenylene diamine. With living systems, this strategy cannot be used, so that it is important to minimize the excitation intensity by reducing the laser power or inserting neutral density filters into the illumination path, and to avoid unnecessary exposure of the specimen. If a three-dimensional data set is to be collected, the serial sections should be collected from front to back, so that the inevitable progressive fading generates an appropriate depth cue, by causing the rear of the object to appear fainter than the front. Prior to any subsequent quantitative measurements from such images, however, correction of the individual pixel intensities in each section both for depth attenuation and bleaching must be correctly undertaken, as detailed by Rigaut and Vassy [1991].

In comparison with conventional fluorescence microscopy, the instantaneous light intensity during confocal laser scanning microscopy is far higher, sufficient in many cases to saturate the fluorochromes present in the illuminated spot, but any one point in the specimen is only illuminated for a brief fraction of the time. The exact relationship between photobleaching rate, light intensity, and scanning dwell time is controversial, and for a fuller discussion of this and many other practical aspects of confocal imaging, the reader is referred to The Handbook of Biological Confocal Microscopy (2nd ed.), edited by James Pawley [1990] and to Chapters 10 to 14 in this volume.

F. Recent Technical Developments

Although the first single beam confocal scanning system was patented by Minsky in 1957 [Minsky, 1957, 1988], and the original Nipkow disk tandem scanning microscope was described by Egger and Petran in 1967 [Egger and Petran, 1967; Petran et al., 1968], the research community was slow to realize the significance of confocal microscopy, particularly its usefulness for noninvasive optical sectioning during fluorescence imaging. The development of commercial confocal laser scanning systems was also dependent upon the advent of affordable digital image processing hardware and microcomputers. It was thus not until about 12 years ago that the first biological applications of confocal microscopy were published, and the real growth of confocal fluorescence microscopy as a tool for biological research can be dated from the marketing of the first confocal laser scanning systems in 1987. A bibliography of the early development of confocal microscopy may be found in Shotton [1989].

The present-day situation is entirely different. The importance of confocal imaging is universally recognized, and new technical developments are being made at an ever-increasing rate, a tribute to the creative imaginations of those working in this field, the best of these ideas reaching the marketplace with extraordinary rapidity. These developments fall into four principal categories: (1) new laser wavelengths, contrast generation systems or imaging modes that extend the functionality of the present generation of single beam and Nipkow disk scanning systems; (2) new confocal microscope configurations for "real-time" confocal imaging; (3) novel super-resolving microscopes that achieve significant advances in terms of image spatial resolution, both in-plane and axial, breaching the traditional resolution limitations of the conventional light microscope; and (4) developments in digital image processing and display procedures, particularly designed to handle the massive amounts of three-dimensional image data generated by confocal optical sectioning. Because it is important for would-be confocal microscopists to be aware of these opportunities, the latter category of developments is discussed in Chapter 2, while the first three, concerned with image acquisition, are briefly summarized below.

1. New functionality. a. Lasers. The laser most commonly supplied with a single beam scanning confocal system is a low-cost air-cooled argon ion laser, which has two strong laser lines at 488 and 514 nm. The former is very well suited for exciting fluorescein and similar blue-sensitive fluorochromes such as bodipy (Molecular Probes, Eugene, OR, USA). The 514 nm line, although a long way from the excitation peak for the longer-wavelength dyes such as rhodamine which are commonly used with fluorescein or bodipy in double immunolabeling protocols, may be used for simultaneous excitation of fluorescein and rhodamine. This has the advantage that, since a single excitatory wavelength is employed, there is no risk, as with separate excitation of two dyes using 488 nm and 568 nm light (see below), that chromatic aberration has caused the excitation to occur at different depths in the specimen. Few ratio imaging dyes can be excited by the 488 nm and 514 nm wavelengths. There is thus a present tendency to equip confocal microscopes with additional or alternative lasers, permitting a wider wavelength choice (Table IV). These include an air-cooled argon/krypton mixed-gas laser giving lines at 488 nm, 568 nm, and 647 nm, well suited for multiple labeling studies using fluorescein or bodipy, a rhodamine dye or Texas Red, and one of the red-excited dyes such as Ultralite T680 or Cy-5 (Table V). Such multispectral imaging

TABLE IV. Principal Emission Lines of Low-Power Air-Cooled Gas Lasers Useful for Confocal Scanning Light Microscopy

Laser	Wavelength (nm)			
	UV	Blue	Green	Red
Helium cadmium	322	442		
Helium cadmium (RGB)		442	534, 538	636
Argon ion		488	514	
Argon-krypton mixed gas		488	568	647
Helium neon (green)			543	
Helium neon (red)				633

TABLE V. Excitation and Emission Data for Some Commonly Used Fluorochromes Useful for Confocal Scanning Light Microscopy

Fluorochrome	Excitation Maximum (nm)	Emission Maximum (nm)
UV excitation, blue emission		
Cascade blue	375, 398	424
Blue excitation, yellow-green emission		
Chromomycin A$_3$[1]	458	590
Fluorescein[2]	494	520
Bodipy[3]	505	512
Lucifer yellow	428	540
Green excitation, red emission		
7-Actinomycin D[4]	523	647
Tetramethyl rhodamine	541	572
Lissamine rhodamine B[5]	567	584
Rhodamine X	578	604
Texas Red	596	615
CY3.18	554	568
Red excitation, far red emission		
CY5.18	649	666
Ultralite T680	656, 675	678

[1]May be used to label DNA, employing the 488 nm line of an argon ion laser for excitation.
[2]The long red tail of the fluorescein emission may cause bleedthrough into the red detection channel. Bodipy is superior in this respect.
[3]Very small Stokes shift, requiring appropriate filters and dichroic mirror.
[4]Selectively stains GC-rich double-stranded DNA, giving chromosome banding.
[5]The excitation spectrum of lissamine rhodamine B is very broad, with a secondary peak at 523 nm, making it far superior to tetramethyl rhodamine for single-wavelength excitation at 514 nm, when used for double labeling in conjunction with fluorescein.

Sources: Chromamycin A$_3$: Sigma Chemical Co. Ltd., Fancy Road, Poole BH17 7BR, UK. Phone (0202) 733114.

Lissamine conjugates: Jackson Immunoresearch Labs, Inc., 872 West Baltimore Pike, P.O. Box 9, West Grove, PA 19390, USA. Phone (800) 367-5296.

CY3 and CY5: Biological Detection Systems Inc., 4617 Winthrop Street, Pittsburgh, PA 15213, USA. Phone (412) 621-3143.

Ultralite T680: Ultra Diagnostic Corp., 4526 11th Ave. NE, Seattle, WA 98105, USA. Phone (206) 545-7807.

Other fluorophors: Molecular Probes Inc., P.O. Box 22010, 4849 Pitchford Ave., Eugene, OR 97402, USA. Phone (503) 344-3007.

is particularly convenient when high-performance double or triple bandpass excitation and emission filters and a double or triple dichroic mirror (from Omega Optical, Brattleboro, VT, USA) are used to avoid the necessity to change filter blocks, and should give good results provided that the objective employed is well corrected for chromatic aberration at all six wavelengths involved, the three excitation wavelengths and the three associated emission wavelengths (see Section IID, above). Alternatively, there is a low-cost helium neon laser that emits green light at 543 nm, in addition to the more usual 633 nm red helium neon laser. Blue (442 nm) and ultraviolet (325 nm) wavelengths, which may be obtained separately or simultaneously from an air-cooled helium cadmium laser, permit excitation of other dyes, including the common DNA stains such as DAPI. The 325 nm line is ideally suited for single-wavelength excitation of the calcium emission ratio imaging dye INDO-1, the calcium-bound form of which has an absorption maximum at 332 nm and a maximal emission ratio when imaged at 405 and 480 nm (Table V). To avoid the potential chromatic aberration problems that may be experienced when using a multiline laser, several single-line lasers may be combined on the same microscope by beam steering mirrors and appropriate dichroic filters, their separate optical paths providing opportunity to insert correction lenses where necessary, particularly to compensate for the chromatic path length difference between UV and visible light using conventional optics [Cogswell et al., 1990a; Fricker and White, 1992]. Lasers are generally used for single beam scanning confocal microscopy primarily because of their brightness and suitability as effective point sources, which results in a bright focused scanned spot and strong fluorescence excitation. However, Hell et al. [1991] have recently reported the use of a short-arc xenon lamp with an extremely high source brightness in the illumination source for a single beam scanning white-light confocal microscopy, thereby overcoming the wavelength limitations of using lasers.

b. Combined confocal transmission and reflection imaging. Confocal transmission imaging has been practiced for many years using stage-scanning laboratory instruments built upon optical benches [reviewed by Shotton, 1989], but as yet no commercial instruments offering confocal transmission imaging have been produced. This is largely because of the difficulties of maintaining the alignment of two objective lenses along the same optical axis and focused upon the same point, and of accommodating such an imaging system within a conventional widefield compound microscope. However, this optical arrangement has advantages, particularly for imaging colloidal gold particles (see Section IF above). Recently, Dixon et al. [1991] published a design for a beam scanning confocal microscope in which, by using matched infinity-corrected objectives and by folding the transmission optical back into the illumination/reflection pathway, reflection images may be obtained from both upper and lower surfaces of the specimen, and interesting image combinations may be made of the transmitted and reflected beams.

c. Differential interference contrast imaging. Many varieties of confocal differential contrast generation techniques have been developed for stage scanning single beam confocal microscopes (see Shotton [1989] and papers in Elder [1990] for discussion), although none of these are conveniently available on commercial beam scanning systems. Corle and Kino [1990] have reported an elegant adaption of the Nipkow disk confocal scanning microscope detailed in Chapter 14 (Masters and Kino) to permit reflection Nomarski differential interference contrast (DIC) imaging. There is no reason why the single beam confocal transmission microscope of Dixon et al. [1991] mentioned above could not be equipped with Wollaston prisms to permit true Nomarski DIC transmission imaging.

d. Confocal spectroscopy. Two groups of workers [Benedetti et al., 1990; Bowron et al., 1991] have reported confocal designs in which the spectral response of the specimen can be measured. Benedetti's instrument has a slit illuminator, using either arc lamp or laser light, and a slit detector [see Section IIF2 below], and performs a one-dimensional scan. By reflecting the resultant line of light off a polychromator grating and imaging it using a two-dimensional CCD array, a rectangular image of the original line is obtained by spectral dispersion in the orthogonal direction, thus yielding an x,λ image. This may be obtainable either by transmission or epi-illumination of the specimen. Bowron et al.'s microscope is a two-dimensional point-scanning instrument in which the photoluminescence or fluorescence from the laser-excited specimen, having passed a blocking filter to remove stray laser light, is reflected from a diffraction grating to the confocal detector aperture. The microscope collects a single x,y image at a fixed wavelength, and performs spectral analysis by repeated imaging at different wavelengths, thus generating a series of wavelength-specific images that comprise an x,y,λ three-dimensional data set for each focal z-plane.

e. Two-photon excitation. Watt Webb and his colleagues [Webb, 1990] have pioneered the use of pulsed red (630 nm) light of extremely high intensity from a mode-locked dye laser during single beam scanning to bring about the simultaneous absorption of two photons by a chromophore normally excited by a single photon of much shorter wavelength. Because the two-photon excitation rate is proportional to the square of the incident intensity, the fluorescence emissions excited by this process occur only at the plane of focus where the waist of the excitation beam is smallest (Fig. 1B). This results in axial resolution equivalent to that of the best confocal microscopy, without the need for a confocal detector aperture.

2. New microscope configurations for "real-time" confocal imaging. In the Nip-

kow disk confocal scanning systems mentioned in Section IIA above, the rapid scanning of the specimen by an array of illuminating light beams results in an apparently continuous optical image being formed, visible through the eyepieces by the human eye, or detectable by an attached video or CCD array camera. Chapters 13 and 14 detail the two principal types of such Nipkow disk confocal microscopes. Several varieties of this type of microscope are now on the market, their chief advantages being their ability to deliver an apparently continuous image and their use of white-light illumination to excite the full range of fluorochromes and to observe natural colors. Recently, a variety of other "real-time" CSLMs have been developed, falling into four classes.

a. Fast beam scanning using mirrors. The first, developed by Roger Tsien specifically for video-rate UV laser confocal calcium ratio imaging [Tsien, 1990], is a classical single beam scanning instrument, in which high-speed scanning and descanning[2] has been achieved by the use of a resonant galvanometer mirror. The data acquisition electronics include a cosinusoidally modulated pixel clock to generate equally spaced pixels during the sinusoidal scan of the mirror, and a first-in last-out data buffer to reverse the direction of every other image line, thus permitting faster bidirectional data collection by utilizing both the forward and the return swing of the mirror, with a first-in first-out data buffer being used to generate an equal time delay for the data collected during the forward swing. This instrument has the advantage that, by using point confocal laser scanning, optical sectioning performance is not sacrificed. Correction for UV chromatic

---

[2]Descanning in a single beam scanning microscope is the passage from the specimen of the fluorescence emission or reflection excited by the scanned beam, back along the same optical path as the scanned beam to the scanning optics, whose action brings this return beam to a stationary on-axis path coincident with the original illuminating beam (see Fig. 8).

aberration may be made easily by adjusting the axial position of the confocal detector aperture.

b. Fast beam scanning using acousto-optic deflectors. Another means of rapid single beam laser scanning uses an acousto-optic deflector (AOD) to scan the fastest moving *x*-axis of the two-dimensional raster scan, the *y*-axis being scanned by an oscillating mirror. This design, developed by Draaijer and Houpt [1988] and now available from Tracor Northern/Noran, is detailed by them in Chapter 12 of this volume, and thus will not be elaborated upon here. Suffice it to say that when used in reflection mode, the same AOD is used to descan the beam to a confocal detector aperture, resulting in good two-dimensional confocal imaging. For fluorescence this is not possible, since the AOD is dispersive and hence the descanning optical path is wavelength-specific. Instead, the beam, still oscillating in *x* after being descanned by the *y*-scanning mirror, is allowed to pass through a confocal detector slit before falling onto a PMT detector. Thus confocal out-of-focus blur rejection is incomplete, since the slit passes some of the light that would normally be excluded by a small circular detector aperture. Nevertheless, reasonably good optical sectioning is obtained. A similar optical system, also employing an AOD and descanning along one axis only, but using a linear CCD array rather than a slit and a PMT for data acquisition, has been developed by Lasertec, with similar confocal performance.

Goldstein and his colleagues [Goldstein et al., 1990; Goldstein and Hubin, 1990] have taken this use of AODs one step further, using two AODs oriented with their deflection axes at right angles to achieve scanning in both *x* and *y*, and employing an image-dissector tube to accomplish descanning in both directions and hence achieve true confocal imaging in a "no-moving-parts" instrument, now in its second-generation design [Goldstein and Hubin, 1991].

c. Full objective aperture one-dimensional slit scanning. Brakenhoff and Visscher [1990,

1991] have developed a rapid scan rate confocal microscope operating on a quite different principle (Fig. 9). Here, rather than using AODs to rapidly scan a point of light through the specimen in a two-dimensional raster, a single oscillating mirror is used to scan a focused slit of laser light through the focal plane of the specimen in a single direction. The resulting linear fluorescence or reflection signal obtained is descanned by the same mirror and is focused to a slit detector aperture where confocal out-of-focus blur rejection occurs. This now-stationary line image is then reflected off the back surface of the same mirror and focused onto the two-dimensional faceplate of a CCD array camera. The result of this arrangement is that the optical image is scanned over the CCD array, from which it can subsequently be read out in electronic form, in exact register with the specimen scanning. Because the CCD is an integrating device, the specimen scan rate can be independent of the CCD readout rate, enabling very high optical scan rates to be achieved. The CCD has additional advantages of possessing higher quantum efficiency and better red sensitivity than photomultiplier tubes. By substituting the human eye at the position of the CCD, a "direct-view" confocal image can be observed. This form of the instrument is being marketed by Meridian Instruments as a slot-in accessory for a standard compound microscope. The same basic design, but with circular illuminating and confocal detector apertures rather than slits, and with the double-sided mirror being used for scanning along both the x- and y-axes, forms a conventional slow-scan single laser beam scanning confocal microscope, albeit still with the advantages of CCD imaging [Brakenhoff and Visscher, 1990].

d. Divided objective aperture one-dimensional slit scanning. Koester [1990, 1991] has developed a similar slit scanning "direct-view" confocal microscope for ophthalmological investigations, using coupled reflections from an oscillating three-sided

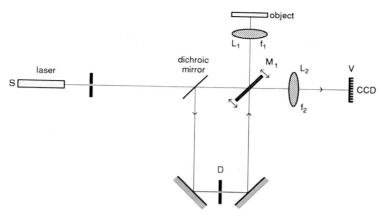

Fig. 9. Schematic layout of a CCD-based confocal microscope. Light originating from light source S, either a circular aperture, a slit, or some other illumination pattern, is scanned over the specimen by the front surface of mirror M_1. The emitted fluorescence light is descanned by the same mirror surface and is made to pass through the confocal detection aperture D, placed in a plane optically conjugate with S. This detection aperture D is subsequently imaged by lens L_2 onto the CCD array. In the case of uniaxial slit scanning, the confocal image is generated as a charge pattern on the CCD by scanning this pattern over the array using the reverse side of scan mirror M_1. The optical system is designed such that the respective mirror surfaces of M_1 are located in pupil planes of the lenses L_1 and L_2, respectively (or at conjugate planes thereof). The use of a single two-sided mirror for both specimen scanning and guidance of the detected light to the CCD array ensures perfect geometrical registration between the specimen and the CCD image. (Reproduced with permission from Brakenhoff and Visscher [1990].)

mirror, in which the microscope objective is divided, a stationary illuminating slit being imaged through one half of the objective, while the reflected rays pass through the other. An equivalent optical arrangement, but based on the rotating disk principle, was first patented by Baer in 1970 and is described by him in two brief publications [Baer, 1989, 1991]. This instrument has, in place of a Nipkow disk, a rotating assembly consisting of an illuminating slit, a mirror reflecting the illuminating light into one half of the objective only, and an imaging slit. Both use white-light illumination and produce an apparently continuous image of high intensity. While this divided objective aperture illumination arrangement sacrifices lateral resolution anisotropically by halving the NA of the objective in one dimension, it permits confocal rejection of out-of-focus blur by an entirely new principle. In all the microscope configurations hitherto considered, the full aperture objective passes both in-focus and out-of-focus information, which is only differentiated at the confocal detector aperture in the primary image plane. Upon moving progressively out of focus, light from a point source is gradually defocused at the image plane, so that progressively less and less passes through the confocal detector aperture (whether this be a circular aperture or a slit) to reach the detector. In contrast, in these divided objective aperture slit scanning confocal microscopes, the specimen is illuminated by an inclined semicone of light that comes to a focused line, which is the demagnified image of the illuminating slit at the focal plane (Fig. 10). Reflections or fluorescence emissions excited along this line of illumination can only be imaged by escaping along another inclined semicone, mirror-symmetrical with the first, into the opposite half of the objective. The restricted three-dimensional volume within the specimen where these two focused semicones inter-

A   Vertical section showing light paths

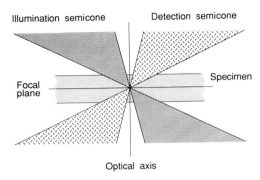

B   Enlargement of the focal diamond
         within the specimen

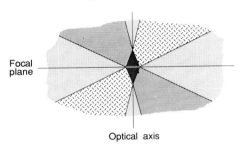

 Diamond of intersection between illumination and
detection semicones from divided objective lens

sect is the only region from which light can be collected (unless it undergoes secondary scattering or reflection). No light from illuminated regions of the specimen which are not in the focal plane is able to reach the detector, since these regions lie completely outside the detection semicone. Thus there is no need of a confocal detector aperture. The axial resolution in these microscopes is defined by the geometry of the optics, namely the width of the illuminated line in the focal plane and the angle of inclination of the least off-axis rays in the semicones with respect to the optical axis of the microscope, and is significantly better than in an equivalent full objective aperture confocal microscope. The images Koester [1990] has obtained from the very weakly reflecting corneal epithelium using this system are remarkable for their clarity. Although it has not yet been implemented, there is no reason in principle why similar semiconical illumination could not be implemented on other geometries of CSLM, by the use of appropriately positioned half-apertures at illuminating and imaging planes conjugate with the back focal plane of the objective.

In Wilson [1990b], Wilson and Hewlett [1990], and Chapter 10 of this volume, Tony

Fig. 10. The principle of confocal imaging by slit scanning through a divided high numerical aperture objective lens. **A:** The diagram shows schematically a vertical section through the specimen at right angles to the orientation of the demagnified image of the scanning slit in the focal plane, which is rendered in this diagram as the intersection point in the center of the specimen. The specimen is illuminated by light from an illuminating slit, transmitted by one half of the divided objective (not shown), which forms an almost complete semicone of light (darkly shaded) that comes to a focus as the line image of the scanning slit. Fluorescence emissions or reflections from the specimen can contribute to the image if they occupy the detection semicone (lightly shaded) entering the opposite half of the divided objective lens. **B:** A close-up of the focal plane, showing the intersection of the illuminating and detection semicones. The only region of the specimen which is both illuminated and in the detection light path is the illuminated line in the focal plane, of width determined by the reso-

lution of the objective lens and the width of the scanning slit (shown as a white line), and regions within the small intersection diamond above and below (shown darkly cross-hatched). Fluorescence emissions or reflections from elsewhere in the illumination semicone have no possibility of coinciding with the detection light path and thus contribute nothing to the image. Confocality is thus achieved without the need for a confocal detector aperture, but at the expense of some anisotropic loss of lateral spatial resolution due to the use of only half the objective lens for each light path. The axial resolution is determined by the shape of the overlap zone (diamond of intersection) between the illuminating and detection semicones, which is a function both of the width of the demagnified focused image of the scanning slit at the focal plane and the angle of inclination of the least off-axis rays in the semicones with respect to the optical axis of the microscope. For clarity, objective, slide, coverglass, and refraction effects are not shown (compare with Fig. 1B and Fig. 7).

Wilson discusses the degree to which the use of a slit rather than circular confocal detector aperture in a full objective aperture scanning microscope compromises the confocal performance and optical sectioning capabilities of slit-scanning and slit-detection systems. Fortunately, the short answer is that this compromise degrades the confocal axial resolution by no more than a factor of 2, although resulting in some anisotropy in the confocal response, as might be expected. Such instruments are thus well suited for obtaining reasonable blur-free images where the ability to view a continuous image directly, or to video record it, is more important than obtaining the highest possible confocal optical performance.

3. Confocal super-resolution. Even greater super-resolution advantages than those achievable by imaging with a single small on-axis confocal detector aperture may be obtained using a variety of innovative novel microscope designs, still in prototype. In the first, the light reflected from a single laser beam is imaged by not one but an array of detectors which record both the on-axis illumination, as is normally recorded, and also the off-axis light that is normally excluded by the confocal detector aperture at the image plane. The additional information obtained is processed by new algorithms, leading to a theoretical increase in lateral resolution to twice the Rayleigh criterion for coherent illumination, and of four times the Rayleigh criterion for incoherent illumination. While this approach is still in its formative stages, such a resolution enhancement has very recently been demonstrated experimentally for a low numerical aperture system [Bertero et al., 1990; Pike et al., 1990; Walker et al., 1991]. In other recent developments, Kino and Chim [1990] and Hobbs and Kino [1990] have reported new microscope configurations which permit independent phase and amplitude data to be obtained from the specimen. This allows subsequent digital filtering of the signals, permitting a doubling of the spatial resolution of the system, while retaining good

axial resolution. Such digital filtering is considerably more accurate and convenient than attempts to increase resolution by apodizing the pupil function of the microscope with physical filters or annular apertures [Hegedus, 1990].

4. Conclusions. Future developments in confocal microscopy are hard to predict, being limited only by the imagination and resourcefulness of research microscopists, and the ability of vendors to develop their ideas into timely and affordable products. However, it seems to me that confocal microscopes for biological research will, in the immediate future, fall into one of two classes. On one hand, there will be a variety of low-cost analogue "direct-view" or "real-time" confocal accessories for conventional broad-field research microscopes, utilizing slit scanning or Nipkow disk scanning to achieve moderately good confocal optical sectioning performance. These will be used primarily for direct observation or video recording of dynamic events in living systems, where confocality will be used primarily to rid the image of out-of-focus blur, thus enabling the users to obtain two-dimensional views of the living world that were previously unobtainable. I believe such microscopes are likely to become as common in research environments as are wide-field epifluorescence microscopes today. Separately, there will be a smaller number of dedicated high-resolution slow-scan single beam point-scanning confocal systems, equipped with a variety of lasers and incorporating or attached to dedicated compound microscopes, that will be used to collect accurate three-dimensional data sets by serial optical sectioning. These will be interfaced to high-performance workstations equipped with software for state-of-the-art scientific visualization of these 3D images, capable of interactive processing, deconvolution, analysis, rendering, animation, and stereoscopic display of volume data, as described in Chapters 2 and 11 of this volume.

REFERENCES

Aikens RS, Agard DA, Sedat JW (1989): Solid-state imagers for microscopy. In Wang Y-L, Taylor DL (eds): "Fluorescence Microscopy of Living Cells in Culture. Part A: Fluorescent Analogs, Labeling Cells and Basic Microscopy." Methods Cell Biol 29:291–314.

Allen RD (1985): New observations on cell architecture and dynamics by video-enhanced contrast optical microscopy. Ann Rev Biophys Chem 14:265–290.

Arndt-Jovin DJ, Robert-Nicoud M, Kaufman SJ, Jovin TM (1985): Fluorescence digital imaging microscopy in cell biology. Science 230:247–256.

Baer SC (1970): Optical apparatus providing focal-plane-specific illumination. US Patent 3,547,512.

Baer SC (1989): Tandem scanning slit microscope. Proc SPIE 1139:99–101.

Baer SC (1991): The divided aperture moving slit confocal microscope. Scanning 13 (Suppl 1):111.

Benedetti PA, Evangelista V, Guidarini D, Vestri S (1990): Spectral imaging in confocal-line microscopy. Trans R Microsc Soc 1:271–274.

Bertero M, Boccacci P, Brakenhoff GJ, Malfanti F, van der Voort HTM (1990): Three-dimensional image restoration and super-resolution in fluorescence confocal microscopy. J Microsc 157:2–20.

Bowron JW, Damaskinos S, Dixon AE (1991): A new spectrally resolved confocal scanning laser microscope. Scanning 13 (Suppl 1):76–77.

Brakenhoff GJ, Visscher K (1990): Novel confocal imaging and visualization techniques. Trans R Microsc Soc 1:247–250.

Brakenhoff GJ, Visscher K (1991): Confocal imaging with bilateral scanning and array detectors. Scanning 13 (Suppl 1):65–66.

Bright GR, Fisher GW, Rogowska J, Taylor DL (1989): Fluorescence ratio imaging microscopy. In Taylor DL, Wang Y-L (eds): "Fluorescence Microscopy of Living Cells in Culture. Part B: Quantitative Fluorescence Microscopy—Imaging and Spectroscopy." Methods Cell Biol 30:157–192.

Castleman KR (1979): "Digital Image Processing." Englewood Cliffs, NJ: Prentice-Hall.

Castleman KR (1987): Spatial and photometric resolution and calibration requirements for cell image analysis instruments. Appl Optics 26:3338–3342.

Cogswell CJ, Hamilton DK, Sheppard CJR (1990a): Full-colour confocal reflection imaging using red, green and blue lasers with an on-axis scanning microscope. Trans R Microsc Soc 1:267–270.

Cogswell CJ, Sheppard CJR, Moss MC, Howard CV (1990b): A method for evaluating microscope objectives to optimize performance of confocal systems. J Microsc 158:177–185.

Corle TR, Kino GS (1990): Differential interference contrast imaging on a real time confocal scanning optical microscope. Appl Optics 29:3769–3774.

DeBiasio R, Bright GR, Ernst LA, Waggoner AS, Taylor DL (1987): Five-parameter fluorescence imaging: Wound healing of living Swiss 3T3 cells. J Cell Biol 105:1613–1622.

De Brabander M, Geuens G, Nuydens R, Moeremans M, De Mey J (1985): Probing microtubule-dependent intracellular motility with nanometer particle video ultramicroscopy. Cytobios 43:273–283.

De Brabander M, Nuydens R, Geuens G, Moeremans M, De Mey J (1986): The use of submicroscopic gold particles combined with video contrast enhancement as a simple molecular probe for the living cell. Cell Motil Cytoskeleton 6:105–113.

De Brabander M, Nuydens R, Geerts H, Nuyens R (1989): Detection and use of gold probes with video-enhanced contrast slight microscopy. In Verkleij A, Leunissen J (eds): "Immunogold Staining in Cell Biology." Baton Route, FL: CRC Press, pp 217–232.

De Mey J (1983): Colloidal gold probes in immunocytochemistry. In Polak JM, Van Noorden S (eds): "Immunocytochemistry." Bristol, London, Boston: Wright PSG, pp 82–112.

De Mey J, Lambert AM, Bajer AS, Moeremans M, De Brabander M (1982): Visualization of *Haemanthus* endosperm with the immuno-gold staining method. Proc Natl Acad Sci USA 79:1898–1902.

Dixon AE, Damaskinos S, Atkinson MR (1991): A scanning confocal microscope for transmission and reflection imaging. Nature 351:551–553.

Draaijer A, Houpt PM (1988): A standard video-rate confocal laser-scanning reflection and fluorescence microscope. Scanning 10:139–145.

Eggar MD, Petran M (1967): New reflected-light microscopy for viewing unstained brain and ganglion cells. Science 157:305–307.

Elder HY (1990) (ed): "MICRO '90." Trans R Microsc Soc, Vol 1. Bristol, Philadelphia, New York: Adam Hilger.

Ellis GW (1985): Microscope illuminator with fiber optic source integrator. J Cell Biol 101:83a.

Fricker MD, White NS (1992): Wavelength considerations in confocal microscopy of botanical specimens. J Microscopy 166:29–42.

Goldstein SR, Hubin T (1990): Improved acousto-optic laser scanning system for a video-rate confocal microscope with no moving parts. Trans R Microsc Soc 1:255–258.

Goldstein SR, Hubin T (1991): Second generation video-rate no-moving-parts confocal microscope. Scanning 13 (Suppl 1):77–78.

Goldstein SR, Hubin T, Rosenthal S, Washburn C (1990): A confocal video-rate laser-beam scanning reflected light microscope with no moving parts. J Microsc 157:29–38.

Hansen EW (1986): Modulation transfer function analysis in video microscopy. In Inoué S (ed): "Video Microscopy." New York: Plenum Press, pp 467–475.

Hegedus, ZS (1990): Pupil filters in confocal imaging. In Wilson T (ed): "Confocal Microscopy." London: Academic Press, pp 171–183.

Hell S, Witting S, v. Schickfus M, Wijnaendts van Resandt RW, Hunklinger S, Smolka E, Neiger M (1991): A confocal beam scanning white-light microscope. J Microsc 163:179–187.

Herman B, Jacobson K (1990) (eds): "Optical Microscopy for Biologists." New York: Wiley-Liss.

Hobbs PCD, Kino GS (1990): Generalizing the confocal microscope via heterodyne interferometry and digital filtering. J Microsc 160:245–264.

Inoué S (1986): "Video Microscopy." New York: Plenum.

Kino GS, Chim SSC (1990): Mirau correlation microscope. Appl Optics 29:3775.

Koester CJ (1990): High efficiency optical sectioning with confocal slits. Trans R Microsc Soc 1:327–332.

Koester CJ (1991): The scanning slit confocal microscope. Scanning 13 (Suppl 1):37.

McNeil PL (1989): Incorporation of macromolecules into living cells. In Wang Y-L, Taylor DL (eds): "Fluorescence Microscopy of Living Cells in Culture. Part A: Fluorescent Analogs, Labeling Cells and Basic Microscopy." Methods Cell Biol 29: 153–174.

Minsky M (1957): US Patent 3013467. Microscopy apparatus, Dec 19, 1961 (Filed Nov 7, 1957).

Minsky M (1988): Memoir on inventing the confocal scanning microscope. Scanning 10:128–138.

Minta A, Kao JPY, Tsien RY (1989): Fluorescent indicators for cytosolic calcium based on rhodamine and fluorescein chromophores. J Biol Chem 264:8171–8178.

Nederlof PM, Robinson D, Abuknesha R, Wiegant J, Hopman AHN, Tanke HJ, Raap AK (1989): Three colour fluorescence in situ hybridization for the simultaneous detection of multiple nucleic acid sequences. Cytometry 42:87.

Nederlof PM, van der Flier S, Wiegant J, Raap AK, Tanke HJ, Ploem JS (1990): Multiple fluorescence in situ hybridization. Cytometry 11:126.

Pawley J (ed) (1990): "Handbook of Biological Confocal Microscopy" (2nd ed). New York: Plenum Press.

Petran M, Hadravsky M, Eggar MD, Galambos R (1968): Tandem-scanning reflected-light microscopy. J Opt Soc Am 58:661–664.

Pike ER, Davies RE, Walker JG, Young MR (1990): An introduction to singular systems with applications to confocal scanning microscopy. J Microsc 160:107–114.

Ploem JS, Tanke HJ (1987): "Introduction to Fluorescence Microscopy." Royal Microscopical Society Microscopy Handbook 10. Oxford: Oxford University Press.

Reynolds GT (1968): Image intensification applied to microscope systems. Adv Optical Electron Microsc 2:1–40.

Reynolds GT (1972): Image intensification applied to biological problems. Q Rev Biophys 5:295–347.

Reynolds GT, Taylor DL (1980): Image intensification applied to light microscopy. BioScience 30:586–592.

Rigaut JP, Vassy J (1991): High-resolution three-dimensional images from confocal scanning laser microscopy. Anal Quant Cytol Histol 13:223–232.

Schnapp BJ (1987): Viewing single microtubules by video light microscopy. Methods Enzymol 134: 561–573.

Sheppard CJR, Cogswell CJ (1990): Three-dimensional image formation in confocal microscopy. J Microsc 159:179–194.

Shotton DM (1988a): Video enhanced light microscopy and its applications in cell biology. J Cell Sci 89:129–150.

Shotton DM (1988b): The current renaissance in light microscopy. II. Blur-free optical sectioning of biological specimens in confocal scanning fluorescence microscopy. Proc R Microsc Soc 23:289–297.

Shotton DM (1989): Confocal scanning optical microscopy and its applications for biological specimens. J Cell Sci 94:175–206.

Shotton DM (1991): Video and opto-digital imaging microscopy. In Cherry RJ (ed): "New Techniques of Optical Microscopy and Microspectrometry." London: Macmillan, pp 1–47.

Song L, Vrolijk J, Verwoerd NP, Bonnet J, Beverloo HB, Nederlof PM, Tanke HJ (1990): An imaging system for time-resolved fluorescence microscopy. Trans R Microsc Soc 1:467–470.

Spencer M (1982): "Fundamentals of Light Microscopy." Cambridge: Cambridge University Press.

Spring KR, Lowy RJ (1989): Characteristics of low light level television cameras. In Wang Y-L, Taylor DL (eds): "Fluorescence Microscopy of Living Cells in Culture. Part A: Fluorescent Analogs, Labeling Cells and Basic Microscopy." Methods Cell Biol 29:270–291.

Taylor DL, Wang Y-L (1978): Molecular cytochemistry: Incorporation of fluorescently labelled actin into living cells. Proc Natl Acad Sci USA 75:857–861.

Taylor DL, Wang Y-L (1980): Fluorescently labelled molecules as probes of the structure and function of living cells. Nature 284:405–410.

Taylor DL, Wang Y-L (1989): "Fluorescence Microscopy of Living Cells in Culture. Part B: Quantitative Fluorescence Microscopy—Imaging and Spectroscopy." Methods Cell Biol 30. London: Academic Press.

Tsay T-T, Inman R, Wray B, Herman B, Jacobson K (1990): Characterization of low-light-level cameras for digitized video microscopy. J Microsc 160:141–159.

Tsien RY (1990): Laser scanning confocal fluorescence microscopy at video rate (30 frames/sec) with dual-wavelength emission ratioing for quantitative imaging of intracellular messages. Proc R Microsc Soc 25(4):S53.

Tsien RY, Poenie M (1986): Fluorescence ratio imaging: A new window into intracellular ionic signaling. Trends Biochem Sci 11:450–455.

Waggoner A, DeBiasio R, Conrad P, Bright GR, Ernst L, Ryan K, Nederlof N, Taylor D (1989): Multiple spectral parameter imaging. In Taylor DL, Wang Y-L (eds): "Fluorescence Microscopy of Living Cells in Culture. Part B: Quantitative Fluorescence Microscopy—Imaging and Spectroscopy." Methods Cell Biol 30:449–478.

Walker JG, Pike ER, Davies RE, Young MR, Bertero M (1991): Superresolving scanning optical microscopy: The practical application of singular system theory. Scanning 13 (Suppl 1):62–63.

Wang Y-L, Heiple J, Taylor DL (1982): Fluorescence analog cytochemistry of contractile proteins. Methods Cell Biol 25:1–11.

Wang Y-L, Taylor DL (1989): "Fluorescence Microscopy of Living Cells in Culture. Part A: Fluorescent Analogs, Labeling Cells and Basic Microscopy." Methods Cell Biol 29. London: Academic Press.

Webb WW (1990): Two photon excitation in laser scanning fluorescence microscopy. Trans R Microsc Soc 1:445–450.

Wehland J, Weber K (1980): Distribution of fluorescently labelled actin and tropomyosin after microinjection in living tissue culture cells as observed with TV image intensification. Exper Cell Res 127:397–408.

Weiss DG, Maile W, Wick RA (1989): Video microscopy. In Lacey AJ (ed): "Light Microscopy in Biology. A Practical Approach." Oxford: IRL Press, pp 221–278.

White N, Lackie PM, Shotton DM (1989): Imaging of immunogold labelled antigens on capping thymocytes by confocal reflection contrast scanning optical microscopy. Cell Biol Int Reports 13:941–948.

Willingham MC, Pastan IH (1978): The visualization of fluorescent proteins in living cells by video intensification microscopy (VIM). Cell 13:501–507.

Willingham MC, Pastan IH (1983): Image intensification techniques for detection of proteins in cultured cells. Methods Enzymol 98:266–283 and 635.

Wilson T (1990a): "Confocal Microscopy." London: Academic Press.

Wilson, T (1990b): The role of detector geometry in confocal imaging. J Microsc 158:133–144.

Wilson T, Hewlett SJ (1990): Imaging in scanning microscopes with slit-shaped detectors. J Microsc 160:115–139.

Wilson T, Sheppard CJR (1984): "Theory and Practice of Scanning Optical Microscopy." London: Academic Press.

An Introduction to Digital Image Processing and Image Display in Electronic Light Microscopy

David Shotton

I. THE DIGITAL IMAGE PROCESSOR

A. Basic Structure

Digital image processors suitable for video-rate enhancement of electronic light microscopic images consist of hardwired electronic components dedicated to storing and manipulating the vast volumes of digital image data derived from a live video signal at high speeds, controlled by a micropro-

Electronic Light Microscopy, pages 39–70 © 1993 Wiley-Liss, Inc.

cessor responsible for the selection and synchronization of the processing functions. Simple image processors are configured as stand-alone units, the operator activating pre-programmed image processing instructions from a menu. Such image processors have the advantages of being easy to use and relatively inexpensive, and are quite adequate for straightforward image-enhancement procedures such as digital contrast optimization, background subtraction or image averaging, where only a video output of the enhanced image is required. However, these lack the flexibility to undertake some of the more sophisticated image processing procedures discussed below, and are usually limited in their ability to store images in digital form for subsequent processing, or to transmit such digital images to other computers for further processing or archival storage. In more sophisticated systems, the image processing hardware is incorporated within, or is connected by a high-speed direct memory access (DMA) link directly to, a host microcomputer or workstation, whose central processor can process the digital image using software algorithms that complement the image processing hardware, and whose magnetic or optical disk storage facilities can be employed to archive digital images. Processing of three-dimensional confocal images is computationally more intensive than that of two-dimensional video images, and for these it is usual to employ image processors of the second type or high-performance workstations capable of executing software-encoded image processing algorithms at significant speeds.

A "real-time" digital image processor capable of operating on an oncoming video signal at 25 or 30 frames per second typically comprises the following basic hardware components: an analogue-to-digital converter, capable of transforming the incoming video signal into a series of binary numbers; a high-speed random-access image memory dedicated to storing the digital images; an arithmetic logic unit capable of performing an image processing operation on a single image or between two images (for example, a stored image and an incoming live image); input and output look-up tables, by means of which the intensity values of the image are modified according to preassigned functions; and an output digital-to-analogue converter, whereby the processed digital image may be converted back to video format for display on a suitable monitor. The functioning of each of these components is briefly considered in turn.

B. The Analogue-to-Digital and Digital-to-Analogue Converters

For processing video signals, the incoming analogue video image is first converted to its digital equivalent by a high-speed "flash" analogue-to-digital converter (ADC), which samples and quantizes the continuous temporally varying analogue signal into a predetermined number of discrete digital measurements. The image thus becomes represented as a two-dimensional array of discrete digitized points—picture elements or *pixels*—each assigned a defined numerical x and y coordinate corresponding to a location in the original analogue image, and a particular numerical brightness or intensity *gray level* that results from the digitization of the analogue image brightness at that location (Figs. 1, 2). Ideally the ADC should operate at a digitizing frequency (the Nyquist frequency) at least twice the limiting horizontal frequency of the video signal, to ensure that the analogue signal is sufficiently sampled to be accurately represented by its digitized samples [Castleman, 1979], as discussed in Chapter 3 of this volume. In practice, many image processors for convenience digitize each incoming video line into a digital image line only 512 (i.e., 2^9) pixels wide, thus somewhat undersampling the video images produced by high-resolution video cameras (see Chapter 1, Shotton, this volume).

The instantaneous analogue voltage values of the video signal [varying from $+0.3$

volt (black) to $+1.0$ volt (white)][1] correspond temporally to the Cartesian locations of particular pixels, and upon digitization each value is normally encoded as a single 8-bit byte[2] of data. Each of the integer values possible for each byte specifies one of 256 possible intensity gray levels for the pixel; successive gray levels thus correspond to an analogue voltage increment of 2.73 millivolts. To reduce digital storage requirements, the x,y coordinates are not specifically encoded, but, since the pixels are recorded sequentially, these may be determined at any time by incremental counting along the raster from the first pixel, obeying the x and y dimensions of the image which are recorded with the image pixel intensity data as part of the image file.

Usually, to accommodate the video image within a 512^2 array of pixels, only the central 512 lines are digitized out of the 575 video lines actually displayed in the raw CCIR video image. In addition, in order to keep the pixels geometrically square, only the central ¾ of each of these lines is digitized (Fig. 2), giving a digitized image with a 1:1 aspect ratio corresponding to a large central square area within the original TV image. The flash ADC has 39 μsec (¾ of 52 μsec) to digitize the central ¾ of a single video line into 512 pixels, and consequently works at a digitizing frequency a little in excess of 13 MHz, digitizing a new pixel every 76.17 nsec. In the American RS-170 (NTSC) standard, where only 485 lines are displayed out of the 525 total lines per frame, all the displayed lines can be digitized into a 512^2 array, albeit with 27 blank

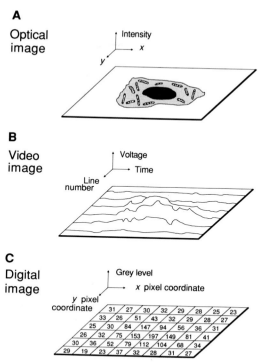

Fig. 1. An illustration of the relationship between an optical image **(A)** and the video image **(B)** and digital image **(C)** that may be derived from it. In the optical image, the intensity varies continuously in the x,y plane of focus. In the video image, this continuity has been sampled in the y direction by the video scan lines. For clarity, only six of the 575 video lines normally displayed in a full video frame (see Chapter 1) are illustrated. To form the digital image, the continuously varying voltage fluctuations of each video line are regularly sampled and quantized, as described in the text and in Figure 2, to yield integer intensity values (gray levels) that correspond to the instantaneous voltage of the video signal at each pixel clock sampling point, and that hence are proportional to the original intensity of the optical image. For clarity, only 8 pixels are shown for each of the six video lines, although in practice each video line is normally digitized into 512 pixels. The illustrated image thus contains only 48 pixels, rather than the 262,144 pixels present in a conventional 512^2 digital image.

---

[1]As in the previous chapter (q.v.), numerical values for video signals refer to the CCIR video standard. Unless otherwise specified, numerical values for digital images refer to 2D 512^2 pixel images, digitized to 8 bits per pixel.

[2]For computational convenience, units of computer memory, each capable of storing one binary unit of information (one *bit*), numerically a zero or a one, are grouped in eights, called *bytes*. One byte (8 bits) is capable of storing 2^8 integer values from 0 to 255. Two bytes ($= 16$ bits $=$ one *word*) can store 2^{16} integer numbers, from 0 to 65,535. 1 kilobyte (Kbyte) $= 1,024$ bytes $= 2^{10}$ bytes $= 2^{13}$ bits of digital storage. 1 Mbyte $= 1,024$ Kbytes $= 1,048,576$ bytes $= 2^{20}$ bytes $= 2^{23}$ bits.

pixel rows. Some image processor designers have chosen to digitize entire video lines into 512 pixels per line, rather than just digitizing the central ¾, giving processed images which fill the screen in an aesthetically pleasing fashion, but in which the pixels are

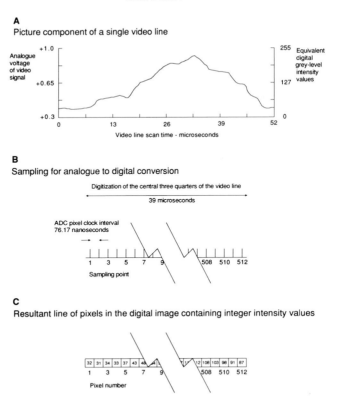

A
Picture component of a single video line

B
Sampling for analogue to digital conversion

C
Resultant line of pixels in the digital image containing integer intensity values

Fig. 2. An illustration of the process of analogue to digital conversion. The information in a single video line is sampled and quantized to form a line of pixels in the digitized image. **A:** The picture component of a single video line, omitting the horizontal blanking interval (compare with Fig. 5 of Chapter 1), showing the intensity information of the original optical image encoded as an analogue voltage signal that fluctuates with time between +0.3 volt (black) and +1.0 volt (white). **B:** The central three-quarters of this video line is sampled by the analogue-to-digital converter at 512 time points each separated by 76.17 nanoseconds. At each of these sampling time points, the voltage of the video line is measured and converted to a corresponding integer gray level between 0 (black) and 255 (white), yielding a line of pixel intensities as shown in **C.** For clarity, only the first and last few sampling points and pixels in the line are shown in B and C, with an exaggerated horizontal scale in comparison with A. When this process is repeated on every video line in the image, a complete two-dimensional digital image is obtained.

rectangular rather than square. More sophisticated image processors employ a slightly larger 768 × 576 pixel array having a 4:3 aspect ratio, in order to give a screen-filling digital image while maintaining square pixels. Rectangular pixels have the disadvantage of complicating any subsequent digital processing procedure which involves image rotations or whole image Fourier transforms (see below), but for many straightforward electronic light microscopic applications such procedures are not required.

In order to output a digitized image from an image processor as a conventional video signal for video cassette recorder (VCR) recording or display on a monitor, a flash digital-to-analogue converter (DAC) reverses the above procedure, any nondigitized area surrounding the processed video image usually being arbitrarily set to black. In Chapter 3, Kenneth Castleman discusses the problems of displaying the digital image on an analogue video monitor without loss of resolution.

The digitizing frequency required to digitize analogue confocal and CCD images is usually less than for video images, since the image readout frame rate is generally less than video rate. On the other hand, the greater dynamic range of the image acquisition hardware frequently permits the intensity value for each pixel to be digitized to an accuracy of 12 or 14 bits [4,096 or 16,384 gray levels, stored for computational convenience as a 2 byte (16 bit) integer]. Such 2D pixel images often comprise part of a three-dimensional image, acquired by serial optical sectioning. In such a three-dimensional spatial (x,y,z) image, the image element is not a pixel but a volume element termed the *voxel*. 3D images collected from living cells at successive *time* points form a four-dimensional (x,y,z,t) image in which the 4D image element is termed the *tixel* [Kriete, 1990].

C. Digital Image Memory

The heart of any video-rate digital image processor is a large amount of dedicated high-speed dynamic random-access memory (RAM), into which an entire incoming video frame can be stored, separately accessed for processing by other components of the image processor *and* output to a monitor, all within the 40 msec cycle time before the arrival of the next video frame, which upon arrival automatically overwrites the preceding frame in memory and thus updates the stored video image. A quarter of a megabyte (256 Kbytes) of such memory is sufficient to store one digitized image of 512^2 pixels, digitized to 8 bits per pixel, while two Mbytes would be required to store a single $1,024^2$ 16-bit image. Such a block of memory is variously termed an *image memory*, a *frame memory*, or a *frame store*, and video-rate image processors require at least two image memories to permit mathematical and logical operations to be performed in real time between the current video frame in memory A and a previously stored image in memory B, the resulting image either being stored in memory C (if a third memory exists) or overwriting the previous contents of memory A or B.

D. Look-up Tables and Pseudocolor Images

En route to and from the image memory, the digitized video signal passes through a high-speed 256-element RAM (usually of 8 bits' depth, but see below) serving as an *input look-up table* (LUT), which can be used in real time to alter the intensity values of the data. To accomplish this, the integer gray level of each incoming pixel is used as an *address* specifying one of the 256 elements in the LUT, and the memory *content* of that element, a variable between 0 and 255 previously loaded into the LUT from the controlling microprocessor or host computer, supplants the input value and becomes the output gray level of that pixel (Fig. 3A). Digital contrast enhancement may thus be achieved "on the fly" at this stage by linear or nonlinear gray-level optimization, in which the input image is modified according to a preassigned function stored in the input look-up table, for instance to selectively enhance the contrast of a particular gray-level zone, or to more efficiently occupy the full dynamic range from black to white. The contrast of a processed image may be similarly optimized by passage through an *output look-up table* (see below). The resulting final image is then converted back to analogue monochrome video signal by the output DAC for display.

Alternatively, a pseudocolor image, in which each monochrome image gray level may be assigned an arbitrarily different color, may be generated via three such look-up tables and three DACs, one for each 8-bit color channel, the resultant video image being displayed on an RGB color monitor.[3] Since the human eye is much bet-

---

[3] An RGB monitor is one which receives a color video signal as three separate "monochrome" inputs, each specifying the luminance of one of the three color guns, red (R), green (G), and blue (B), rather than as a single composite color video signal typical of broadcast TV. Thus, for RGB display of a processed image, three separate output channels are required from the image processor.

A Processing a single pixel grey level through a 256-element look-up table containing a contrast reversal mapping function

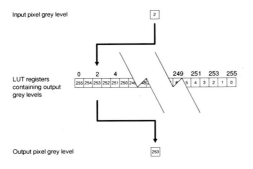

B   A contrast reversal look-up table map

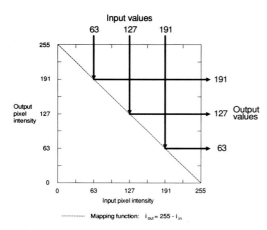

Mapping function: $i_{out} = 255 - i_{in}$

C   A threshold look-up table map

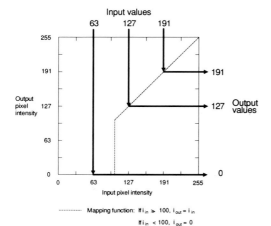

Mapping function: If $i_{in} \geqslant 100$, $i_{out} = i_{in}$

If $i_{in} < 100$, $i_{out} = 0$

ter at discriminating different shades of color than slightly varying gray levels, use of this procedure is extremely useful for highlighting particular regions of a gray-level image, or, by using a suitable juxtaposition of contrasting colors specified in adjacent elements of the LUT, for aiding the interpretation of low-contrast images. However, it is widely acknowledged that indiscriminate use of pseudocolor, while producing aesthetically pleasing images, may confuse, distort, or disguise the real nature of the data, and thus its usage should be undertaken with caution and restricted to situations where it is strictly necessary.

Fig. 3. A: Diagrammatic illustration of the operation of a look-up table. In this example, the LUT has been preloaded with the gray levels corresponding to a contrast reversal mapping function, the gray-level content of each location in the 256-element RAM thus being the binary inverse or complement of that element's address. The gray level of each input pixel, in turn, is used to specify the address of the LUT element whose content provides the gray level of the output pixel, as illustrated for an input gray level of 2. **B,C:** Examples of look-up table maps which specify the manner in which look-up tables are used to modify the pixel gray levels of a digital image. The look-up table map is a square with the input image pixel intensities plotted horizontally and the output image pixel intensities plotted vertically. The mapping function is shown by the dashed line, and specifies the output gray level to be loaded into each element of the 256-element LUT RAM. To use the map to determine the output gray level for any input gray level, extend a vertical line from the input value to the mapping function, and read off the output gray level horizontally from this point of intersection, as shown in both examples. **B:** A contrast reversal look-up table map. **C:** A threshold look-up table map that changes all pixel gray levels below a threshold of 100 to black (0), leaving other intensities unalterd. While these simple examples show look-up table maps in which the mapping functions are composed of linear elements, these functions may be of any shape. Example of curved mapping functions are those for logarithmic conversion of an image, with the mapping function $i_{out} = \log_e i_{in}$, and its antilog equivalent.

E. The Arithmetic Logic Unit

Mathematical and logical operations between two images (in the case of video microscopy, usually between the incoming live video signal and a previously stored image) require a further basic component of the digital image processor, the arithmetic logic unit or ALU, whose operations are outlined in Section IIB below. To avoid truncation errors, if the intensities of the input and output images are stored to a precision of 8 bits, the ALUs operate to a precision of 16 bits, thus allowing multiplication of two image pixel intensities without danger of overflow. Eight-bit input LUTs operate on the data entering the ALU as required, and the 16-bit output LUT may be suitably programmed, for instance, to rescale the output image after a 16-bit multiplication of two 8-bit images into an 8-bit output image which can be accommodated by the 8-bit dynamic range available for the output video signal, or to return absolute values after an image-from-image subtraction. High-resolution systems, in which pixel intensities are stored as 16-bit integers, use 16-bit input LUTs, and 32-bit ALUs and output LUTs.

II. BASIC CONCEPTS IN TWO-DIMENSIONAL DIGITAL IMAGE PROCESSING

A. Single-Image Pixel Point Processing

Pixel point processes are those that apply an identical processing procedure to all the individual pixels in an image, without reference to the intensity values of the neighboring pixels. *Single-image pixel point processes* are those that have a single input image and generate a single output image.

1. LUT mapping functions. Using an LUT, the relationship between the gray levels of the input and the output images is entirely under the operator's control, and may be specified by the general equation $O(x,y) = M[I(x,y)]$, where $I(x,y)$ represents the input gray level of the pixel at the image location specified by coordinates x and y, $O(x,y)$ rep-

resents the output gray level of that pixel, and M is the look-up table mapping function. This relationship can be diagrammatically represented by a graph of input (ordinate) against output (abscissa) gray levels, known as the *LUT map* (Fig. 3B,C). By this simple but ingenious device, a wide variety of real-time alterations of the incoming image may be accomplished by manipulating individual pixels. For instance, if the contents of elements 0, 1, 2, . . . , 255 of the LUT were 255, 254, 253, . . . , 0, respectively (Fig. 3A,B), passage of the image through the LUT would result in contrast reversal, producing a *negative, inversion,* or *complement image*. Alternatively, if the content of all LUT elements from 0 to 99 were zero, and the remaining elements 100, 101, . . . , 255 had contents equal to their address value (i.e., 100, 101, . . . , 255) (Fig. 3C), the output image would be unchanged except that all the darker areas, with original gray levels of 99 or less, would now be set to black, an example of a process known as *thresholding*. LUT may also be used for *contouring*, whereby pixels with intensities of certain increments (e.g., 32, 64, 96, 128, . . . , etc.) corresponding to regular contour intervals are set to white or a bright pseudocolor.

2. Gray-level histograms and contrast enhancement. Among many other single-image pixel point processes that are possible, one of the most important for light microscopic images is that of digital contrast enhancement. For this it is helpful to calculate the *gray-level histogram* of an image. For a 512^2 pixel 8-bit image, the gray-level histogram is a display of the population distribution of the 262,144 pixels in the image among the 256 possible gray-level values (Fig. 4). This graphical representation makes it easy to appreciate the contrast and dynamic range of an image. For instance, the gray-level histogram of a hypothetical image of fairly low contrast and with limited dynamic range (Fig. 4A) shows the pixels to be clustered about a central mid-gray value of 100, with none in the darkest quadrant (<63) or brightest quadrant (>191) of the range. By

A Grey-level histogram of original image

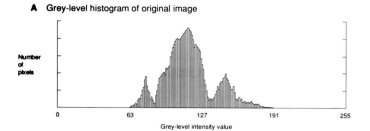

B The look-up table map for contrast stretching

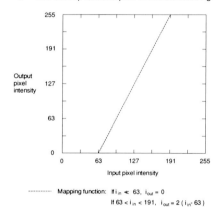

------ Mapping function: If $i_{in} \leqslant 63$, $i_{out} = 0$

If $63 < i_{in} < 191$, $i_{out} = 2\,(\,i_{in}\text{-}\,63\,)$

If $i_{in} \geqslant 191$, $i_{out} = 255$

C Grey-level histogram of enhanced image

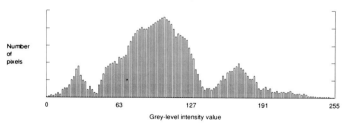

Fig. 4. Illustration of gray-level histograms, and of the use of a look-up table to expand the dynamic range of an image, thereby enhancing its contrast. **A:** A gray-level histogram of an input image, which shows the number of pixels in the complete image that have each of the 256 possible gray levels. In this image, there are no pixels having gray levels below 63 or above 191. The dynamic range of the image intensities is thus restricted to 128 possible values in the central mid-gray half of the range. **B:** A look-up table map designed to stretch this central half of the gray-level range to fill the entire dynamic range of the digital image from black (0) to white (255). **C:** The gray-level histogram of the resultant image produced by processing the input image using the look-up table map shown in B. The contrast of the image has been doubled by this expansion of the dynamic range. Note that the number of pixels in each group has remained unchanged. Only the gray level assigned to the group has been altered, in accord with the specified mapping function.

histogram stretching, using the LUT map given in Figure 4B to spread these gray levels over the full dynamic range of brightness, one can generate an image with enhanced contrast (Fig. 4C).

Histogram equalization is a related procedure in which the original pixel intensities are reassigned absolute values, without changing their relative positions in the histogram, until the resulting gray-level histogram shows, as nearly as can be achieved, equal pixel occupancy in all the intensity ranges. This results in reduction or compression of contrast (i.e., of differences in gray level) in regions of the image that originally had high contrast, and enhancement of contrast in previously low-contrast areas, thus optimizing usage of the available contrast range. The mechanism whereby this is achieved is as follows: From the gray-level histogram of the original images is calculated the definite integral of this histogram, i.e., the *cumulative gray-level histogram,* a plot of the total number of pixels having an intensity *less than or equal to* each gray level. This curve, after rescaling the abscissa between 0 and 255, is termed the *histogram transfer function,* and is used directly as the look-up table map for histogram equalization of the original image. Using this LUT map, pixels originally crowded into a particular intensity range of the original image are redistributed over a wider range of gray levels, to an extent proportional to their original crowding, thus enhancing the contrast of this region at the expense of the contrast of less populated intensity ranges in the image. As would be expected, a cumulative gray-level histogram calculated from the resultant equalized image approximates a straight line with a slope of unity.

B. Dual-Image Pixel Point Processing

Dual-image pixel point processing, as the name implies, involves pixel point processes between a pair of input images to generate a single output image, and takes the general form $O(x,y) = a \, I_1(x,y) \, \& \, b \, I_2(x,y)$,

where O, I_1, and I_2 represent the intensities of the output and two input pixels, respectively, a and b are definable constants, and the symbol & represents a combination function. Combination functions include the mathematical operators $+$, $-$, $\times$, and $\div$, and the logical operators AND, OR, NOT, and EXCLUSIVE OR (XOR). The most commonly used forms of dual-image pixel point processing involve image addition and subtraction.

1. Digital image-averaging procedures. Image addition is important for the temporal averaging of noisy images, such as are obtained from intensifying video cameras, in order to improve the signal-to-noise ratio (SNR). In its simplest form, applicable only to static specimens, a straightforward average may be achieved by instructing the arithmetic logic unit of the digital image processor to perform the recursive operation $y_i = y_{i-1} + (x_i/n)$, where y_{i-1} is the old value of the stored image, x_i is the ith incoming video frame, n is the total number of video frames to be averaged, and y_i is the new value of the stored image, y_0 being initially set to zero. The SNR enhancement achieved in this way is proportional to the square root of n, the number of averaged frames, and a new averaged image is generated every $0.04\,n$ seconds.

Such simple averaging over a large number of video frames is inappropriate for specimens in which motion is present, since blur will result. However, a compromise between blurring and noise reduction may be achieved by dynamic filtering, namely the calculation of a running average. Conventionally this is done by a first order recursive filter of the form $y_i = k \cdot y_{i-1} + (1 - k) \cdot x_i$, where k, the filter constant, has a value between zero and one and the other terms are defined as above. The output from this filter is a weighted sum of all previous frames. The smaller the value of k, the greater the weight given to the most recently acquired frame, and thus the shorter the effective time constant over which the running average is made, a situation suitable for more rapidly moving specimens. Such recursive filtering has the great advantage of producing a continuously updated output image, in contrast to the discontinuous images produced by the simple averaging procedure, and is similar in effect to a very long persistence phosphor on a video screen.

However, Erasmus [1982] showed that the improvement in SNR achieved by such recursive filtering is always less than that obtained by simple averaging, reaching a plateau value after a finite number of frames proportional to the value of k, as one might expect. He suggested that if one's objective is to produce the best possible continuous averaged display, one should use instead a recursive version of the true averaging procedure, named the *Kalman filter*, of the form $y_i = [(i - 1) \cdot y_{i-1}/i] + [x_i/i]$, in which the filter parameters are not constant but vary with the frame number i in such a way that the latest averaged image y_n always equals $(x_1 + x_2 + \ldots + x_n)/n$. This filter thus gives exactly the same result as a straightforward average over n frames, achieving the maximum theoretical SNR improvement, with the two advantages that the number of frames, n, over which the average is to be made does not have to be specified in advance, and that an open-ended continuous display of the true average is available. The averaging procedure may be stopped at any time, when the desired SNR has been obtained. Unfortunately, many image processors are insufficiently versatile to permit Kalman filtering, since it demands that the value of one of the constants in the algorithm, namely i, is incremented every frame.

As mentioned in Chapter 1, cooled charge coupled device (CCD) array cameras are able to integrate incoming light on the chip during an extended exposure, and thereby achieve true temporal averaging prior to image readout. In practice, for all types of electronic image acquisition system—CCD, confocal, or video—mechanical vibrations, specimen drift, and other experimental factors usually limit the period over which temporal averaging may usefully be employed to some 10 sec (256 video frames). Image addi-

tion is also important for spatial averaging, for example, the calculation of an x,y 2D projection through an x,y,z three-dimensional confocal image collected by serial optical sectioning, as detailed in Section V below.

2. Other dual-image pixel point operations. One other use of dual-image processing, particularly relevant for video-enhanced contrast microscopy (VECM), is image subtraction, which permits both background noise removal, by subtraction of a previously stored "background" or "mottle" image from an incoming live image, and also motion detection, by subtraction between two images of the same specimen recorded at different times, as discussed in Chapters 1 (Shotton) and 5 (Weiss and Maile, this volume). This latter procedure, after addition to each pixel in the difference image of a constant equal to the mean intensity of the two input images, yields an output image in which all stationary features are removed, while the old position of each moving object is revealed in negative contrast and its new position in positive contrast upon an otherwise featureless mid-gray background. However, when Nomarski optics are used to generate the initial optical images, in which objects are already seen as shadowcast, with one side brighter than the background and the opposite side darker, the old position of a moving object in the output difference image is seen with reversed contrast, as shown in Figure 7 of Chapter 5.

Multiplication and division of an image pixel by pixel by another image (as opposed to multiplication or division of all the pixels in an image by a numerical constant) have more specialized applications. One is to correct for variations of quantum efficiency between one pixel and the next on a CCD array, a process known as *flat fielding*. In this procedure, several "dark" images are first collected with the shutter closed, each with an exposure time equal to that to be used subsequently for imaging the biological specimen. These are then averaged and saved as the *averaged dark image, D,* which is a record of the averaged dark current and readout noise of each pixel in the CCD array under the conditions to be used for subsequent imaging. Next, several "flat-field" images are collected with the same exposure time, but with the shutter open, the microscope adjusted to observe a uniform specimen (for example, a solution of a fluorochrome sandwiched between slide and coverslip) using the optics to be used for the subsequent imaging of the biological specimen, and with the illumination source intensity adjusted to give approximately the same mean image intensity as that of the subsequent specimen images. These flat-field images, in turn, are averaged to yield the *averaged raw flat-field image*, F_r. From this, a *corrected flat-field image*, F_c, is then calculated by subtracting the averaged dark image from the averaged raw flat-field image: $F_c = F_r - D$. This digital image varies in intensity from pixel to pixel in proportion to the relative quantum efficiencies of the individual pixels on the CCD array, and also as a result of any field inhomogeneities (shading and mottle) in the illumination and imaging optics. The mean pixel intensity of this corrected flat-field image, M, is also calculated and stored. Finally, each incoming raw CCD image, I_r, obtained using the same illumination conditions and exposure time as were employed to determine F_c, is then simultaneously corrected, pixel by pixel, for the dark current, the readout noise, and the variations in the individual pixel quantum efficiencies of the imaging device, and also for shading and mottle in the microscope optics, to give corrected image, I_c, as follows: $I_c = (I_r - D) \cdot M/F_c$, the multiplication being calculated before the division to avoid the inaccuracies that would otherwise occur by integer truncation when dividing one pixel by another of similar gray level. Individual pixels in I_c are constrained to have gray levels between 0 and 255, as a precaution against sign inversion should the gray level of that pixel in D exceed that in I_r.

Perhaps the most important use of image division is in the determination of intracellular physiological parameters such as free

ion concentrations by fluorescence ratio imaging. In this procedure, a cell loaded with the ratio reporter dye is excited and imaged sequentially (or, in the case of dual emission wavelength confocal laser scanning images, simultaneously) at two selected wavelengths. Division of one image by the other yields the ion concentration image. This procedure, detailed by Michael Whitaker and his colleagues in Chapter 8 of this volume, has the particular advantage of simultaneously correcting both for variations in cell thickness and for inhomogeneities in dye concentration throughout the cytoplasm.

An increase in processing speed for image divisions may be obtained by passing both images through logarithmic LUTs, from which the output intensity values are the natural logarithm of the input intensities. The two images may then be subtracted—a computationally rapid operation—subtraction of logarithms being equivalent to division. The ratio image is then generated by passing through an antilog output LUT. The limitation of this procedure, which otherwise nicely illustrates the usefulness and versatility of look-up tables, is the compression of gray levels that occurs at the upper end of the dynamic range upon logarithmic conversion, with resulting loss of precision.

Dual-image combinations employing the Boolian logical operators are generally useful in postprocessing, for overlaying and combining images. Some examples of this type of processing are mentioned in Section III below.

C. Pixel Group Processing

The pixel point operations considered above allow gray-level modifications and image combinations individual pixel by individual pixel, with no reference to the character of adjoining pixels. *Pixel group processes* are more sophisticated context-dependent procedures which modify the intensity of each pixel in the image according to some function of the intensity values of their immediate neighbors. Such processes involve

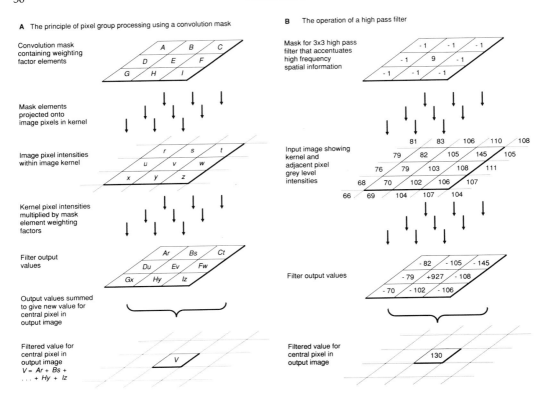

A The principle of pixel group processing using a convolution mask

Convolution mask containing weighting factor elements

Mask elements projected onto image pixels in kernel

Image pixel intensities within image kernel

Kernel pixel intensities multiplied by mask element weighting factors

Filter output values

Output values summed to give new value for central pixel in output image

Filtered value for central pixel in output image
$V = Ar + Bs + \ldots + Hy + Iz$

B The operation of a high pass filter

Mask for 3x3 high pass filter that accentuates high frequency spatial information

Input image showing kernel and adjacent pixel grey level intensities

Filter output values

Filtered value for central pixel in output image

Fig. 5. A: The process of digital image filtering by pixel group processing using a 3 × 3 convolution mask is illustrated. The convolution mask, containing weighting factors A to I, is centered over one pixel in the image, covering a kernel of 3 × 3 image pixels that contain integer gray levels r to z. The operation of the filter involves first calculating the product of each filter weighting element and its underlying image gray level, and then summing these products to obtain the gray level for the central pixel in the output image. **B:** The same process using numerical values for a high-pass filter that enhances high-frequency spatial information while leaving unchanged the gray levels in areas of uniform intensity, operating on a portion of a simulated digitized image.

a group or *kernel* of adjacent image pixels, comprising a central pixel and its surrounding neighbors, of dimensions $M \times N$. M and N are both odd numbers to give mirror symmetry around the central pixel, and are usually equal (e.g., 3 × 3, giving a 9-pixel kernel, or 5 × 5, giving a 25-pixel kernel). Each pixel group process also requires the definition of a *convolution mask* or *filter*, a two-dimensional array of dimensions equal to those of the image kernel, containing numerical weighting factors (Figs. 5, 6).

Normally, the weighting factors in the filter sum either to zero or to unity. Those filters in which the weighting factors sum to unity produce an output image in which the mean image intensity is unchanged, while those in which the weighting factors sum to zero produce an output image which is black everywhere except where the filter generates a positive result. If application of the filter produces a negative pixel gray-level intensity value, this pixel is set to a gray level of zero (black). Conversely, if it produces a pixel gray level in excess of 255, this is reduced to 255 (peak white).

The manner in which a convolution mask is used is illustrated in Figure 5. The mask is centered sequentially over each pixel in the image to be processed. At each position, the gray level of each of the pixels in the image kernel underlying the filter is multiplied by the weighting factor of the respective overlying filter element. The resulting values are then summed over the whole kernel to yield a single value, which becomes the gray level of the central pixel in the output image. This value does not overwrite the original intensity value of the pixel in the input image, which remains unchanged, but rather is written into a separate memory location in another frame store, in which the filtered output image is being built up. When this process has been applied in turn to all of the pixels in the input image, the filtered output image is complete, and the input image is said to have undergone *spatial convolution* with the filter, or *spatial filtering*. To avoid edge artifacts due to incomplete overlap of the filter with the peripheral image pixels, pixel values beyond the edge of the image are assumed, for the purposes of calculating the filter sum, to be equal to those of their edge neighbors.

When a digital image contains a large amount of detail, for example an image of a tree with bare branches, the gray levels of adjacent pixels are likely to differ considerably from one another. The content of such an image thus varies spatially at high frequency, and the image detail is consequently referred to as high-frequency spatial information. On the other hand, an image lacking fine detail, for example an image of a cloudy sky in which large areas of pixels show similar intensities and the spatial information varies at low frequency, is said to exhibit a lack of high spatial frequencies.

Depending upon the filter size and the weighting factors loaded into the filter, pixel group processing produces a wide variety of results. A *high-pass* filter can be used for high-frequency spatial filtration to enhance details. A *low-pass* filter may be used for low-frequency spatial filtration to smooth the appearance of the image by removing high-frequency noise. Other filters result in isotropic or anisotropic edge enhancement.

The operations of a few of the many possible filters that one can use are illustrated in Figure 6. Recombination, by image addition, of an isotropic edge image produced by application of the Laplacian filter (Fig. 6F) with the original image, or, alternatively, subtraction of a low-pass filtered image in which detail has been suppressed from the original image (the latter procedure termed *unsharp masking*), are both particularly effective methods of enhancing feature visibility in images of low contrast.

The intensities of the surrounding pixels in the kernel may also be used in different ways. In the *median filter*, the output pixel intensity is not a mathematical combination of the pixel intensities of the kernel pixels, but rather the median value when these intensities are ranked in order (Fig. 6C). As shown, this filter is particularly good for removing shot noise, i.e., an individual pixel within the kernel with a spuriously high or low intensity, mathematical inclusion of which locally biases the output of the other filters described above (see Fig. 6A,B). Another filter, the *top hat filter* [Bright and Steel, 1987], is particularly useful for identifying small, compact, round features (e.g., spots in the 2D diffraction pattern of a periodic object) and extracting them from 2D digital images in which the background intensity is not uniform. The form of this filter is shown in Figure 7. The top radius R_t is set equal to the maximum expected radius of the spots or particles to be selected, while the radius of the brim R_b is such as to include a suitable surrounding neighborhood, often being set just smaller than the expected minimum distance between two particles. The height of the top hat H defines the amount by which the maximum intensity of the top pixels must exceed the maximum intensity of the brim pixels if the central pixel is to be recorded with its original intensity in the output image. If this criterion is not fulfilled, the intensity of the central pixel is set to zero. Figure 8 illustrates the action of the top hat filter on a model image. To detect dark spots on a light back-

A   High pass filter

Mask for 3x3 high pass
filter that accentuates
high frequency
spatial information

convolved with

Input image showing
kernel and
adjacent pixel
grey level
intensities

yields

Kernel pixel intensities
in high pass
filtered output
image

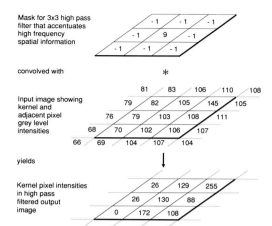

B   Low pass filter

Mask for 3x3 low pass
filter that attenuates
high frequency
spatial information

convolved with

Input image showing
kernel and
adjacent pixel
grey level
intensities

yields

Kernel pixel intensities
in low pass
filtered output
image

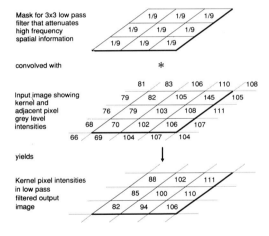

C   Median filter

Medial filter mask
contains no weighting
factors, but simply
defines the size of
the image kernel

Input image showing
kernel and
adjacent pixel
grey level
intensities

Select median values

Median value of pixel
intensities in each
input kernel defines
value of each
central pixel in
output image

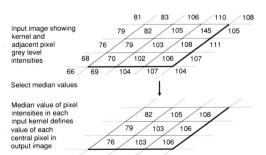

D   East gradient filter

Mask for east
gradient edge
detection filter

convolved with

Input image showing
kernel and
adjacent pixel
grey level
intensities

yields

Kernel pixel intensities
in east gradient
filtered output
image

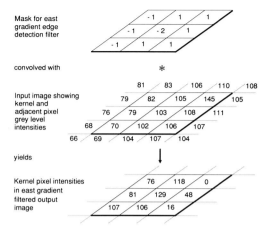

E   South gradient filter

Mask for south
gradient edge
detection filter

convolved with

Input image showing
kernel and
adjacent pixel
grey level
intensities

yields

Kernel pixel intensities
in south gradient
filtered output
image

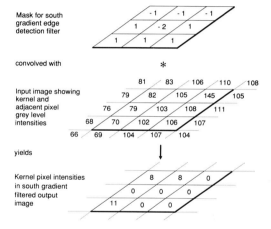

F   Laplacian filter

Mask for 3x3 Laplacian
filter that detects edges
isotropically

convolved with

Input image showing
kernel and
adjacent pixel
grey level
intensities

yields

Kernel pixel intensities
in Laplacian
filtered output
image

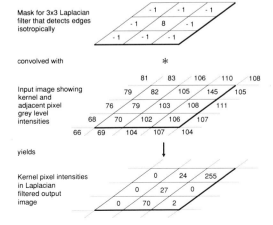

Fig. 6. Digital image filtering with six different 3 × 3 convolution masks acting upon the same numerical image data as given in Figure 5B. In contrast to Figure 5, this figure omits display of the products obtained by multiplying the individual weighting factors with their underlying kernel pixel gray levels, but illustrates the output intensities not only of the central pixel, but also of the surrounding eight pixels in the image kernel, these values being determined by calculating the effect of the convolution mask when centered over each of the kernel pixels in turn, using the intensity information of the additional pixels surrounding the kernel in the input image. The output thus illustrates in miniature the effect of convolution of the mask with an entire digital image. The input image gray levels are chosen to simulate an area of the image in which pixel intensities change abruptly (i.e., with high spatial frequency) along the left–right (west–east) x-axis, the left part of the image portion shown being distinctly lower than the rest, with a clear intensity step between the second and third columns of pixels. Along the north–south y-axis there is no such step. In addition, the data contain one spuriously high "noise" gray level of 145 in the top right corner of the kernel, to illustrate the effect of such noise on the output of the various filters. Filter output values that exceed 255 are set to 255 (peak white), while those which are negative are set to 0 (black).

A: A high-pass filter accentuates both the image intensity step and the noise pixel. **B:** A low-pass filter almost halves the step size and effectively smooths out the noise pixel. **C:** A median filter removes the anomalous intensity of the noise pixel, replacing it by the median value (108) of its neighbors, while leaving the rest of the kernel virtually unchanged. **D:** An east gradient edge detection filter, which will detect the increase in intensity in the direction from west to east (left to right) across a vertically oriented edge, responds very strongly to the edge in the data but completely suppresses the single noise pixel which does not constitute an edge. **E:** A south gradient edge detection filter, which will detect an increase in intensity in the direction from north to south (top to bottom) across a horizontally oriented edge, gives an essentially black output image in response to the absence of horizontal edges in the input data. **F:** A Laplacian edge detection filter, which detects edges isotropically, gives a moderate positive response to the edge in the test image and a very strong response to the noise pixel. Note that the form of this filter closely resembles that of the high pass filter, except that the weighting factors sum to zero rather than one. It will thus give a zero (black) output in regions where the pixel intensity do not vary, and a value close to this where the variation is slight.

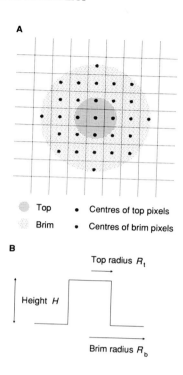

Top   •   Centres of top pixels
Brim   •   Centres of brim pixels

Fig. 7. The form of the top hat filter described by Bright and Steel [1987] for the selection of bright spots from a digital image. **A:** A plan view of the filter, composed of two concentric circles overlying an array of image pixels. The "top" of the hat, the darkly shaded inner circle, covers the centers of the central five pixels (black dots), while the "brim" of the hat, the lightly shaded area between the inner and outer circles, covers the centers of 24 of the surrounding pixels (stippled dots). **B:** A side view of the top hat shape, showing the top and brim radii, R_t and R_b, and the height of the hat, H, measured in gray levels. R_t is set to the maximum expected size of the spots to be detected, while R_b is set just smaller than the expected nearest neighbor distance between spot centers. The height H defines the minimum gray level by which the spot must exceed its local background to be selected by the filter. Its operation is as follows: If the maximum intensity of any of the "top" pixels exceeds the maximum intensity of any of the "brim" pixels by at least H gray levels, the original intensity of the central pixel is passed to the output image. If this condition is not met, the gray level of the central pixel is set to 0 (black) in the output image. This process is repeated in turn for each pixel in the digital image, except at the edge of the image, where a border of width R_b is not processed but is set to black.

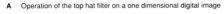

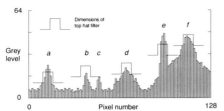

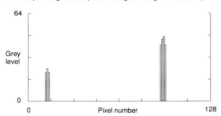

Fig. 8. The operation of the top hat filter illustrated using a one-dimensional digital image of 128 pixels having a gray-level range from 0 to 63. The top hat can be thought of as sliding in a horizontal orientation along the top surface of the pixel gray levels with its brim just touching, usually at a single point. Only those peaks that are sufficiently large to "penetrate" the top of the hat are recorded in the output image. **A:** Input image. **B:** Output image. In this simulation, the top hat filter selects from the input image the central pixels of peak *a* (on a flat background) and peak *e* (on a steeply rising background), but peak *b* (too close to peak *c*), peak *c* (too small), and peaks *d* and *f* (too broad) are not passed by the filter.

ground, for instance images of immunogold particles obtained by video-enhanced bright-field microscopy, the top hat filter is inverted. A more sophisticated version of this procedure, using a Gaussian curve rather than an inverted top hat shape, which more closely resembles the image of a single subresolution immunogold particle, is described by De Brabander et al. in Chapter 6 for locating and tracking gold-labeled antigens.

Since pixel group processing using the smallest (3 × 3) kernel involves nine multiplications and nine additions per pixel, only the fastest digital image processors can accomplish this within a single frame time, permitting video-rate pixel group process-

ing, since this requires some 120 million hardware calculations per second. Such pixel group processing procedures are thus usually employed subsequent to initial image capture and pixel point processing, in order to enhance the appearance of selected recorded images.

D. Frame Processing

Certain processing operations do not fall within the confines of point processing applied to individual pixels or group processing of small neighborhoods of pixels, but involve manipulation of entire images. These diverse operations, termed *frame processes*, are usually handled by software algorithms within the microprocessor or the host computer, and can only infrequently be performed in real time.

1. Geometric transformations. Those geometric transformations most frequently used and most easy to implement as interactive procedures include the ability to translate, rotate, and scale the image.

Image movement (*panning* laterally and/or *scrolling* up and down) is achieved by simply redefining pixel coordinates: $x_{new} = x_{old} + A$; $y_{new} = y_{old} + B$, where A and B are the required movement distances measured in whole pixels. Depending upon the processor, regions of the image which are moved off one edge of the screen will either disappear entirely or will reappear at the opposite edge, a phenomenon known as *wrap-around*.

Rotation of images involves a similar redefinition of coordinates, by the application of the coordinate transform $x_{new} = x_{old} \cos\theta - y_{old} \sin\theta$; $y_{new} = y_{old} \cos\theta + x_{old} \sin\theta$, where θ is the required clockwise rotation angle of the image within a stationary viewing frame. Since this transform rotates the image with respect to the coordinate origin $x = 0$, $y = 0$ (conventionally the upper left corner for digitized video images), it results in the image being progressively rotated out of the frame of view! To achieve rotation about the center of the screen, a temporary redefinition of the image coordinates is made during the calculation:

$$x_{new} = (x_{old} - \tfrac{1}{2}x_{max}) \cos\theta -$$
$$(y_{old} - \tfrac{1}{2}y_{max}) \sin\theta + \tfrac{1}{2}x_{max};$$
$$y_{new} = (y_{old} - \tfrac{1}{2}y_{max}) \cos\theta +$$
$$(x_{old} - \tfrac{1}{2}x_{max}) \sin\theta + \tfrac{1}{2}y_{max}.$$

Obviously, rotation about any other point on the screen is equally easy to define. If the rotational angle required is other than 90° or 180°, the rotated pixels will not exactly overlie the pixel locations in the frame of view. In these cases, this geometric operation requires bilinear interpolation of the pixel intensity values, achieved by assigning to the new displayed pixel a weighted average of the intensities of the nearest rotated image pixels in the x and y directions. For correct image geometry to be preserved, the image must be composed of square pixels. For images composed of rectangular pixels, special processing is required to accommodate for the pixel shape during rotation. To avoid curious wraparound effects during rotation, it may be desirable to first shrink the image so that after rotation it still fits entirely within the frame of view, or to treat it as the central part of a larger image in which the surrounding pixels are set to black.

Magnifying the image by powers of two (*zooming*) may be achieved either by pixel duplication, so that each original pixel becomes a square of 2×2, 4×4, or 8×8 identical pixels, clearly visible as square "superpixels" on the screen, or by bilinear interpolation of the original pixel values to obtain the intensity values for the intervening pixels added to bring about the expansion, which gives a smoother appearance. Shrinking the image by powers of two, useful for generating image montages, is similarly accomplished by pixel omission or, better, pixel averaging. For all magnifications and demagnifications other than by powers of two, interpolation is also required, but in all cases the coordinate transformation used is $x_{new} = S\, x_{old};\ y_{new} = S\, y_{old}$,

where S is the scaling factor. S thus acts to modify the coordinates of the image in a manner analogous to that in which the look-up table mapping function, M, modifies the intensity values of the image. Unless otherwise specified, the origin for the magnification or shrinkage will be the image coordinate origin $(0, 0)$. However, as for image rotation, relocation of the zoom origin to the center of the screen or to any other coordinate point m, n may simply be achieved by redefining the coordinate origin during the calculation: $x_{new} = S \cdot (x_{old} - m) + m;$ $y_{new} = S \cdot (y_{old} - n) + n$. Areas of the image that are expanded beyond the field of view are not displayed. However, interactive variation of m and n, most conveniently by use of a mouse, enables one (apparently) to roam around by panning and scrolling within the scaled image.

More complex geometric operations, termed *rubber sheet transformations*, are able to change the shape of the image in nonlinear ways. While useful, for example, when producing apparently overhead orthogonal views from obliquely photographed satellite images, and while potentially useful for correcting geometric distortions in video camera images, they have as yet found little use in electronic light microscopy.

2. Fourier transforms and spatial filtering. Image transformations such as the Fourier transformation are computationally much more intensive. They generate a pictoral representation of the distribution of spatial frequencies within the two-dimensional image, either in the form of the two-dimensional Fourier transform itself (also known as the *diffraction pattern* or the *spatial frequency image*) or as a one-dimensional graphical representation of these data, the *power spectrum histogram*, in which the amplitudes of particular frequency components (abscissa) are plotted against frequency (ordinate). Spatial filtering using Fourier transforms is both intuitively more straightforward than by using pixel group convolutions on the image, and infinitely more flexible. To remove (or modify) a particular spatial frequency component from a real image, one simply calculates the Fourier transform of that image, sets the corresponding region of the two-dimensional spatial frequency image to zero (or some other value), and then computes the reverse Fourier transform of the spatial frequency image to generate a "filtered" real image of the original object.

Such image Fourier transformations are widely used in X-ray protein crystallography and in the processing of digitized electron micrographs of periodic biological specimens (for instance, two-dimensional crystalline arrays of membrane proteins). Here the periodic nature of the specimen permits image averaging by reciprocal space image filtering, with a significant improvement in the signal-to-noise ratio. An excellent introduction to Fourier transforms and image filtering of periodic objects is to be found in Holmes and Blow [1965]. The use of Fourier transforms in conjunction with electronic light microscopy has to date largely been limited to out-of-focus blur deconvolution procedures associated with three-dimensional reconstruction studies from nonconfocal optical sections of fluorescently labelled cells, discussed in Section IV below, and to determination of the precise coordinates of moving objects in video-enhanced contrast images by image cross-correlation techniques, which has been useful in studies of the molecular motors involved in organelle motility, and of the lateral movement of cell surface membrane protein molecules [Gelles et al., 1988; Sheetz et al., 1989].

3. Image compression. The final type of frame processing procedures are those concerned with data compression, removing redundancy or irrelevance within the information content of images to permit speedier transmission, denser storage, or faster computation of the data. Of particular importance, for example, in the transmission of satellite images from deep space probes back to earth, data compression processing has as yet seen little use in electronic light microscopy, but will become increasingly important as the transition is made from analogue (videotape) recording to digital image storage devices to archive more and larger images. Image compression algorithms are of three major types: those that allow complete restoration (decompression) of all the image element intensity values of the original image from the compressed image with absolute fidelity (*lossless compression*), those that allow only partial restoration (*controlled quality compression*), and those, particularly employed for three-dimensional images, in which certain image data are completely discarded, preventing restoration of the original image.

One of the simplest of the lossless image compression algorithms that allow complete image restoration, both to understand and to implement, is linear run-length encoding. In this, each line in the image, originally recorded as a 512-element array of 8-bit integers representing the intensities of successive pixels in the image line, is recoded as a series of *pairs* of 8-bit numbers. The first pair records the intensity of the first pixel in the line and the number of adjacent pixels, including this first one, sharing that intensity. Traveling along the line to where the pixel intensity changes, the second pair of numbers records the new intensity value and the number of pixels in this second intensity group, and so on. For images with large areas of uniform intensity lacking high-frequency components, such as sparsely populated fluorescence images with large areas of low-intensity background, this simple procedure can easily achieve fivefold data compression. At the other extreme, if every pixel had a intensity different from those of its neighbors in the line, the amount of memory required to encode the image would be doubled. Various other image-compression algorithms of increasing sophistication, operating in one, two, or more dimensions, achieve greater compression while still allowing complete image restoration, at the expense of being computationally more intensive to implement.

Controlled quality image compression permits greater compression but only partial restoration of the original image. While useful, for example, for video conferencing over the restricted data bandwidth of telephone lines, it is not suitable for electronic light microscopy, since having gone to the great effort and expense of obtaining a high-resolution image, the last thing the research worker wishes is to intentionally and irrevocably degrade that resolution.

Image compression of the third type is a form of data reduction undertaken to permit the rapid rendering or display of certain features selected from within an original 3D image, and is associated with a change in image format from *raster* to *vector*. Raster images are those considered hitherto, in which each voxel is represented by a single intensity value, the coordinates of that voxel being encoded only implicitly, by the position of that byte in the image record and by prior knowledge of the x, y, and z dimensions of the entire image. Vector images, in contrast, are those in which the spatial coordinates as well as the intensity of each image element are explicitly defined and recorded. If the dimensions of the image do not exceed $256 \times 256 \times 256$ voxels, each coordinate value can be stored as a single-byte integer. However, for images of a larger size, each image coordinate requires 9, 10, or more bits to encode it, and is normally stored as a 2-byte word. Unless special encoding is used to save space, a large image thus requires 7 bytes per voxel, as opposed to only 1 byte per image element for an 8-bit intensity raster image. Transformation of an entire raster image into vector format would obviously increase the volume of data to be stored. However, when only a small subpopulation of image elements is selected from an entire raster image, for example, those voxels defining the boundary of a cell from within a 3D confocal image, the vector image format becomes a much more efficient means of recording the data.

Two processes are necessary for the generation of such a vector image from a raster image: (1) the application of image analysis procedures, primarily thresholding (discussed in Section III below), to identify the voxels defining the surface of the object to be displayed; and (2) the discarding of all other voxels of the image, to produce a much smaller data set of perhaps 2^9 voxels, which, if stored in 7-byte vector format, would occupy 3.5 Kbytes, selected (let us suppose) from the original 2^{25} 1-byte voxels (32 Mbytes of data) of a complete 512^2 (x,y) $\times$ 128 (z) 3D confocal data set. This reduces the number of vector data points to be computed and displayed to a scale which can be handled by a graphics workstation at a sufficiently rapid computational speed to allow interactive processing. Such display is discussed in Section VD below. Vectorization is thus not an image compression procedure *per se*, but its use leads to efficient data storage of images in which data selection has reduced the number of image elements to be further processed. Because one has discarded most of the voxel data, one forfeits the ability to restore the original raster image from its vector derivative.

Detailed considerations of the theory of digital image processing, of the image manipulation algorithms themselves, and of the microelectronic architecture and data flow timing of digital image processors which make it possible, are beyond the scope and purpose of this chapter. Those interested in gaining a deeper understanding are referred to Gregory Baxes's excellent, although now somewhat dated, text *Digital Image Processing: A Practical Primer* [Baxes, 1984], to Kenneth Castleman's authoritative and readable classic work *Digital Image Processing* [Castleman, 1979], and to Gonzales and Wintz's more recently revised text with the same title [Gonzales and Wintz, 1987].

III. IMAGE ANALYSIS

An operational distinction may be made between digital image processing, discussed above, and digital image analysis. The purpose of image processing is to improve the clarity or visual appearance of an image, thereby aiding its comprehension by the human eye/brain, and the operation may be typified by the slogan *image in–image out*. In contrast, the primary purpose of image analysis is to abstract meaningful numerical data from an image, which concisely describe the salient features of the image and facilitates, if desired, a significant reduction in the volume of data stored: *Image in–data out* describes the image analysis operation. However, this distinction is not an absolute one, since many of the early stages of image analysis involve standard image processing routines, and, conversely, image processing of multidimensional images for display on a two-dimensional video monitor, discussed in Section VB below, involves image analysis procedures such as thresholding, feature selection, and data reduction.

While it is not my intention to describe image analysis in any depth, the following brief outline exemplifies a typical simple image analysis operation performed on a two-dimensional digitized image. Such an analysis can be undertaken satisfactorily using a digital image processor by running compatible image analysis software on the host computer. Alternatively, and usually at considerably greater expense, a purpose-designed digital image analysis system may be employed, in which a very large range of image analysis routines are pre-programmed for ease of execution.

Let us suppose that the original image is a low-magnification VECM image of a small cluster of fibroblasts growing on a glass coverslip, and that the purpose of the analysis is to determine the mean area covered by a single cell (Original image [say] 512^2 pixels; intensity range of the image already optimized to fill all 256 gray levels [8 bits]; image data size 262,144 bytes [0.25 Mbyte]). The following image analysis operations would be undertaken:

1. (Optional) Definition of an *area of interest* (AOI): The operator marks on the image that region which he/she wishes to analyze, by defining a rectangular box within the full image or by drawing around the features of interest using a cursor or mouse. This reduces the time taken for subsequent processing by restricting the amount of raw image data subjected to analysis. (Area of interest [say] 256 × 128 pixels; intensity range of each pixel unchanged at 256 gray levels [8 bits]; AOI gray-level image data size 32,768 bytes; data reduction factor from original gray-level image × 8).

2. Definition of a *threshold* intensity value to separate features of interest from the rest of the image, this threshold value being displayed as a contour superimposed upon the gray-level image by setting that particular intensity value to white (or to a bright pseudocolor) using a look-up table. This process is normally done interactively, by adjusting the selected threshold intensity value until the displayed contour accurately circumscribes the edges of the objects of interest, in this case the cell perimeters. (Threshold set [say] at an intensity value of 95. Image data size unchanged.)

For a VECM image previously subjected to background correction by mottle subtraction, the background intensity should be reasonably uniform, and thresholding may be simply accomplished by selecting a single gray level midway between the mean gray level of the background and the mean gray level of the flattened cell margins. Calculation of the gray-level histogram is often useful for deciding upon this threshold, it being set to a gray level in the trough between the peak of low-intensity background pixels and the usually smaller peak of higher-intensity object pixels. However, for images in which the background is non-uniform, either a preliminary background shading correction needs to be performed, or a local *adaptive thresholding* procedure should be applied in which the threshold is automatically adjusted, region by region across the image, according to the local mean background intensity.

3. Binary *segmentation* of the intensity image to generate a *binary image*, in which pixel intensities are constrained to one of

two values, black or white: The AOI image data is passed through a look-up table in which all pixel intensities below the threshold value are set to black, and all intensities at or above this value are set to white. The selected cells will thus appear as uniform white shapes on a black background. (Area of interest 256×128 pixels; intensity range of each pixel 2 gray levels, white (1) or black (0): Note that only 1 bit is needed to encode each pixel of the binary image; AOI binary image data size 4,096 bytes; data reduction factor from gray-level AOI image $\times 8$; data reduction factor from original image $\times 64$.)

Sophisticated digital image analyzers will at this stage uniquely number each discrete object in the binary image, and will assign adjacent objects different pseudocolors for ease of recognition. Additionally, multiple thresholds may be set. Two might be used to define two separate classes of objects; for example, one threshold to distinguish the boundaries of cells from their background, and another to differentiate between the cytoplasm and the darker nucleus within each of these cells. The nucleus and the cytoplasm may then be displayed with different pseudocolors.

4. Separation of touching objects: If two or more cells were touching in the original image, thresholding should not have separated them. Either an interactive or an automatic procedure may be used to achieve the separation.

The interactive method relies upon the operator drawing a dividing line with a cursor or mouse along the line of contact between each pair of touching cells, and thus requires viewing of the original intensity data for these cells. For this purpose, a gray-level image of the selected objects upon a uniform background may easily be generated by combining the binary image, which acts as an *image mask*, with the original gray-level image, using the Boolian logical AND operator. Wherever the binary image is white, the resultant image will show the original intensities of the cells.

Wherever it is black, the background of the resultant image will be set to black (or, if desired, to a background pseudocolor). After separation, the image analyzer will recognize the two parts of the original binary object as two separate objects in the binary image.

Automatic objective separation, which works best on objects that are touching only over very small areas, requires two binary image operators, *erosion* and *dilation*. Erosion removes the outermost layer of pixels from objects in the binary image, while dilation adds a single layer of pixels around the perimeter of such objects. One or more rounds of erosion are used until all of the cells are separated from one another. These are then identified and numbered as distinct objects. An equal number of rounds of dilation are then employed to restore the cells to (approximately) their original size and shape, while maintaining their separate identities. The combined operation of erosion and dilation is known as an *opening* operation. It has the additional property of totally removing very small objects, which disappear during the erosion phase, and of removing filamentous extensions of larger objects, which cannot be restored by dilation. The reverse procedure, *closing*, which involves dilation followed by erosion, may be used to connect separate but adjacent fragments of a single object. These procedures are discussed more fully by van der Voort and Smeulders in Chapter 11.

5. Other geometric operations on the binary image: Although not of relevance to the example being considered, other geometric operations may at this stage be performed on the binary images of the selected objects. One of the most useful for extended objects is *skeletonization*, thinning by erosion until the object is everywhere only 1 pixel wide, at which point no further thinning occurs, thereby defining the central axes and branch points of the object.

6. Categorization and measurement of objects: Once the desired objects have all been successfully identified and isolated,

one is free to categorize them (for instance, by area) into groups, to count the number of objects in each group, to calculate statistics from these data, and to determine the properties of the individual objects. In this example, the area of each cell may be determined by counting the number of pixels comprising each cell image. From these data, the required mean cell area ± standard deviation may be calculated. (Number of cells [say] 15; data stored: identifying number and area of each cell, plus mean area and SD; total data recorded, 32 bytes; data reduction factor from the binary image: × 128; data reduction factor from the original gray-level image: × 8,192.)

Of course, having identified the objects, many other measurements may be made upon them. Digital image analyzers will automatically calculate many of the following parameters for all the selected objects, and will generate statistical analyses of the population data thereby generated: *size factors* (area, diameter, center of gravity, etc.); *edge factors* (roughness, fractal dimension, etc.); *shape factors* (aspect ratio, maximum and minimum diameters, perimeter-to-area ratio, roundness or lack of it, width and branching topology of extended objects, etc.); *intensity factors* (maximum, minimum, and mean object pixel intensity, variance, contrast range); *texture factors* calculated by analysis of the spatial frequencies within the original intensity data within each object; and *spatial distribution parameters* for the population of objects (mean nearest neighbor distance, percentage area occupied, porosity, etc.).

The determination of the area of a digitized image of a discrete object is trivial, being achieved, as mentioned above, simply by counting the number of pixels comprising the object. However, the process of digital sampling converts the smoothly changing boundary of the original specimen into a jagged stepped boundary formed by the edge pixels of a digital image of the object, whose positions only approximate the original edge. This makes the measurement of the perimeter of digitized images open to systematic overestimation if simple perimeter-determining algorithms are employed. Much effort, discussion of which is beyond the scope of this chapter, has thus been invested in designing more accurate algorithms for perimeter measurement, and the interested reader is referred to Dorst and Smeulders [1987] and to Chapter 11 of this volume by Hand van der Voort and Arnold Smeulders for further details. The extent to which such sophisticated perimeter-measuring algorithms are employed by commercial image analyzers is rarely made clear by their suppliers.

The subsequent sections of this chapter discuss how the image processing and analysis processes outlined above may be applied to images of more than two dimensions.

IV. COMPUTATIONAL DEBLURRING OF FLUORESCENCE IMAGES

Conventional wide-field fluorescence microscopy suffers from one major disadvantage, namely, out-of-focus blur. The illumination of the entire field of view of the specimen with intense light at the excitatory wavelength excites fluorescence emissions throughout the whole depth of the specimen, rather than just at the focal plane (see Chapter 1, Fig. 1). Much of the emitted light coming from regions of the specimen above and below the focal plane is collected by the objective lens and thus contributes as out-of-focus blur to the final image of the specimen at that focal plane, seriously degrading it by reducing the contrast and sharpness of the image.

Fluorescence images free from such blur may be obtained either optically, by confocal microscopy, as discussed in Chapter 1 and Chapters 10 to 14, or by the computational removal of out-of-focus blur from conventional wide-field fluorescence images. The advantages of the latter approach in comparison with confocal scanning microscopy include (1) the ability to employ arc lamp illumination, and thus to excite the full range of conventional chromophores, including those excited by ultraviolet illumination, rather than being restricted to those excitable by certain laser lines; (2) simultaneous parallel data acquisition from all regions of the specimen, rather than serial data collection during single beam scanning; and (3) collection of all of the light from the sample that enters the objective lens rather than just that proportion of it from the focal plane, discarding the rest at the confocal imaging aperture. Both (2) and (3) result in more rapid image acquisition. While point (3) might seem counterproductive at first sight, the subsequent process of computational image deblurring may be seen as an image restoration procedure that seeks to remedy the effects of out-of-focus blurring by returning the "stray light" to its rightful place in the three-dimensional image.

This approach, pioneered particularly by David Agard, John Sedat, and their colleagues at the University of California, San Francisco [Agard and Sedat, 1983; Agard, 1984; Agard et al., 1989], is detailed in Chapter 9 of this vollume by their collaborator Peter Shaw. Here I wish to give a brief introduction to this approach, since it forms one of the most important and exciting areas of digital image processing applied to 3D electronic light microscopic images.

The relationship between an ideal point object and the blurred image obtained from it by conventional microscopy using a high numerical aperture objective at any chosen focal plane (Fig. 9) is defined by the image-forming characteristics of the objective lens, namely the lens three-dimensional *point spread function* (psf), $s(x, y, z)$. In practice, subsequent lenses in the imaging pathway (oculars, transfer optics, etc.) do little to modify this objective lens psf, which thus defines the resolution of the complete microscope imaging system. Castleman [1979] demonstrated that the imaging of an entire three-dimensional object may be described as the convolution of the three-dimensional object with the three-dimensional psf: $i(x', y', z') = o(x, y, z) \otimes s(x, y, z)$, where $o(x,$

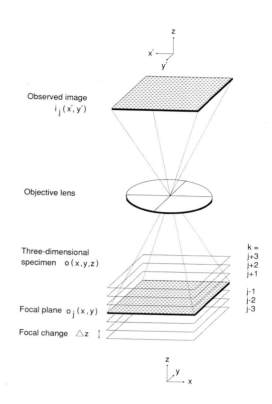

Fig. 9. A schematic representation of the geometric relationship between a fluorescently labeled three-dimensional specimen and its observed blurred two-dimensional image, to illustrate the methods employed for computational blur removal from images obtained using a conventional fluorescence microscope. Each observed image, $i_j(x', y')$, is the sum of the in-focus information from the selected focal plane, $o_j(x, y)$, and out-of-focus contributions from the remainder of the specimen, $o(x, y, z)$, inverted and magnified by the objective lens. For clarity in this diagram, a magnification factor of $\times 1$ is shown. From a set of such observed images, separated from one another by changes of Σz in the focal plane, and with knowledge of the three-dimensional point spread function of the objective lens, it is possible, as discussed in the text, to calculate to a reasonable approximation the original in-focus information of the specimen sampled at these planes, and hence to obtain a complete three-dimensional description of the fluorescence emission of the specimen free from out-of-focus blur. (Reproduced with permission from Shotton [1991]. Adapted from an original diagram by Castleman [1979].)

y, z) is the three-dimensional *object*, i (x', y', z') is its three-dimensional *image*, and $\otimes$ is the convolution symbol. The in-focus psf s_i (x, y) is a central section through the three-dimensional psf at $z = 0$, and describes the blurring present in a single 2D image i (x', y') of an infinitely thin two-dimensional object o (x, y) situated at the focal plane (when $z = 0$), thus being the mathematical description of the familiar Airy disk image. In this notation, the relationship between the dimensions in the original object (x, y, z) and those in the image (x', y', z') is defined by the image rotation and magnification caused by the objective lens.

According to optics theory, this three-dimensional psf should, for an ideal lens, be symmetrical above and below the focal plane, and thus the image should look identical for similar degrees of overfocus and underfocus. In practice, using the best high-NA objectives to image subresolution (100 nm diameter) fluorescent beads, this has been shown to be far from the case [Hiraoka et al., 1987]. Not only is the psf strikingly asymmetric along the z-axis, but it also fails to show the expected circular symmetry in planes perpendicular to z. Consequently, to achieve accurate computational image deblurring, it is clear that the three-dimensional psf should first be determined experimentally using subresolution point test objects (e.g., fluorescent beads), employing the same objective lens and imaging conditions as will be used to collect the specimen data.

The lens *contrast transfer function* (CTF) is the Fourier transform of the point spread function, and defines the way in which the various spatial frequency components of the object are altered by passage through the objective. With conventional wide-field light microscopic imaging, certain image spatial frequencies occupying conical regions above and below the origin of the three-dimensional CTF (the "missing cones") have zero intensity, resulting in severely impaired spatial resolution along the optical axis. Thus when optical section images from a conventional

optical microscope taken at different focal planes through a three-dimensional object are combined, even after computational deblurring, the z-axis resolution is always inferior to the in-plane (x,y) resolution, causing the image to appear blurred and elongated along the optical axis. A method for overcoming this problem is described below.

In principle, the magnitudes of the out-of-focus contributions of other regions of a 3D specimen to any image plane of a 3D image may be calculated, to a reasonable approximation, from in-focus images at other focal planes, if the psf of the objective lens is known, as detailed by Castleman [1979], Agard [1984] and Agard et al. [1989]. If one has collected a through-focal series of fluorescence images by uniform optical sectioning through a single selectively labeled cell or nucleus which has a true fluorescence emission intensity distribution o (x, y, z), it is possible to estimate, for each image plane, the out-of-focus contributions made by the other planes in the object (Fig. 9). Thus the observed fluorescence image i_j (x', y') at a particular focal plane j within this specimen will be composed of the in-focus information arising from that focal plane within the object o_j plus the sum of each of the individual contributions from the pairs of adjacent out-of-focus planes k (where $k = j + 1, j - 1, j + 2, j - 2$, etc.). Each of these out-of-focus contributions is itself the true in-focus emission intensity for that particular plane within the object o_k (x, y) convolved with the appropriate out-of-focus plane psf s_o selected from the complete three-dimensional psf s, the exact value of s_o used for each plane being determined by its distance z from the focal plane. By subtracting the sum of these contributions from each observed image, one could in principle derive a set of "deblurred" optical section images of enhanced clarity.

Unfortunately, life is not always that easy. The in-focus images of the other planes in the three-dimensional specimen required for this calculation are themselves unknown, since these planes too can only be imaged in the presence of out-of-focus blur. Thus it is impossible to calculate an exact solution for the desired deblurred image plane, even with accurate knowledge of the psf. Various procedures have been developed to overcome this dilemma, all of necessity using the observed blurred images at the various focal planes as initial approximations for the desired in-focus images to be used as input data for the deblurring algorithms. They include a method for the rapid calculation from a set of blurred section images of a deblurred *projection* of the three-dimensional specimen (i.e., a two-dimensional image obtained by parallel projection of the 3D image intensity down the optical axis onto a plane); simple approximate methods for deblurring images using only information from the two adjacent sections (one above and one below), which have the benefit of computational ease and speed; and more complex computationally intensive three-dimensional procedures involving the entire set of optical sections comprising the 3D image that yield more exact solutions [Agard et al., 1989].

The projection method relies upon the fact that the Fourier transform of a projection of a three-dimensional object is a central section through the object's three-dimensional reciprocal-space diffraction pattern. Thus a projection image i_p of a stack of n blurred observed section images i_k (where $i_p = \Sigma\ i_k/n$; $k = 1$ to n) may be deblurred simply by using the in-focus psf alone, since $i_p = o_p \otimes s_i$. This projection method is elegantly simple, and has the additional advantage that, by employing stereoscopic pairs of projections made by projecting down slightly inclined axes (as described in Section VC below), deblurred stereoscopic images may be obtained quickly and easily [Agard et al., 1989]. However, it does not yield a complete deblurred three-dimensional data set. Those procedures that do yield a complete deblurred three-dimensional data set are detailed by Peter Shaw in Chapter 9 of this volume, and will not be elaborated upon here.

Because conventional microscopy results in the irrecoverable loss of spatial frequencies in the missing cones, the resolution along the z-axis of all such deblurred 3D reconstructions calculated from a single stack of optical sections will always be low. Agard [1984] described the improvements in z-resolution that may be expected by combining different views of the same specimen, for instance, two sets of optical sections collected along orthogonal axes, obtained by specimen rotation by 90° after collecting the first data set, since the second set of sections will contain spatial frequencies absent from the first. Such data sets from *Drosophila* embryo nuclei have been successfully collected and combined, with the expected improvement in resolution [Shaw et al., 1989; Shaw, 1990].

Confocal images, for all their advantages over wide-field images, are not themselves perfect. A particularly promising avenue for future work, presently being pioneered by various groups, including Carrington et al. [1990], Conchello and Hansen [1990], and Peter Shaw and his colleagues [Shaw and Rawlins, 1991; Shaw, Chapter 9, this volume), is the application of these computational deblurring algorithms to remove residual distortions from confocal images. It has been found that deblurred conventional images closely resemble raw confocal images in image quality, while deblurring of the confocal images leads to a further small but significant increase in image clarity.

V. MULTIDIMENSIONAL IMAGE DISPLAY AND ANIMATION

Once a series of blur-free noninvasive optical sections has been obtained, either by computational deblurring of conventional fluorescence images, by Nomarski differential interference microscopy, in which the depth of field is intrinsically very small [Inoué, 1989], or by confocal microscopy, the three-dimensional image so formed may be manipulated in many ways, in common with similar data obtained from X-ray or magnetic resonance tomographic imaging of the human body, or electron-density maps of protein structures calculated from X-ray crystallographic analyses, to take two familiar nonmicroscopic examples. This is a particularly interesting and rapidly expanding field for biological light microscopy. The display of such 3D and higher-dimensional images on the two-dimensional screen of a conventional video monitor requires either the selection of a 2D display plane, or a reduction in the dimensionality of the data to be presented. In Sections VA and B I briefly consider these options. Stereoscopic images having the appearance of three dimensions may be presented by the combination of pairs of 2D images representing the left-eye and right-eye views, using procedures outlined in Section VC.

A. Image Plane Selection

One advantage of having a complete x, y, z image is that it is trivial to view planes other than the x, y planes in which the image was originally collected. This is particularly fast if the entire image can be accommodated in computer memory at one time. Thus x, z planes selected at different y positions may be displayed to show side views of the specimen, impossible to obtain (without remounting the specimen) by conventional microscopy. Similarly, oblique views may be selected in any desired plane, although all such nonorthogonal image plane selections and projections (discussed below) require interpolation of the image element intensities, a computationally intensive task. Multidimensional images involving time can either be restructured to show time along one axis of the display, or may be animated, as described in Section VC below, to show a two-dimensional spatial image changing over time.

B. Dimensional Reduction

Frequently it is desirable to reduce the dimensionality of the image, both for display purposes and in order to save space during image storage. Ideally, this should be done without sacrificing essential data about the specimen. The simplest procedure that may be employed to compute an orthogonal 2D projection from a 3D image is to average the intensity values in each column of voxels parallel to the projection axis, generating an *AVERAGE* projection. Alternatively, one may record the value of the brightest voxel in each column, to generate a *MAXIMUM* projection. Despite its inability to remove hidden lines and surfaces, this procedure works surprisingly well for many specimens, particularly those which have relatively few fluorescent features distributed in a nonfluorescent space [van der Voort et al., 1989; Shotton, 1989; White and Shotton, 1990; White et al., 1991b]. Images thus generated contain no intensity-based depth clues, objects at the front and rear of the image being displayed equally. Similarly, the *MINIMUM* projection records the minimum voxel intensity encountered along each projection column.

Such projection images need not be used to record intensity information. A *HEIGHT* projection is one in which one projects down a particular axis (usually z) and records the *distance* (z coordinate) from the top of the object at which one first encounters a voxel above (or below) a particular specified threshold intensity, or the *distance* to the voxel having the maximum or minimum brightness [Wilson and Sheppard, 1984; Shotton, 1989; White and Shotton, 1990; White et al., 1991a,b]. Analogously a *TIME* projection is one in which projection is down the time axis, and one records the *time* at which one first encounters an image element above (or below) a particular specified threshold intensity, or the *time* at which the intensity reaches a maximum or minimum. This latter procedure is particularly effective for recording in a single 2D image the time at which the intracellular free calcium concentration peaks in different regions of a cell during a physiological experiment. The x, y, t data set from which the projection is calculated is first obtained as a time series of

2D calcium ratio images, and the resultant TIME projection is displayed with the different times of arrival of the calcium wave shown in different pseudocolors.

It is particularly useful for the rotational animation (see below) of certain types of 3D fluorescence confocal images to calculate both the MAXIMUM intensity projection *and* the HEIGHT projection of the 3D image at each viewpoint, since the former accurately records the maximum intensities of the objects in the image, while the latter retains the x, y, and z coordinates of these objects [Bennett et al., 1990; White et al., 1991a,b]. Almost all of the pertinent information about the original image is thus encoded in these two projections, and can be numerically extracted from them subsequently. If the pair of projections for each animation display viewpoint are loaded into two image memories in the display framestore, it becomes easy to interactively switch between them during display of the rotating animation, and thus to extract the original x, y, and z coordinate information *and* the intensity data from any maximum-intensity voxel in the image.

While the examples given above have been of dimensional reduction from 3D to 2D, identical projections may, of course, be made from 4D to 3D, giving, for example, a resultant x, y, z TIME projection from an x, y, z, t tixel data set collected by 3D confocal calcium ratio imaging over time, which would show the time at which the free calcium concentration peaks in the cell in three dimensions, rather than in a particular place. The problem of displaying the resultant 3D image remains.

C. The Perception of Three Dimensions in Microscopic Images

In normal life, the three-dimensional nature of our surroundings is revealed by a variety of clues. In addition to stereoscopic vision, perspective, familiarity with the relative sizes, shapes, and appearances of objects, surface shading, texture and shadowing, haziness and a loss of contrast at large distances, hidden surface obscuration by intervening objects, and motion parallax are all potent indicators of the relative dispositions and shapes of objects in the environment, so that people who lack the ability to see stereoscopically are still able to function normally, for example, driving cars without serious risk of collision.

In the unfamiliar microscopic world, many of these additional clues are absent. While stereoscopic images may indeed be seen directly using a compound microscope, by arranging the optics so that the left eye receives only the left half side-band of the Fourier components of the image in the back focal plane of the objective, and the right eye only the right half, such optical tricks are not conveniently combined with high-resolution electronic imaging. The following discussion therefore assumes that conventional full-aperture back focal plane images have initially been collected, with the consequence that only those clues that can be generated *post hoc* from the static three-dimensional images formed from them are available for a perception of the 3D nature of our specimens. These are *stereopsis* achieved by viewing computer-generated left- and right-eyed views of the specimen, *motion parallax* generated by computational animation of conventional or stereoscopic image projections, *depth coding* in which regions toward the front of the object are made to appear brighter than, or in a different pseudocolor from, regions further back, *hidden surface obscuration*, and *computer-generated shading*.

1. Stereoscopic display. Despite the fact that many people (about 10% of the population) are unable to perceive depth by stereopsis, one of the most straightforward and useful image manipulations that may be applied to a 3D image is the generation of a stereoscopic pair of image projections, permitting the image to be directly inspected in three dimensions while being displayed only in two. Such stereo pairs may be generated "on the fly" during image acquisition, or subsequently, by the mathematical projection of the complete 3D data set down two selected inclined axes corresponding to the required left- and right-eye views. This is achieved by combining all the image planes, voxel by voxel, after small incremental shifts of successive sections by a constant displacement to right or left. For shifts involving integral numbers of voxels, this is straightforward and rapid to achieve using a simple image processor, although this method places restrictions on the possible stereo angles obtainable between the left- and right-eye images. Complete flexibility, at the expense of speed, may be obtained by nonintegral shifts involving voxel gray-level interpolation. To obtain the correct geometric depth perspective during stereoscopic display, attention must be paid to the relationship between the stereo viewing angle and the interplane spacing. As discussed in Chapter 1, refractive index differences between the specimen and the mounting medium will cause the actual optical distance between adjacent optical sections to differ from the physical distance by which the microscope focus was stepped.

Surprisingly few image planes (as few as 12) are required to generate a stereoscopic image in which the individual planes are not perceivable, and inclusion of all the sections from a large 3D image may indeed result in confusing stereoscopic images, because of superposition of detail at different depths. This reflects the fact that the human eye is less sensitive to variations in depth than in lateral directions. Such stereo image pairs may be viewed stereoscopically in a number of ways designed to ensure that the right image is not seen by the left eye and vice versa, the most common of which, in order of increasing sophistication (and cost!), involve their display as follows: (1) as side-by-side photographic prints or images on a single video monitor, requiring parallel or cross-eyed stereoscopic fusion, either by "free-viewing" or by using a stereoscope, a simple optical aid incorporating lenses and/or mirrors that direct distinct images to

one eye or the other; (2) as a red/green anaglyph photograph or image on a color monitor viewed through red/green spectacles (only possible for monochrome images, and not suitable for people with red/green color blindness); (3) as separate images on two monitors fitted with polarizing screens of opposite polarity, viewed superimposed by means of a half-silvered mirror while wearing appropriately polarizing spectacles [one video image being laterally inverted (left-right) by reversing the polarity of the horizontal scan driver on one of the monitors]; (4) as full-screen superimposed color video images projected from two aligned video-projectors through left and right polarizing filters onto a nondepolarizing screen, viewed using polarizing spectacles; or (5) as full-screen color video images occupying alternate video fields on a single high field rate monitor, viewed either through liquid crystal shutter spectacles, in which the left and right shutters are alternatively made opaque in synchrony with the video field display frequency, or through circularly polarizing spectacles and a liquid crystal screen fitted to the monitor which switches the state of circular polarization of the viewed image from clockwise to counterclockwise and back again in synchrony with the change in displayed image.

The manner in which the individual sections should be combined for 3D viewing is to some extent a matter for subjective evaluation, depending upon the nature of the specimen being imaged. In addition to the use of AVERAGE or MAXIMUM projections, convincing *DEPTH-CODED INTENSITY images* may be obtained by weighting successive sections before averaging, such that the front of the specimen is made to appear brighter or in a different pseudocolor than the rear, thereby adding an extra depth cue to that of parallax.

Various workers have explored the use of image processing procedures to improve the clarity of confocal stereo pairs by enhancing the contrast and edges of the individual images before their use in stereo pair generation. For this purpose, several of the common two-dimensional image-processing algorithms, including edge-enhancement filters, have now been extended to work isotropically on 3D images, as described by van der Voort and Smeulders in Chapter 11 of this volume.

2. Image animation. The ultimate development of cellular tomography is to display not just a single image, albeit stereoscopically, but an incremental series of images in rapid succession, thus forming an animation sequence [Shotton and White, 1989]. If successive projections are calculated at small incremental circumferential angles, the specimen will appear to rotate before one's eyes. If, alternatively, successive images are calculated at increasing distances along one axis—for instance, down the z-axis—then one will have the illlusion of moving progressively through the specimen. For this latter approach to work most effectively, perspective rather than orthogonal stereoscopic projections are desirable, giving the viewer the impression of actually floating through the structure, like on of Christian de Duve's "cytonauts" [de Duve, 1984]. This technique can be further extended by displaying a series of rotationally animated images of the same space which change over time. Such a display of a four-dimensional image collected as a series of 3D spatial images from a live three-dimensional specimen enables one to study cellular phenomena with a clarity and facility hitherto impossible to contemplate.

While the AVERAGE and MAXIMUM intensity image projections discussed above may be used to generate animation sequences, these, because of their inherent transparency, often result in an ambiguity concerning the handedness of rotation if the animation is viewed nonstereoscopically. This ambiguity may be resolved if opacity and hidden object obscuration can be introduced into the display. This may be achieved conveniently by using a THRESHOLDED MAXIMUM *projection*, in which the frontmost voxel exceeding a given threshold is selected to generate the projection image [Shotton, 1989; White et al., 1991b]. This results in a surface image in which hidden objects lying further back in the specimen are automatically occluded, even if of higher intensity.

Apparent lighting may be applied to any of the above projections by the further calculation of a SHADOWED *projection*, in which a simulation of shadows generated by the "absorption" of oblique illumination is made to fall upon a surface behind the object. This is achieved by the calculation of an AVERAGE projection after rotation through a small angle (say 15°), its negation (contrast reversal) to simulate the shadows cast by the objects illuminated from this oblique lighting direction, and then its recombination with the chosen projection of the original image. This recombination is done in such a way that the shadow of one object feature is displayed in the final image only if it falls on the background, but not if it falls upon or behind other object features, thus preserving the ability to retrieve accurate intensity data from all the object features in the SHADOWED projection [Bennett et al., 1990; White et al., 1991a,b].

D. Surface and Volume Rendering

An alternative image display mode for three-dimensional images, particularly useful for those of compact fluorescent specimens, is that of displaying them as if surface-shadowed or surface-luminant. Two distinct ways exist for doing this, both of which involve preselection of the desired features to be displayed, usually by a thresholding operation, and the conversion of the selected data from raster to vector image format, with a consequent reduction in data volume and hence an increase in subsequent computational rendering and display speed.

The first method is to make use of *surface-rendering* procedures that are now standard on graphics workstations used for displaying computer-generated vector images. Here the primary problem is that of partitioning the object surface information required for

these procedures from the original continuous three-dimensional gray-level voxel
image, using the image analysis thresholding
procedures discussed above. Once the required surface has been defined, as contours
within each image plane, the coordinates
(only) of the voxels comprising this surface
are stored in vector format, defining a series
of polygons covering the surface. A wide
variety of surface-rendering procedures may
then be applied to tile such polygons with
triangles, to smooth the appearance of the
resultant tiled surface, and to apply apparent lighting from a variety of specified directions, complete with accurate reflections and
shadows, using ray-tracing algorithms. By
altering the constants that model the transparency or reflectivity of the surfaces, the
objects can be displayed as opaque or semitransparent, so that it is possible to see structures within structures, for instance the
nucleus within a cell. It is further possible
to use an "electronic knife" to section the
model at any arbitrary angle and look within.
The parameters used to control the rendering can usually be varied interactively and
adjusted subjectively until the most pleasing or informative appearance is obtained.
While extremely useful for the nonstereo-
scopic presentation of 3D data, such images
may also be made stereoscopic.

Alternatively, the objects of surfaces selected from within the entire image may be
rendered using the original intensity values
of the selected voxels. This procedure,
known as *volume rendering*, is relatively new,
and few software products are commercially
available for this purpose. It is significantly
more computationally demanding than surface rendering. However, its advantages are
that the entire raster image is not initially
discarded, but may be resampled interactively to optimize the selection parameters
for the desired features, and that the original intensity values of the selected voxels,
as well as their coordinates, are used in the
rendering process. As for surface rendering,
lighting, color, and transparency may be
modeled and adjusted in volume-rendered

images, and the rendered object may be
rotationally animated and displayed stereoscopically. In Chapter 11 of this volume, van
der Voort and Smeulders give an example
of one such voxel-rendering procedure,
known as *simulated fluorescence emission*, first
described by van der Voort et al. [1989].

In contrast to the projection methods
described in Sections VB and C above, the
surface brightness of such rendered images
has no simple correspondence with the original image intensity, preventing one from
retrieving accurate quantitative data from
them. One further disadvantage of these
rendering procedures over the more straightforward image projections discussed above
is that they are less appropriate for handling
images of dispersed or fragmented objects,
or of diffuse fluorescence staining patterns,
since in such images extended surfaces are
harder to define. Nevertheless, they are very
useful for generating striking representations of the shapes of more compact objects.

E. Computational Requirements

Given the appropriate software, for instance *ThruView*, written for the display and
animation of multidimensional digital confocal images for BioRad Microscience Ltd.
by my colleague Nicholas White, it is quite
possible to undertake the sort of image visualization described in Sections VB and C
above with great success using nothing
more powerful than a personal computer
with a 80386 processor and a 80387 math
coprocessor [White and Shotton, 1990; White
et al., 1991b]. The required projections are
individually precalculated and recorded digitally onto a magnetic or optical disk, or
time-lapsed onto videotape, and are then
played back sequentially to produce the
desired animation.

However, the requirement for sophisticated interactive postprocessing of large
three-dimensional confocal images may
soon outstrip the capabilities of a simple personal computer, and in my opinion this type
of work is best handled by off-loading these
tasks to a more powerful computational

workstation, freeing the microcomputer for its primary task of controlling microscopic image acquisition. Fortunately, the tumbling costs of computer workstations with powerful digital image-processing capabilities, and the availability of optical digital storage and stereoscopic image display hardware, now makes possible such sophisticated processing, animation, and stereoscopic display of digital light microscopic images.

To handle the variety of image processing and image display computational procedures which may be required, this workstation should be a general-purpose image processing engine of significant power (a computational speed of at least 10 million instructions per second), memory allocation (minimally 20 Mbytes), and disk storage capacity (at least 300 Mbytes), with fast graphics display hardware, a high-resolution screen display (at least 1,024 by 768 pixels, with a 60 hz refresh rate) and, ideally, special hardware for stereoscopic image display. The amount of image memory required for the display of animation sequences may be appreciated by considering a simple rotational animation at 5° intervals (72 views) of projections through a 512^3 voxel image. To provide the ability to switch between the animation of either of two of the three possible orthogonal rotational viewpoints [*tumble* (x), *roll* (y), and *spin* (z)], 36 Mbytes of available image display memory would be required. Four-dimensional animation, in which spatial animation of 3D objects is combined with display of temporal changes in the objects themselves, may also be achieved, and minimally requires the same amount of image memory as for two spatial animation sequences.

Electronic light microscopy, coupled with these digital image processing and display techniques, now makes possible a new era of cellular visualization, in which one can acquire, quantify, animate, stereoscopically view, and transform images of cellular ultrastructure and intracellular physiological parameters with a clarity previously unobtainable. The subsequent chapters in this book illustrate these possibilities.

REFERENCES

Agard DA (1984): Optical sectioning microscopy: Cellular architecture in three dimensions. Ann Rev Biophys Bioeng 13:191–219.

Agard DA, Hiraoka Y, Shaw PJ, Sedat JW (1989): Fluorescence microscopy in three dimensions. In Taylor DL, Wang Y-L (eds): "Fluorescence Microscopy of Living Cells in Culture: B. Quantitative Fluorescence Microscopy—Imaging and Spectroscopy." Methods in Cell Biology 30:353–377.

Agard DA, Sedat JW (1983): Three-dimensional architecture of a polytene nucleus. Nature 302:676–681.

Baxes GR (1984): "Digital Image Processing: A Practical Primer." Englewood Cliffs, NJ: Prentice-Hall.

Bennett ST, Fricker MD, Bennett MD, White N (1990): The 3-D localization of chromosomes using confocal microscopy. Trans R Microsc Soc 1:441–444.

Bright DS, Steel EB (1987): Two-dimensional top hat filter for extracting spots and spheres from digital images. J Microsc 146:191–200.

Carrington WA, Fogarty KE, Lifschitz L, Fay FS (1990): Three-dimensional imaging on confocal and wide-field microscopes. In Pawley JB (ed): "Handbook of Biological Confocal Microscopy." New York, London: Plenum Press.

Castleman KR (1979): "Digital Image Processing." Englewood Cliffs, NJ: Prentice-Hall.

Conchello J-A, Hansen EW (1990): Enhanced 3-D reconstruction from confocal scanning microscope images. 1. Deterministic and maximum likelihood reconstructions. Appl Optics 29:3795.

De Duve (1984): "A Guided Tour of the Living Cell," Vols 1, 2. New York: WH Freeman Co.

Dorst L, Smeulders AWM (1987): Length estimators for digitized contours. Computer Vision, Graphics, and Image Processing 40:311–333.

Erasmus SJ (1982): Reduction of noise in TV rate electron microscope images by digital filtering. J Microsc 127:29–37.

Gelles J, Schnapp BJ, Sheetz MP (1988): Tracking kinesin-driven movements with nanometre-scale precision. Nature 331:450–453.

Gonzalez RC, Wintz P (1987): "Digital Image Processing," 2nd ed. Reading, MA: Addison-Wesley.

Hiraoka Y, Sedat JW, Agard DA (1987): The use of a charge-coupled device for quantitative optical microscopy of biological structures. Science 238:36–41.

Holmes K, Blow DM (1965): The use of X-rays in the study of protein and nucleic acid structure. Methods Biochem Anal 13:113–239.

Inoué S (1989): Imaging of unresolved objects, super-resolution, and precision of distance measurement, with video microscopy. In Taylor DL, Wang Y-L (eds): "Fluorescence Microscopy of Living Cells in Culture. Part B: Quantitative Fluorescence Microscopy—Imaging and Spectroscopy." Methods in Cell Biology 30:85–104.

Kriete A (1990): 4D data acquisition and visualization methods in computer assisted microscopy. Trans R Microsc Soc 1:323–326.

Shaw, PJ (1990): Three-dimensional optical microscopy using tilted views. J Microsc 158:165–172.

Shaw PJ, Agard DA, Hiraoka Y, Sedat JW (1989): Tilted view reconstruction in optical microscopy: Three dimensional reconstruction of Drosophila melanogaster embryo nuclei. Biophysics J 55:101–110.

Shaw PJ, Rawlins DJ (1991): The point-spread function of a confocal microscope: Its measurement and use in deconvolution of 3-D data. J Microsc 163:151–165.

Sheetz MP, Turney S, Qian H, Elson EL (1989): Nanometre-level analysis demonstrates that lipid flow does not drive membrane glycoprotein movements. Nature 340:284–288.

Shotton DM (1989): Confocal scanning optical microscopy and its applications for biological specimens. J Cell Sci 94:175–206.

Shotton DM (1991): Video and opto-digital imaging microscopy. In Cherry RJ (ed): "New Techniques of Optical Microscopy and Microspectrometry." London: Macmillan Press, pp 1–47.

Shotton DM, White N (1989): Confocal scanning microscopy: Three-dimensional biological imaging. Trends Biochem Sci 14:435–439.

van der Voort HTM, Brakenhoff GJ, Baarslag MW (1989): Three-dimensional visualization methods for confocal microscopy. J Microsc 153:23–132.

White NS, Bennett ST, Kenton AY, Callimassia MA, Fricker MD (1991a): Characterizing plant chromosomes and their 3-D organisation using CLSM. Scanning 13(Suppl 1):128–131.

White N, Fricker MD, Shotton DM (1991b): Quantitative 3D visualisation of biological CLSM images. Scanning 13 (Suppl 1):51–53.

White N, Shotton DM (1990): Processing and visualization of 3D confocal images. Trans R Microsc Soc 1:319–322.

Wilson T, Sheppard CJR (1984): Theory and practice of scanning optical microscopy. London: Academic Press.

CHAPTER 3

Resolution and Sampling Requirements for Digital Image Processing, Analysis, and Display

Kenneth R. Castleman

Electronic Light Microscopy, pages 71–93 © 1993 Wiley-Liss, Inc.

I. INTRODUCTION

Before a particular digital imaging problem can be approached properly, one must determine that the instrumentation in use is adequate for the task. If it is not, then success becomes less likely, and failure is rendered inconclusive in that one does not know if a better system might have produced a more favorable result.

In particular, the resolution, magnification, number of pixels, pixel size, and pixel spacing must be properly selected for the task at hand. In this selection there should be a match between the optics (microscope), the image sensor (camera), the image digitization, storage, and display hardware, and the algorithms for processing and quantitative anlaysis of the digital images. Applicable theory from the fields of optics, electronics, statistics, and signal processing allows development of workable rules for establishing such a balance.

The question we address in this chapter is: How can one analyze a system and determine if it is both adequate and cost-effective for carrying out the image processing and quantitative image analysis projects of interest on cells or other biological specimens? We seek a method to establish a balance among the various components in the imaging chain so that the overall performance is adequate for the task, and none of the components represents overkill compared with what is required to do the job.

We address the topics of spatial resolution and image sampling with the goal of establishing a balance between the performance of each of the system components and that of the instrument overall. We are aided in this endeavor by a wealth of well-developed and applicable theory in the several disciplines that are involved in quantitative digital image analysis. We can use this to relate the performance of the various components to that of the system as a whole.

While it is possible to analyze and specify spatial resolution parameters with great accuracy, to do so makes the analysis unnecessarily complex. It better serves our purposes to make some well-founded assumptions that then lead us to simple and broadly applicable rules of thumb. From there a margin of safety can be added to protect against error in the underlying assumptions. In most cases of practical interest, the resulting loss of accuracy is acceptable.

A. The Resolution Question

Considerable confusion often arises around the concept of resolution. To avoid this, one needs a clear definition of resolution and a clear understanding of the goal of any analysis of the resolving power of an imaging instrument. For our purposes, the key resolution question is: Will the system adequately reproduce the small detail of interest in the specimen? If such detail is lost in the imaging process, success is endangered. This question can be answered readily if we first have a concise, quantitative answer to the question: How well does the system reproduce objects of different sizes? Then, assuming we know the size of the objects of interest, we can obtain a yes or no answer to the key question of resolution.

B. The Sampling Question

Another key question regards the parameters of the sampling process. It can be stated as: How many pixels are required to ensure that the digitized image adequately represents the content of the microscope image? This involves an entirely different set of considerations from those associated with resolution. Any analysis that fails to keep the sampling and resolution considerations separate is doomed to create confusion and end in failure.

C. The Display Question

Yet a third key question relates to the conversion of the digital image back into visible form. It can be stated as: Will the displayed image adequately represent the objects of interest? In some applications, usually those involving only quantitative analysis, image display is less important or even unnecessary. In others, usually those involving image processing and human interpretation, it is a vital element. As before, image display involves different considerations from those of resolution and sampling, and it must be dealt with separately.

D. The Approach

A workable approach that answers the key questions quantitatively is outlined here and developed in this chapter.

1. Resolution. To approach the resolution question, we apply the tools of linear system theory to those components of the system that precede the actual sampling (conversion from analogue to digital form). These components can be assumed to behave as linear, shift-invariant system components, and thus linear system theory applies.

In particular, we shall analyze the image-forming optics and the image sensor (camera) to determine the effective scanning spot size and shape. From this comes the imaging system point spread function (psf) and its equivalent, the imaging system modulation transfer function (MTF). This MTF forms the quantitative specification of resolution we require for this portion of the analysis.

2. Sampling. To approach the sampling question, we apply the Shannon sampling theorem to the analogue-to-digital conversion step. This yields a simple way to determine if the pixel spacing is small enough, and it describes what happens if it is not.

3. Display. To approach the display question, we view the image display process as an interpolation step and again apply the Shannon sampling theorem. This yields a way to determine if the display process is adequate, and gives guidelines for improvement if it is not.

4. Practical considerations. Once each of the basic processes mentioned above has been analyzed, one can combine the three results to determine if the overall system design is balanced and adequate for the specific applications for which it is intended. Finally, one must assess the effect of each of the assumptions and approximations that are made in the analysis and the effect that noise in the system will have.

II. RESOLUTION

While most people have a concept of what resolution is, system analyses often fail to produce definitive results. In this section we develop an approach that can be used in practical situations.

A. Definitions

Before we can develop a coherent approach to specifying the resolution of an imaging system, we must state some definitions. These are by no means standard in the field, but they form a workable basis upon which we can proceed.

By *resolution* we mean the ability of the imaging system to reproduce the contrast of objects of various sizes. Of particular interest are the smaller objects, since they are usually the most troublesome.

By *contrast* we mean the intensity differences within an object or between it and the surrounding background. If an object were to lose contrast as a result of the imaging process, it would appear fainter in the image than it actually was in the specimen. If its contrast were reduced all the way to zero, it would disappear.

1. The sine wave bar target. We need to establish a standard type of test object that can take on different sizes. Then we can specify what happens to the contrast of these differently sized test objects as they pass through the imaging system.

For our purposes, the best choice of these test objects is alternating dark and light bars. Such a pattern is called a *bar target*. For reasons to be pointed out later, the best form for the variation from light to dark is that of the sine function from trigonometry. This pattern is called a *sine wave bar target* (Fig. 1).

The objects in the test image are actually pairs of adjacent bars, one light and one dark. These are referred to as *line pairs*. The size of the line pairs can be specified by their spatial width. This is the actual dimension of the repeat pattern. A more useful size specification for our purposes is the *frequency* in line pairs per unit of length. Thus large bars have a low frequency, while small bars have a higher frequency.

2. The modulation transfer function. For our purposes, the most convenient expression of the resolution of an imaging system takes the form of the *modulation transfer function (MTF)*. The MTF is a graph that shows how the contrast of differently sized bars is affected while passing through the imaging system. It is a plot of the ratio of output to input contrast for the line pairs in a sine wave test target as a function of the frequency of the line pairs.

The MTF is a real-valued function that accounts only for the loss of contrast of the line pairs during the imaging process, and not for any positional (sidewise) shift (called

phase shift) that may occur, whereas the complex-valued *transfer function*, commonly used in linear system theory, can account for both. The MTF can, however, go to zero at some frequencies, indicating a total loss of the line pairs at those frequencies. It can even go negative at some frequencies, indicating a black-for-white reversal of the line pairs at those frequencies.

3. Linearity and shift invariance. Many of the components of an imaging system can be assumed to behave as *linear system* components. This means that an increase in the contrast of the input (specimen) produces a proportional increase in the contrast of the output (image).

Many of the components of an imaging system also can be assumed to behave as *shift-invariant* components. This means that the relationship between the image of an object and the object itself is independent of where that object is located within the field of view of the imaging system.

It is a well-known result of linear system theory that a shift-invariant linear system, when presented with a sine wave input, can only produce an output that is likewise a sine wave having the same frequency [Castleman, 1979]. Such a system can change the amplitude of the sinusoidal input and shift it right or left, but it cannot change it into something other than a sine wave, nor can it change its frequency. It is also well known from linear system theory that a shift-invariant linear system is completely specified by its complex-valued transfer function [Castleman, 1979].

Commonly the components of a digital imaging system are, or can reasonably be assumed to be, *phaseless, shift-invariant linear systems*. Phaselessness means they introduce no phase shift in the image, as discussed above. The transfer function of a phaseless component is real-valued (rather than complex-valued), and is simply the MTF. Thus the MTF tells all there is to know about a phaseless, shift-invariant linear system, and becomes a very convenient specification for our purposes.

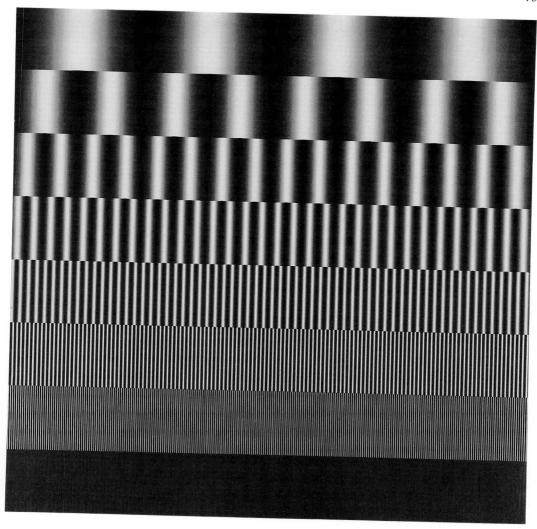

Fig. 1. Sine wave bar target. This type of test pattern is useful for measuring the resolution of an imaging system. It is made up of alternating light and dark bars having a sinusoidal variation of intensity. Each row contains bars of a different frequency. In this orientation, the target would aid measurement of resolution in the horizontal direction. It can be rotated to measure resolution in other directions.

It is fortunate that the commonly used components of electronic light microscopes (lenses, cameras, amplifiers, display monitors, etc.) have, at least approximately, these three characteristics of linearity, shift invariance, and phaselessness defined above. It allows us to use the MTF as a complete specification of how each component performs during the imaging process.

B. Component MTFs and the Overall System MTF

Most imaging systems are composed of a series of components through which the image passes sequentially. When components do occur in cascade in the imaging chain, their MTFs combine by multiplication to form the overall MTF of the system. Thus if the MTFs of the individual components are known, the system MTF can be determined by multiplying them together.

For most imaging systems encountered in biological research, the components critical to the resolution question are those located between the specimen and the analogue-to-digital converter. For these reasons we shall use the imaging system MTF as the basic specification for the resolution of the electronic light microscope system.

The imaging system MTF of a digital imaging microscope is determined principally by two components, the microscope objective lens and the image sensor (camera). In general there may be several lenses, filters, mirrors, and beam splitters in the optical path from specimen to sensor. The microscope objective lens, however, is ordinarily the primary optical component that limits resolution and, together with the camera, determines the overall image quality.

C. Optical Image Formation

In this section we discuss the process of image formation by a lens or a system of optical components. Our primary interest is the MTF of the microscope objective which forms an image of the specimen on the image sensor.

For an initial approach, it is sufficient to analyze the simple case of an objective lens imaging a specimen directly onto a planar image sensor. This arrangement is more straightforward and simpler than designs involving additional lenses in the optical path. More complicated optical systems can generally be modeled as being equivalent to a single-lens system for purposes of analysis.

1. Imaging geometry. Figure 2 shows the geometrical model we shall use to analyze the image formation process. A lens of diameter a is located a distance d_i from an *image plane*. The image sensor would be located at the image plane, as this is where the optical image of the specimen is formed. A *focal plane* exists a distance d_f on the oppo-

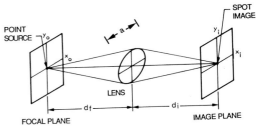

Fig. 2. Geometrical model of the imaging process. A lens of diameter a and focal length f is located a distance d_i in front of an image plane. Then the focal plane exists a distance d_f in front of the lens. A point source in the focal plane produces a small spot in the image plane. In operation, the specimen is placed at the focal plane and the image sensor (camera) at the image plane.

site side of the lens. This is where the object of interest or specimen must reside if it is to be imaged in focus. Thus it can also be called the *specimen plane* or the *object plane*. The focal plane, the lens, and the image plane are all orthogonal to the optical axis.

Some of the light passing through, or emitted from, the point on the specimen at x_o, y_o will intercept the lens. The action of the lens is such as to bend each incoming light ray to make it converge to a corresponding point in the image plane. All rays of light emanating from the point source at x_o, y_o and entering the lens will be converged to the point x_i, y_i in the image. Thus a point source in the focal plane gives rise to a small spot of light in the image plane.

In general, the specimen will contain a continuous distribution of light-absorbing or light-emitting objects. Nevertheless, each point in the focal plane maps to a unique location in the image plane. A point source anywhere in the focal plane will produce a spot at the corresponding location in the image plane. The specimen, then, can be thought of as a distribution of point sources.

In a high-quality imaging system, a point source in the focal plane produces a very small spot in the image plane. In less well-designed systems, the spot image is not so small. Thus the resolution question can be approached by examining how small a spot image is produced by a point source in the focal plane.

The image produced in the image plane will, in general, be larger than the specimen itself, due to the factor of magnification. Ideally the image would be an exact scaled-up copy of the pattern of light emerging from the specimen. Geometrical distortion, if present, introduces a "warping" that renders the image an inexact copy of the specimen. For present purposes, we shall assume that geometric distortion is negligible.

The *field of view* of an objective specifies the size of the area of the specimen which can be imaged at one time.

2. The parameters of an imaging system. In this analysis we make use of the following fundamental parameters of the imaging model in Figure 2: f, focal length of objective lens; a, aperture diameter of objective lens; d_i, lens to image plane distance; d_f, lens to focal plane distance; λ, wavelength of the narrow band incoherent light.

The lens to image plane distance, d_i, is fixed in the design of the microscope. Each microscope manufacturer uses a particular standard value for the mechanical tube length, 160 mm being a popular choice [Bradbury, 1984]. This places d_i at about 190 mm.

The aperture diameter and focal length of the objective vary from one lens to the next. These parameters determine the magnification and resolution of the image.

The focal plane exists at a distance d_f in front of the lens. Objects in the focal plane will be imaged in focus. The distance, d_f, is determined by the focus equation

$$\frac{1}{d_i} + \frac{1}{d_f} = \frac{1}{f} \qquad (1)$$

which implies

$$d_f = \frac{f \cdot d_i}{d_i - f} \qquad (2)$$

The magnification factor that relates the size of objects in the focal plane to their images in the image plane is given by

$$M = \frac{d_i}{d_f} \qquad (3)$$

Ordinarily d_f is much smaller than d_i, and M is thus much greater than 1. M is the familiar "objective power" that is typically engraved on the side of the objective lens. It represents the enlargement factor present in the imaging process.

Although objects appear in the image plane magnified by the factor M, it is their actual size (in the focal plane) that is of interest. For this reason it is convenient to adopt the convention of always referring size parameters (pixel spacing, resolution, specimen size, etc.) back to the focal plane. This is easily done by dividing image plane values by the magnification factor. In complex systems involving a cascade of components, it avoids possible confusion that can result from having a different magnification at each point in the imaging chain.

While the aperture diameter a is, itself, an adequate specification, it is convenient to use an alternative called the *numerical aperture*. For the simplified imaging model shown in Figure 2, this is given by

$$NA = \frac{a}{2 \cdot d_f} \qquad (4)$$

and it expresses the aperture radius in terms of the distance from the lens center to the focal plane. More general discussions of numerical aperture of microscope objectives appear elsewhere [Born and Wolf, 1964; Bradbury, 1984].

3. The point spread function. In the discussion surrounding Figure 2, we noted that a point source in the focal plane produces a small spot of light in the image plane. If the optical system had unlimited resolution, the spot in the image plane would have zero spatial dimension, just like the point source.

In practice, however, the spot image has finite size.

The spot that is produced by an imaging system with a point source input is called its *point spread function* (psf). Techniques developed in the field of optics allow us to derive mathematically the shape and dimension of the point spread function for a particular imaging system.

In the following analysis, we assume that the lens is without aberration. This means it is a theoretically perfectly manufactured lens whose resolution is limited only by the physical properties of light itself. Such an ideal lens is said to be *diffraction-limited*, because the only limitations on its resolving power are those imposed by the diffraction of light.

Assuming a diffraction-limited lens with a circular aperture, operating in narrowband incoherent light, it can be shown [O'Neill, 1963; Born and Wolf, 1964; Harris, 1964; Goodman, 1968; Castleman, 1979] that the point spread function is given by

$$h(r) = \left[2 \cdot \frac{J_1\left[\pi \cdot \left[\dfrac{r}{r_o}\right]\right]}{\pi \cdot \left[\dfrac{r}{r_o}\right]} \right]^2 \quad (5)$$

where $J_1[x]$ is the first order Bessel function of the first kind. The constant r_o, which serves as a dimensional scaling factor, is given by

$$r_o = \frac{\lambda}{2 \cdot NA} \quad (6)$$

and r is the radial distance measured from the optical axis and given by

$$r = \sqrt{x_f^2 + y_f^2} \quad (7)$$

where x_f and y_f are coordinates in the focal plane. The expression in Equation 5 refers to the focal (specimen) plane.

The two-dimensional point spread function is circularly symmetric, and its shape

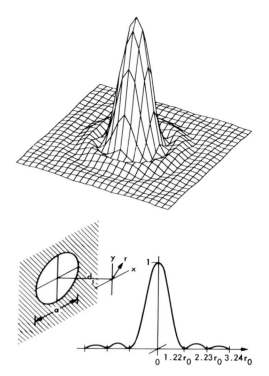

Fig. 3. The point spread function of an ideal lens. A point source of light in the focal plane produces this distribution of light in the image plane. For our purposes, we refer all size calculations back to the focal (specimen) plane. For a 1.2 NA lens in green light, the diameter at half amplitude of the central peak is 0.28 μm, which is also the radius at the first zero. The outlying rings are much smaller in amplitude than the central peak.

is plotted in Figure 3. This is the familiar *Airy disk* pattern, which has a main central peak and a series of small-amplitude side lobes of decreasing intensity.

Ideally (for best resolution) the point spread function would be extremely narrow, and without side lobes. From Equation 5 we see that the scale factor r_o is a specification of the psf width. It becomes smaller with larger numerical aperture, and it takes on smaller values at shorter illumination wavelengths (Eq. 6). The side lobes cannot be eliminated.

To better understand how a larger aperture improves resolution, consider the following. As light enters the specimen parallel to the optical axis, the small structures therein cause the light to be bent by the phenomenon of diffraction. The smaller the structures, the greater the angle of diffraction. For structures below some limiting size, the light diffracted by them will exit the specimen at such an angle that it will not enter the lens aperture, and thus will not contribute to the formation of the image. Increasing the aperture diameter, and thus the NA, will allow light diffracted by smaller structures to contribute to the image.

In Figure 3, the point spread function has its first zero at a radius of 1.22 r_o. According to the *Rayleigh criterion* of resolution, two point sources can just be distinguished if they are separated, in the focal plane, by that distance [Castleman, 1979, 1987; Bradbury, 1984; Inoué, 1986]. Thus a commonly used way to specify the resolution of an imaging system is the Rayleigh criterion given by

$$d = 0.61 \cdot \frac{\lambda}{NA} \tag{8}$$

Notice again that resolution improves (d becomes smaller) at shorter wavelength and higher numerical aperture. For a $100\times$, 1.2 NA objective in green light, the Rayleigh resolution is approximately 0.28 μm.

Another way to specify the resolution of a lens is by the diameter of the central peak of the point spread function. To a good approximation, the equivalent diameter of the psf is also given by Equation 8 [Castleman, 1979, 1987]. Thus the psf diameter also becomes smaller at shorter wavelength and higher numerical aperture.

4. The optical transfer function. The optical transfer function (OTF) is a complex-valued function of two orthogonal spatial frequency variables. It specifies how well the lens can reproduce in the image plane sinusoidal features that occur in the focal plane. Its modulus (magnitude) is the MTF of the lens.

High-quality lenses are designed to introduce a minimum of phase shift, and can be assumed to be phaseless. This means the OTF reduces to a (real-valued) MTF. For our purposes, then, we can use the terms OTF and MTF interchangeably. As stated before, the value of the MTF, at a particular frequency, is the factor by which the contrast of sinusoidal features in the specimen, at that frequency, is multiplied in the imaging process.

The OTF is obtained mathematically as the two-dimensional Fourier transform of the psf [Bracewell, 1965]. It is a property of the two-dimensional Fourier transform that, if the psf is an even function (symmetrical about the x- and y-axes), then the OTF will be real-valued and, likewise, an even function [Bracewell, 1965; Castleman, 1979]. Since the lens is circularly symmetric, the image it forms of a point source is likewise symmetrical. Thus we see that a phaseless system, a real-valued OTF, and a symmetrical psf occur together.

The OTF of a diffraction limited lens is given by O'Neill [1963], Harris [1964], Born and Wolf [1964], Goodman [1968], and Castleman [1979]:

$$H(q) = \frac{2}{\pi - 2} \cdot \left[\cos^{-1}\left[\frac{q}{f_c}\right] - \sin\left[\cos^{-1}\left[\frac{q}{f_c}\right]\right] \right] \tag{9}$$

where q is the spatial frequency variable, measured radially in two-dimensional frequency space. It is given by

$$q = \sqrt{u^2 + v^2} \tag{10}$$

where u and v are spatial frequencies in the x and y directions of the focal plane, respectively.

The parameter f_c is called the *cutoff frequency*, and it is given by

$$f_c = \frac{2 \cdot NA}{\lambda} \tag{11}$$

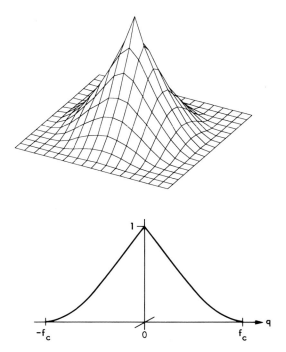

Fig. 4. The optical transfer function of an ideal lens. The lens will not pass image detail corresponding to a sine wave frequency greater than f_c. Detail at frequencies near f_c is attenuated severely. For a 1.2 NA lens in green light, f_c is 4.36 cycles per micron.

The OTF for a lens with circular aperture is plotted in Figure 4. It starts at unity at zero frequency (large objects) and decreases steadily to zero at the cutoff frequency. Sinusoidal structures having frequencies above f_c do not appear in the image. This points out that any lens would be incapable of resolving structures smaller than some limiting size.

In Equation 11 it is evident that the cutoff frequency can be improved (increased) by going to a higher numerical aperture or a shorter wavelength. For a $100\times$, 1.2 NA objective in green light, f_c is 4.36 cycles per micron.

The psf and the MTF are related by the two-dimensional Fourier transform. The Fourier transform of the psf is the MTF, and the inverse Fourier transform of the MTF is the psf. Thus, having either, one can obtain the other.

There is a property of the Fourier transform known as the similarity theorem. This states, loosely, that the broader a function is, the narrower will be its Fourier transform, and vice versa [Bracewell, 1965; Castleman, 1979]. Since the psf and MTF are related by the Fourier transform, we see that a narrow psf and a broad MTF go hand in hand. Clearly, the better imaging system is the one able to form a smaller spot image from a point source. Likewise, the better imaging system has a broader MTF and thus allows higher frequencies to pass through.

D. The Spectrum of the Specimen

An important consideration so far not addressed is how much detail is present in the specimen itself. Clearly the imaging system need only be capable of resolving the smallest objects of interest that actually occur in the specimen. A plot of the amplitude of the sinusoidal components in the specimen itself, as a function of frequency, is referred to as the *spectrum* of the specimen.

An approximation to the spectrum of the specimen can be obtained numerically by applying the two-dimensional Fourier transform to a high-quality image of a typical specimen. In this case, one must account for the fact that the imaging system's MTF has operated on the image. If that MTF is known, one can estimate what the spectrum of the specimen was at the input to the imaging process.

Once the spectrum of the specimen is known, even if only approximately, one has an idea which frequencies are important in the imaging process. When the specimen spectrum and the imaging system MTF are plotted on the same scale, one can see the relationship between what information is present in the specimen and the system's ability to pass that information through to the image.

E. Digital Image Size

We can now address the question of how many sample points (pixels) per scan line and how many scan lines are required to digitize an image from an optical microscope. This involves considerations of the area of the specimen to be covered by the digital image and the pixel spacing at the specimen.

1. Field of view. The diameter in millimeters of the circular field of view at the specimen covered by a typical microscope objective lens is

$$D = \frac{18}{M} \tag{12}$$

where M is the magnification (power) of the objective. Specially designed wide-field objectives have a somewhat larger numerator in Equation 12.

Assuming this diameter is reduced by 20% as a precaution to avoid possible defocus in the corners of the image, the approximate width of the largest square image field that can be placed within the circular field of view of the objective is given by

$$W = \frac{10,000}{M} \tag{13}$$

where W is measured in microns at the specimen.

2. Resolution cells. The value d (Eq. 8) defines circular resolution cells in the image, since two point sources cannot be resolved if they fall within the same resolution cell. It is also approximately the diameter of the point spread function of a perfect lens. For practical purposes we can consider the psf diameter and the resolution cell diameter to be the same and use them interchangeably. Actual objectives can only approach this theoretical upper limit of resolution, and then only if clean and properly aligned and if the NA of the condenser equals or exceeds that of the objective.

Table I shows, for several objective lenses, the maximum width of an inscribed square image in microns (from Eq. 13), the diameter of the resolution cells in microns (from Eq. 8), and the number of resolution cells N across the image computed from

$$N = \frac{W}{d} \tag{14}$$

Table I assumes a theoretically perfect lens with a condenser of at least equal NA and green light ($\lambda = 0.55$ μm) for illumination. For specially designed wide-field objectives, the values in the third and fifth columns would increase by about 40%.

TABLE I. Width of Square Image, W (μm), Diameter of Resolution Cells (μm), and Number of Resolution Cells, N, Across the Image, for Several Objective Lenses[1]

Objective Power (M)	Numerical Aperture (NA)	Image Width (W)	Resolution Cell Diameter (d)	No. of Cells (N)
10	0.4	1,000	0.84	1,190
20	0.6	500	0.56	893
20	0.8	500	0.42	1,190
40	0.6	250	0.56	446
40	0.8	250	0.42	595
40	1.0	250	0.34	735
63	1.0	159	0.34	468
63	1.4	159	0.24	663
100	1.2	100	0.28	357

[1]It assumes a perfect lens, with green ($\lambda = 0.55$ μm) light for illumination. It does not take into account the sensing spot of the electronic image sensor, which would reduce the number of resolution cells across the image (see text).

There are many factors that reduce the actual number of resolution cells below the values presented in Table I. Among these are (1) imperfection and aberration in the objective lens, (2) the NA of the condenser, if less than that of the objective, (3) improper alignment of the microscope, (4) foreign matter on the optical surfaces, (5) any defocus of objective, condenser, or relay lenses, (6) the psf of relay lenses if present, and (7) the size of the scanning spot of the image-sensing element. Thus Table I represents an upper limit on resolution that can only be approached in practice.

The number of resolution cells across the image represents a minimum for the number of pixels per scan line required to digitize the image adequately. In order to resolve adjacent point sources in a digital image, for example, one would have to have a pixel spacing of half the Rayleigh distance so that the pixels would alternatively fall upon and between the point sources in the image. Due to this and other practical considerations discussed later, it might be necessary to increase the sampling frequency by a factor of two or so beyond the minimums specified in Table I.

F. Image Sensor Resolution

The scanning spot size and pixel spacing are actually established by the electronic image sensor that is located at the image plane. These parameters, however, can be scaled down to the focal (specimen) plane simply by dividing by the magnification factor, and it is useful to do this as a matter of course.

The sensing spot of the electronic image sensor combines with the psf of the objective lens, by the process of convolution, to form the overall psf of the image-input side of the system. This specification, called the *imaging system psf*, accounts for all imaging components up to the analogue-to-digital converter, which is where the image is converted to digital form. While relay lenses in

the system contribute similarly to the objective, they typically can be neglected. Their psfs are usually quite narrow compared with that of a high-power objective, and this produces little effect on the imaging system psf.

The imaging system psf will be broader than either the objective lens psf or that of the image sensor. To a reasonable approximation, the width of the imaging system psf is the square root of the sum of the squares of the widths of the lens and sensor psfs.

As a numerical example, consider a microscope with a $100 \times$, 1.2 NA objective lens and a TV camera having a 25 μm diameter scanning spot and a 10 mm square image scanning area on the tube face or charge-coupled device (CCD) array. From Equation 8, the lens psf is 0.28 μm wide at the focal plane. Assuming no additional magnification beyond that of the objective, the scanning spot scales down to 0.25 μm diameter at the specimen, and the scanning area scales down to a 100 μm square. Thus the approximate diameter of the imaging system psf is

$$w = \sqrt{0.28^2 + 0.25^2} \ = 0.38 \, \mu m \quad (15)$$

at the focal plane. It follows that the actual number of resolution cells across the 100 μm-wide image would be 263 rather than the 357 shown in Table I, which ignores the scanning spot of the image sensor.

In this case, a standard 512 by 512 pixel digitizer operating on the TV camera output signal should be adequate. The number of pixels per scan line would be almost twice the number of resolution cells across the image.

III. SAMPLING

Image sampling and quantization are necessary prerequisites for processing an image in a computer. The sampling process can be done well or poorly. If it is done poorly, the image will suffer a loss of information content. In this section we address the factors that influence how well the sampling and quantization process preserves the content of the image.

A. The Basics of Image Sampling

So far in our discussions we have considered an image to exist in continuous form. That is, for example, a smooth, continuously varying pattern of light intensity cast on a surface by a lens. Such an image can be described mathematically by a continuous function of two continuous variables [i.e., $a(x,y)$, where a, x, and y are continuous variables]. A continuous variable is one that can take on any value.

We have considered an image to be something that exists everywhere on a plane, and that, at every point, can take on any of infinitely many possible values.

Electronic light microscopy involves both optical and *digital* imaging—the processing and analysis of images in a digital computer. A digital computer cannot accept or process data in continuous (analogue) form. It can only process data that is entered in digital form.

For our purposes, the definition of the word *digital* is "organized in discrete units, and occurring in integer form." The word *discrete* means organized as separate individual units. *Integer form* means that fractional values (of brightness, for example) are not allowed.

Before we can process an image in a digital computer, it is necessary to convert this continuous intensity function of two continuous spatial variables into a discrete intensity function of two discrete spatial variables. This involves two separate processes, sampling and quantization.

Whereas a continuous image exists everywhere within its borders, a *sampled* image exists only at a predetermined set of discrete sample points (picture element or *pixel* locations). Whereas a continuous image can take on any possible intensity value at any point, a *quantized* image can only take on one of a finite number of integer values.

Image sampling and quantization together comprise the process of digitization, the generation of a discrete function of two discrete variables to represent a particular continuous function of two continuous variables. The approach commonly used is (1) to divide the image into a set of sample points (pixels), thus discretizing the spatial coordinates x and y, and then (2) to measure the local property of the image (e.g., brightness) at each pixel location and represent that value as an integer.

The brightness at each pixel location could be represented, in the computer, as a real data type (floating point variable). Because the number of pixels is large, however, it is a much more effective use of computer memory to represent each brightness value as a single 8-bit binary integer (byte). If the resulting 2^8 or 256 gray levels are not enough to represent adequately the dynamic range of the image, the pixel values can be encoded as 16-bit integers. This allows 2^{16} or 65,536 possible gray levels.

Each pixel has the following three properties: (1) its x,y position on the sampling grid in relation to other pixels, (2) its gray level, which is related to the image brightness or other local property at that point, and (3) its size and shape. The first of these, pixel spacing, is determined by the parameters of the digitizer (i.e., horizontal and vertical pixel spacing at the image sensor, scaled down to the specimen). The third represents the finite area over which the image intensity is averaged (during the digitization process) in order to determine the gray level. The size and shape of this area is specified by the imaging system psf.

1. Sampling for display. In many cases it is necessary to sample an image in such a way that it can be redisplayed after processing. One criterion of how well sampling has been done is whether or not the sampled image can be redisplayed (converted back to continuous form) without a serious loss of resolution or the introduction of distortion or artifact. Clearly, if this can be done, then the sampling process *per se* has not introduced undesirable effects.

As one would expect, the more finely we sample an image during digitization (the more pixels used and the closer they are together), the better we will be able to display the objects within the image later. While it might be tempting always to use a large number of closely spaced pixels, the resulting data volume then becomes so large that practical considerations impose restrictions.

2. Sampling for measurement. Another criterion for adequate sampling would be whether or not the digital image contains sufficient information to permit the extraction of accurate measurement parameters. If the sampling is properly done, measurement parameters reflecting size, shape, and density can be extracted from the objects in the digital image. If the sampling process prohibits accurate measurement of objects in the image, then it is inadequate.

In both display and measurement, the situation reduces to the practical question of how much is enough. The approach commonly used is to determine what pixel spacing is required to produce the desired level of performance and use that value, perhaps increased by a margin of safety. This minimizes the required data volume.

B. The Sampling Theorem

The Shannon sampling theorem offers a very concise way to approach the question of how much is enough with regard to sampling an image for subsequent display. The analysis in this section allows us to determine under which conditions undersampling will produce distortion in the displayed image, and what form that distortion will take.

The steps in the analysis are the following: (1) Assume a functional form for the continuous function that represents the image to be sampled, (2) establish a way to specify the spatial size of the detail in the input image, (3) establish the sample spacing or sampling density with respect to the size of the detail in the input image, (4) establish a "best" method for reconstructing the continuous function from the sample values, and (5) determine how the sampling density affects the accuracy with which one can reconstruct the original function from its sample values using this best method.

In this discussion, the word *sampling* refers to the process of converting the continuous input function into a discrete (sampled) function by picking off the value of the function at specified positions (sample points). The word *interpolation* refers to the process of using the sample values to reconstruct a continuous function. This entails fitting a continuous curve through the sample points.

1. The sine wave input signal. We choose the sine wave as the standard form of the continuous input function for the sampling process. This form permits us to take advantage of the Shannon sampling theorem, a mathematical development that makes very definite statements about whether or not the sampling density is adequate. While the detail within an image is not usually of sinusoidal form, Fourier's theorem points out that any analytic function can be expressed as the sum of a set of shifted sine functions at different frequencies [Bracewell, 1965; Castleman, 1979]. This means that any image, however complex, can be thought of as a summation of different sine waves.

2. Frequency. To describe the spatial size of the continuous input function, we again use the frequency of the sine wave. The period of the wave is the distance between peaks in the uniformly repeating pattern. The frequency is the reciprocal of that spacing. Thus a sine wave having 0.5 mm between adjacent peaks (line pairs) has a frequency of 2 cycles per millimeter. The sampling density can also be specified by its frequency in samples per unit length (e.g., per millimeter).

As one's intuition might suggest, it is actually the relationship between the sine wave frequency and the sampling frequency that is important in determining whether or not the sampling is inadequate. The critical parameter is the number of sample points per cycle of the sine wave.

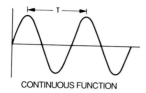

CONTINUOUS FUNCTION

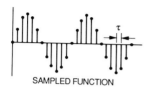

SAMPLED FUNCTION

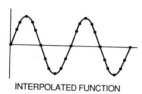

INTERPOLATED FUNCTION

Fig. 5. Sampling and interpolation. Sampling converts a continuous function into a series of sample values by retaining its amplitude only at the sample points. Interpolation converts the sample points back to a continuous function by fitting a smooth curve through the sample points. Here a sine function is sampled at 12 points per cycle.

3. Interpolation. Figure 5 illustrates the process of sampling a sine wave and then reconstructing it by interpolation. The sampling process consists merely of measuring the amplitude of the sine wave at each of the predetermined sample points.

Interpolation is the process of reconstructing a continuous function from a set of sample values. It has the effect of fitting a smooth curve through the sample points. It is done by placing one copy of an *interpolation function* of appropriate amplitude at each of the sample points and then summing these to produce the reconstructed continuous function. This process is seen in Figure 6 and again in Figure 10.

The sampling theorem specifies the mathematical form that the interpolation function should take. That proper form is $\sin(\alpha x)/\alpha x$ (Fig. 6) since this is the only interpolation function that yields error-free reconstruction of a sampled sine function [Castleman, 1979, Chap. 12]. The scale parameter, α, takes on the value $2\pi/2\tau$, where τ is the pixel spacing. This particular value causes the spacing of the zero crossings of the interpolation function to be the same as the pixel spacing. This makes the zero crossings correspond to the locations of the adjacent samples, as shown in Figure 6.

At each sample point, only the copy of the interpolation function that is centered at that point contributes to the summation, since all the other copies have a zero crossing there. Between sample points, however, all copies contribute to the summation.

In Figure 5 we see that reconstruction from samples taken 12 per cycle works well. Indeed, the sampling theorem says that interpolating with the $\sin(\alpha x)/\alpha x$ function will reconstruct the sine wave exactly. It can be recovered from its sample points totally without error. This is very encouraging because it suggests that it may be possible to avoid completely any errors or artifacts due to processing, as we must, images that have been sampled.

Figure 7 shows what happens when we reduce the sampling density to four samples per cycle. Again the reconstructed sine wave is exactly like the original (unsampled) function.

4. Aliasing. In Figure 8, we further reduce the sampling density to 1.5 samples per cycle of the input sine wave. In this case the result is less encouraging. The function reconstructed by interpolation is, as before, a sine wave. Furthermore, it has the proper amplitude. Its frequency, however, is now only half that of the input function.

Figure 8 illustrates the phenomenon of *aliasing*. This is the change of frequency that occurs when a sine wave is reconstructed by interpolation from sample points that were taken too far apart. Aliasing results when the sampling is too sparse compared to the detail in the image. When aliasing

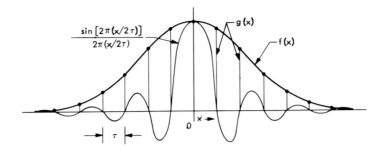

$$\frac{\sin\left[2\pi(x/2\tau)\right]}{2\pi(x/2\tau)}$$

Fig. 6. The interpolation process. The reconstruction of a sampled function involves placing appropriately sized copies of the interpolation function at each sample point and then summing them up. Here $f(x)$ is the result of interpolating the sampled function $g(x)$ with the interpolation function $\sin(\alpha x)/\alpha x$. Although only one is shown, in actual fact one copy of the interpolation function is placed at each sample point.

occurs, there is a change in the frequency (spatial size) of the detail in the interpolated image as compared to the original image.

The phenomenon of aliasing does not come about gradually as the sample points get farther apart. There is a critical value of sample spacing below which no aliasing whatever occurs. This is the value of 2 sample points per cycle. As long as the sampling frequency is more than twice the frequency of the sine wave, that sine wave can be reconstructed perfectly by interpolation with the $\sin(\alpha x)/\alpha x$ function. Whenever the sampling frequency is less than twice the frequency of the sine wave, aliasing will occur, as illustrated in Figure 8.

It is curious that the phenomenon of aliasing does not affect the amplitude of the sine wave, nor does it change its functional form. Aliasing alters only the frequency of the sine wave.

There is a simple rule by which one can predict the frequency change induced by the

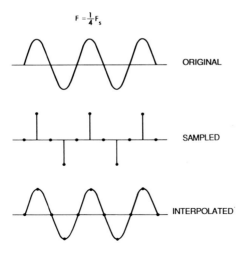

Fig. 7. Sampling at 4 samples per cycle. When a sine wave is sampled with 4 sample points per cycle, it can be recovered exactly by proper interpolation.

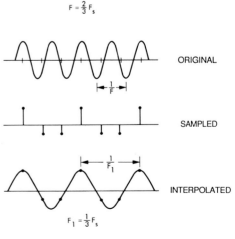

Fig. 8. Sampling at 1.5 samples per cycle. When a sine wave is sampled with 1.5 sample points per cycle, it cannot be recovered exactly. The result after interpolation is as change of frequency (aliasing). In this case, aliasing reduced the frequency from $\frac{2}{3}$ to $\frac{1}{3}$ of the sampling frequency.

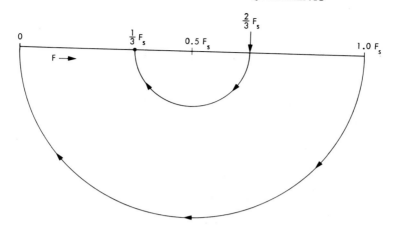

Fig. 9. Frequency folding. When a sine wave whose frequency falls above the folding frequency is sampled and interpolated, aliasing makes the frequency of the reconstructed sine wave fall an equal distance below the folding frequency.

aliasing phenomenon. Figure 9 shows a plot of the frequency axis from zero to F_s, the sampling frequency. If the frequency of an incoming sine wave is less than $0.5 F_s$, there will be no aliasing. If the input function has a frequency between $0.5 F_s$ and $1.0 F_s$, then aliasing will occur. The frequency of the sine wave will be reduced ("aliased down") to a frequency that is equally spaced below the $0.5 F_s$ point. For example, a sine wave at a frequency of two-thirds F_s will be reduced to one-third F_s. Likewise, a sine wave at the sampling frequency $(1.0 F_s)$ will be aliased down to a frequency of zero.

The value of $0.5 F_s$ is sometimes called the *folding frequency* because it acts as a hinge point in Figure 9. This frequency is also called the *Nyquist frequency*.

Under conditions where the sine wave frequency is less than the folding frequency, the sine wave is said to be *oversampled*. When the sine wave frequency is greater than the folding frequency, it is said to be *undersampled*, and aliasing will occur.

In summary, the rules regarding aliasing are: (1) The folding frequency is half the sampling frequency, (2) aliasing will be avoided if the specimen image intrinsically contains no information above the folding frequency, and (3) aliasing will be avoided

if the imaging system MTF eliminates information above the folding frequency. The imaging system MTF is determined primarily by the combination of the objective lens point spread function and the electronic image sensor's sampling spot.

C. Practical Situations

In practice, one can estimate the highest frequencies within an image by looking at the spatial extent of the smallest detail in which one is interested. For estimation purposes, one can locate the smallest structures in the image and assume them to be sinusoidal. If these structures have sharp edges, even higher frequencies (multiples of the fundamental frequency) will be present, but at lower amplitudes. If, however, the fundamental frequency is more than half the cutoff frequency of the OTF (Eq. 11), then these can be ignored, since they are attenuated sharply by the microscope objective. Thus, in most microscope imaging applications, effects of aliasing will not be visible if the size of the smallest structures is greater than twice the sample spacing distance.

As a numerical example, consider a microscope with a $100\times$, 1.2 NA objective lens. From Equation 11, the cutoff frequency of the objective lens OTF is 4.36 cycles per micron at the specimen. Information above that frequency is stopped altogether by the lens. Thus if the sampling frequency is at least twice that, or 8.73 samples per micron, then aliasing cannot occur. The objective will act as an antialiasing filter to stop any fine detail that might otherwise give rise to aliasing. The corresponding pixel spacing at the specimen is the reciprocal of 8.73, or 0.115 μ.

If the image width is 100 μ at the focal plane, as in Table I, then 8.73 times 100, or 873 pixels per line, would be required to ensure against aliasing. If only a 512 by 512 digitizer were available, the magnification could be adjusted so that the width of the digitized image was 512 times 0.115 or 59 μ at the specimen, rather than 100 μ. Alternatively, one could use a high-resolution image sensor such as a 1,024 by 1,024 CCD array or 1,000-line tube-type camera and digitizer.

In this example, we need about three times as many pixels (873 pixels per line) as the number of resolution cells (263 per line). In this example we failed to account for the sensor spot and used the objective lens OTF alone as that of the imaging system. This difference accounts for a factor of about 1.5. In the earlier example we (correctly) combined the objective lens psf with the scanning spot to obtain an estimate of the imaging system psf.

In this example, we sought to use the objective lens MTF as an antialiasing filter. Using the cutoff frequency of the MTF to establish sampling frequency, rather than using the Rayleigh distance accounts for the remaining factor of two.

In actual practice, the truth lies somewhere between these two examples. Using the Rayleigh criterion alone leaves no margin for error in any of our assumptions or in our subsequent measurement algorithms. It also leaves open the possibility of troublesome aliasing if high-contrast detail of

suitably high frequency is present in the specimen, although this is relatively uncommon. On the other hand, ignoring the fact that the sensing spot makes the imaging system psf broader than that of the objective lens alone causes the analysis to overstate the sampling requirement. Requiring that the folding frequency be no lower than the cutoff frequency, as we did here, also imposes a stringent requirement on the digitizer. A technique for resolving this apparent dilemma is presented at the end of this chapter.

IV. IMAGE DISPLAY

Image display is the process of interpolating a digital (sampled) image to form a continuous image which can be viewed (Fig. 6). This process is commonly done by television monitors and by hardcopy image recorder devices.

A. Display with a Gaussian Spot

As stated in the previous section, the ideal function for interpolating a sampled image has the form $\sin(\alpha x)/\alpha x$. The display spot on a television monitor or hardcopy device has a shape that is closer to that of the Gaussian function. Thus one of the underlying assumptions of the sampling theorem is violated when we display an image. Empirically, the result is that fine detail in the image is not reproduced as well as if the proper spot form had been used.

Figure 10 illustrates what happens when a sampled function is interpolated with a Gaussian spot rather than $\sin(\alpha x)/\alpha x$. Even though the sine wave frequency is well below the folding frequency (and aliasing is thus avoided), the interpolation process introduces distortion because the conditions demanded by the sampling theorem have not been met. This distortion becomes worse as the frequency of the sine wave increases toward the folding frequency.

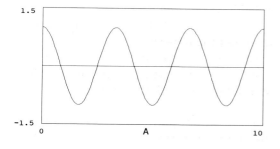

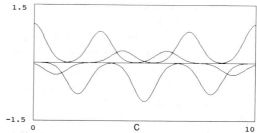

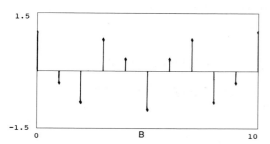

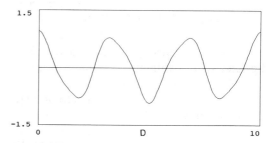

Fig. 10. Interpolation with a Gaussian spot. When a properly sampled sine wave is interpolated with a Gaussian function, rather than the (proper) $\sin(\alpha x)/\alpha x$ function, distortion of the reconstructed function results. Here the frequency of the sine wave in **A** is only 30% of the sampling frequency, and it is thus adequately sampled. The sampled function shown in **B** is interpolated using a Gaussian function of width equal to the sample spacing in **C**, by placing at each of the 10 sample points a Gaussian curve of appropriate amplitude. The interpolated function that results from summing the 10 Gaussians is shown in **D**. It differs from the original (A) in both amplitude and wave shape.

B. Oversampling for Display

One typically must display images using a spot that has some form other than the proper [$\sin(\alpha x)/\alpha x$] shape. One can, however, reduce the distortion that results from improper display spot shape by intentionally oversampling. This increases the folding frequency, placing it well above the highest frequency of the image content. This, in turn, reduces the distortion introduced by the inappropriate display spot shape.

C. Interpolation for Display

While one cannot electronically generate a $\sin(\alpha x)/\alpha x$ display spot, one can use multiple display spots for each input pixel to simulate the ideal situation. By using several display pixels (a 3 × 3 neighborhood, for example) for each pixel in the digital image, one can obtain both a reasonable approximation to the ideal display situation and a satisfactory display. This is accomplished by a digitally implemented interpolation step that increases the size of the digital image prior to display, say from 512 by 512 to 1,024 by 1,024.

When digital interpolation is used to make a digital image larger (and thus oversampled) prior to display, the process is called *resampling*. It moves the folding frequency upward, just as oversampling does. It allows one to approach the theoretically predicted image display quality in spite of a suboptimal display spot shape.

1. Pixel replication. When a neighborhood of several display pixels is to be used to display each single input pixel, some digital interpolation scheme is required. The simplest is pixel replication. This causes each display pixel in the defined neighborhood to have the same gray level as the input pixel being displayed. This, however, is frequently unsatisfactory, since it can give the displayed image a blocky appearance and introduce a stair-step artifact in the vicinity of edges.

2. Bilinear interpolation. Bilinear interpolation [Castleman, 1979, Chap. 8] avoids the blocky appearance of pixel replication, and it produces a smoother overall appearance. When implemented in software, it takes about twice as long as pixel replication. It may produce a barely perceptible loss of image detail, due to the smoothing effect of the local averaging that is inherent in bilinear interpolation.

3. Higher order interpolation. One can use an approximation to the $\sin(\alpha x)/\alpha x$ as a digital interpolation function. This produces a better displayed image, but at the expense of computational overhead. If the image was adequately sampled to begin with (no aliasing, adequate imaging system MTF), such resampling theoretically produces the same result that would have been obtained by oversampling during the original digitization process, but without the need for a high-resolution camera. One would still need a larger frame buffer and a high-resolution display system to accommodate the larger amount of data associated with each digital image.

As an example, suppose one wished to double the size of a digital image prior to display, say from 512^2 to $1,024^2$. The following explanation assumes an understanding of the process of digital convolution [Castleman, 1979, Chap. 9; Inoué, 1986, Chap. 10].

To begin with, a $1,024^2$ image memory is sparsely filled with a 512^2 image and zeros at the pixels in between. Then a new gray-level value must be generated between each pair of original pixels horizontally and vertically. The gray level of each new pixel is computed as a weighted sum of the gray levels of the surrounding original pixels. This is the process of digital convolution.

Theoretically, all of the original pixels would contribute to each new pixel, but in practice only those in a small neighborhood (e.g., 5 by 5 or so) are used. The values of the weights for each neighbor pixel are simply the $\sin(\alpha x)/\alpha x$ located at the neighbor pixel and evaluated at the new pixel position. This is equivalent to simply placing a $\sin(\alpha x)/\alpha x$ interpolating function at the new pixel position. Thus the convolution kernel [Castleman, 1979] becomes a two-dimensional $\sin(\alpha x)/\alpha x$ function, where $\alpha = 2\pi/2\tau$ and τ is the original (prior to resampling) pixel spacing, that is, twice the new pixel spacing.

Since the $\sin(\alpha x)/\alpha x$ interpolating function is of infinite extent, it must be truncated to finite size. This requires that the weights, other than the central weight, must be scaled to sum to unity. Otherwise, the gray levels of new pixels would not match those of the original pixels in areas of constant intensity. The central weight remains at 1.0.

As the kernel moves across the image memory during the process of convolution, the central weight alternately falls upon original pixels and new pixel positions. When it falls on an original pixel, the neighboring original pixels fall at zero crossings of the interpolating function, and thus the central pixel alone contributes to the new gray level. When the central weight falls on a new pixel location, the nonzero values in the kernel fall on neighboring original pixels, which then do contribute to the new gray-level value.

V. THE ANALYSIS OF COMPLETE SYSTEMS

We have covered separate analytical techniques for the separate considerations of resolution, sampling, and display. Now we combine these to specify the complete imaging system.

A. Magnification and Field of View

Each of the spatial parameters, field of view, raster size, scanning spot size, and pixel spacing, can be defined at several points in the imaging chain. Pixel spacing, for example, can be specified either at the face of the image sensor or at the specimen. It is most convenient to refer all measurements to the focal (specimen) plane. This is easily done if the appropriate magnification factors are known. They can be calculated or measured with the aid of a calibration standard such as a stage micrometer or grating slide.

The field of view of the objective, measured at the focal plane, is determined primarily by the microscope design. If the image sensor is properly mounted on the microscope, the scanning raster will be a rectangle inscribed within the circular field of view of the objective.

It is important that the scanning raster fit properly within the field of view. If the magnification factor is too small, the corners of the scanning raster will exceed the limits of the field of view with resulting vignetting (shading) and loss of focus at the periphery. If the magnification factor is too high, one will be using only the central portion of the field of view. In this case it might be better to use a higher power (and consequently higher NA) objective and use more of its field of view.

The magnification from the specimen to the image sensor will typically be the objective power times the magnification factor of any intermediate microscope optics (Optivar, etc.) times another factor which is dependent upon the optical adapter used to mount the image sensor on the microscope. When this latter factor has been determined for one microscope objective, it should remain constant when other objectives are substituted.

It should be noted that the objective power marked on a particular lens is a nominal value and may be subject to errors of 10 to 20% or more. The safest approach is to measure the pixel spacing for each objective configuration and remeasure the pixel spacing each time the configuration is changed.

B. Resolution and Sampling

Determining the psf or MTF of a camera from manufacturer's specifications can be an exercise in frustration. Manufacturers often attempt to specify resolution by a single number (e.g., "700 TV lines"), but this is seldom adequate. They are often reluctant to supply MTF data, but scanning spot diameter is sometimes available. One can assume a Gaussian shape and be approximately correct. The best method is to evaluate the camera under suitable test conditions.

Figure 11 shows the three parameters that should be well matched at the specimen plane. The frequency $F = 1/T$ is the highest spatial frequency of interest that is present in the specimen. T is the period of the smallest detail of interest in the specimen. $F_s = 1/\tau$ is the sampling frequency, where τ is the sample spacing, and f_c is the cutoff frequency of the imaging system MTF.

As a rule of thumb, the diameter of the scanning spot (imaging system psf), w, referred to the focal plane, should be no larger than one-half of T, the period of the smallest image detail. This means one scanning spot would fit within one half-cycle of the highest frequency sine wave. A larger scanning spot is prone to reduce the contrast of the image detail due to its action as a low-pass filter.

In order to avoid aliasing and loss of resolution, the sample spacing at the specimen, τ, should be smaller than both $T/2$ and w. As stated before, however, it is good practice to oversample by some factor such as 2. This means that τ should fall in the range between $T/4$ and $T/2$ and between $w/2$ and w.

In order to obtain good quality of displayed detail (with an inherently Gaussian display spot), it may be necessary to oversample the image by a significant factor or to resample the image digitally prior to display.

In order to obtain accurate measurement data from the objects in the digitized image, it may also be necessary to oversample by

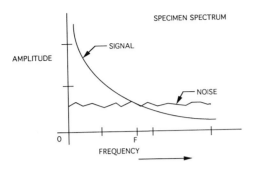

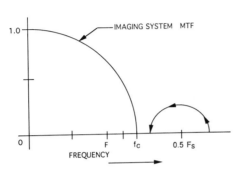

Fig. 11. Combining resolution and sampling considerations in the frequency domain. In this example, the frequency, F, of the smallest detail of interest in the specimen falls below the point at which noise takes control of the image. F also falls well below the cutoff frequency, f_c, of the imaging system MTF, so image detail will not be compromised. Finally, the folding frequency, $F_s/2$, falls above the cutoff frequency, so the imaging system MTF will suppress any information that might cause aliasing if the specimen should contain abnormally fine (high-frequency) detail.

a significant factor, resample the image prior to the analysis, or build resampling into the measurement algorithms.

As an example of the latter, consider the problem of measuring the perimeter of an object in a digitized image of a microscopic specimen. If one programs a straightforward perimeter tracking algorithm that merely sums the center-to-center distances of adjacent pixels on the boundary, one obtains, actually, the perimeter of a polygon that approximates the shape of the object. If the pixel spacing is chosen merely to satisfy the

Rayleigh criterion and the sampling theorem, this approximation may lead to unacceptable errors. This problem can be overcome if the algorithm, instead, fits arcs of curves (e.g., quadratics or cubics) through the boundary points and measures perimeter distance along these.

C. Noise Considerations

In Figure 11, the amplitude of the specimen is shown to fall off with increasing frequency, while that of the noise does not. This is typical of actual specimens and the common noise sources, most of which can be assumed to be "white noise." Thus there are three ways to lose the fine detail in a specimen. One is to have it attenuated by the imaging system MTF (resolution too low), the second is to have it corrupted by aliasing (sampling too coarse), and the third is to have it buried in noise. In the latter case it is necessary to eliminate the offending noise sources before proceeding with the design.

One potentially devastating noise source is the analogue video recorder. The best situation is to avoid recording altogether and digitize the analogue signal as close to the source as possible. A second choice is to use an instrumentation-quality video recorder. Only when absolutely necessary, or when image quality requirements are not critical, should one use a video cassette recorder. These devices alter the video signal in ways that severely degrade the quality of the digital image that can be obtained.

Noise reduction techniques are beyond the scope of this chapter [see, for example, Castleman, 1979]. For present purposes it suffices to say that it serves no purpose to accurately image, sample, and display frequencies at which the signal is buried by noise.

D. System Design

In a well-balanced system, (1) the noise will dominate only at frequencies above the highest frequency of interest in the specimen, (2) the imaging system MTF will pass information at frequencies where the specimen has content of interest and attenuate it at those where the specimen detail is buried by noise. Following that, (3) the sampling frequency will be chosen to be high enough to prevent aliasing.

The above summary, which is stated in the frequency domain, can be restated in the spatial domain. In a well-balanced system, (1) the size scale of the noise will be smaller than the smallest detail of interest in the specimen, (2) the imaging system psf will be smaller than the detail of interest in the specimen but larger than the size scale of the noise. Following that, (3) the sample spacing will be chosen to be small enough to prevent aliasing.

Some of the parameters of a digital imaging system (such as magnification) can be set to any desired value during the design process. Other parameters are set by available components, and only certain specific choices are available (i.e., 500-line versus 1,000-line cameras). The best approach is to become familiar with all candidate components and choose ones that are on the high side of adequate (e.g., modest oversampling), rather than marginal or below requirements.

Often the design of an electronic light microscope system involves trade-offs dictated by cost constraints. One thousand-line cameras and frame buffers cost considerably more than five hundred-line systems.

The well-designed system is the one that accomplishes the task for which it is intended without costing more than is required to do the job. In view of the conflicting constraints, it is clear that no one selection of components is a universal solution.

Each system design exercise is unique. If the criteria discussed in this chapter are met, then the microscope image digitizer system will be well balanced and should produce quality results.

REFERENCES

Born M, Wolf E (1964): "Principles of Optics." Oxford: Pergamon Press.

Bracewell R (1965): "The Fourier Transform and Its Applications." New York: McGraw-Hill.

Bradbury S (1984): "An Introduction to the Optical Microscope." New York: Oxford University Press.

Castleman KR (1979): "Digital Image Processing." Englewood Cliffs, NJ: Prentice Hall, ch 12–13.

Castleman KR (1987): Spatial and photometric resolution and calibration requirements for cell image analysis instruments. Appl Optics 26:3338–3342.

Goodman JW (1968): "Introduction to Fourier Optics." New York: McGraw-Hill.

Harris JL (1964): Diffraction and resolving power. J Opt Soc Am 54:931–936.

Inoué S (1986): "Video Microscopy." New York: Plenum Press.

O'Neill EL (1963): "Introduction to Statistical Optics." Reading, MA: Addison-Wesley.

CHAPTER 4

Image Processing Software for Research Microscopy: Requirements and Design of the User Interface

Theodore Inoué

I. INTRODUCTION

During the last decade, electronic image processing techniques have become widely used in the research laboratory. Starting with simple yet effective analogue contrast-enhancement boxes, researchers have made great progress visualizing structures difficult or impossible to see with conventional microscopic and photomicrographic methods. With a simple flip of a switch and a twist of a knob, analogue image processors greatly improved video images through simple manipulations of electrical signals. Researchers improved contrast, crispened edges, and corrected lighting problems. The simplicity and low cost of these methods allowed numerous people to employ enhancement methods successfully.

Recently, with the advent of more powerful microcomputers, digital image processing systems have become more generally available. For roughly the same price as a good research microscope, these systems provide sophisticated methods of image enhancement, such as image averaging to reduce noise, shading correction to allow accurate photometry, and image sharpening to provide a clearer view of small image details. They also make possible new image analysis functions, letting users measure areas, perimeters, and even intracellular ion concentrations. However, as powerful as they are, the majority of these systems are pitifully underutilized even in today's technically sophisticated research community. Many are used as glorified title makers or for little more than the simple enhancement applications that the original analogue boxes excelled at. Worse yet, digital systems often do not perform even these basic tasks as well as their less expensive analogue counterparts.

Given that computers are capable of performing these tasks quite competently, we are forced to the conclusion that the current generation of digital imaging systems fail because of an absence of good software. The software used is simply far too cumbersome for a majority of the users, rendering these costly systems useless for day-to-day use.

Electronic Light Microscopy, pages 95–103 © 1993 Wiley-Liss, Inc.

Why is it so difficult to find a useful imaging system? To find the answer, we must look at the history of image processing. Many of the problems currently plaguing digital image processing systems can be traced back 20 years to the original satellite imaging systems. These first generation systems were designed for use by highly trained technicians and engineers. For this reason, the developers did not have an overwhelming incentive to make the programs easy to use or truly interactive, as required today in research labs. Even if they did desire to do so, the programming tools needed to develop highly interactive programs were not generally available at that time. Unfortunately, many commercial imaging systems were offshoots of the satellite image processing systems, using only slightly modified versions of the original programs. As a result, they were slow, cumbersome, and generally too inefficient to use for day-to-day lab work. Sadly, these poor programming methods have been perpetuated and are present even in today's software.

Worthy of note are the early commercial systems that were successful in the lab environment. These systems originated in research labs, conceived and developed to extend the capabilities of the useful analogue enhancement boxes. Out of necessity they were designed to be powerful and versatile so as to satisfy the needs of the lab without detracting from experiments. Of paramount importance was the need to develop a system that anyone in the lab could operate with a minimum of training and effort. Unfortunately, very few companies were willing to spend the time and commit the effort needed to develop systems of this sort, and therefore very few commercially available systems are suitable for use in the research environment.

Many of the deficiencies present in imaging software stem from a poor user interface, which makes the system too confusing for the average user. Good user interfaces *are* difficult to design and implement. However, regardless of the difficulty, it is possible to design an imaging system that is easy to use as well as suited to the research environment. This chapter outlines the features and design strategies that should be considered when purchasing or designing such a system. It is also suggested here that the user interface, encompassing all interactions between the system and the user, is *the* major determinant of the ultimate utility of any system.

II. THE USER INTERFACE

An indication of the importance of a program's user interface can be found when one considers those researchers who have made successful use of digital imaging systems. Quite often they utilize numerous computers in their labs, performing tasks as varied as image processing, data analysis, and word processing. Many of these researchers have computer programmers working in their labs writing and operating programs tailored to specific tasks. These tailor-made systems allow the researcher to accomplish a huge amount, albeit at a great expenditure of time and money, because the systems are specifically designed to meet the needs of the researcher. However, when the programmers leave, the owner of the equipment may be stranded with an unusable bank of computers and software. The only researchers who continue to accomplish their goals are those who themselves use the programs daily and have a full understanding of their workings. This is the unfortunate but typical result of relying upon an inadequately designed system. Therefore, regardless of whether one is looking for a commercially available system or one that is custom-designed, the same design problems must be addressed.

Software development is an enigmatic process. With a few simple commands, one can make a computer do many tasks, giving the illusion that it is easy to develop complex software. Unfortunately, upon trying to do so, one soon learns how easily software development efforts get out of control. Programs quickly evolve into complex monstrosities with hundreds or even thousands of functions. Accessing functions becomes so difficult that the program is rendered useless; and yet every year, countless programmers rework poorly implemented software without understanding that they should be designing new software, software that makes the user feel that it was written specifically for his or her research needs, while retaining ease of use and flexibility.

A good user interface is key to a *functional* system. Yet not only is the user interface usually the last portion of a program that is designed, it is often poorly designed. Under ideal conditions, the user interface allows one to interact with the program in a totally natural and intuitive manner; the ultimate program lets one realize the solution to any problem with little more effort than that required to verbalize a desire. This obviously requires that the user interface be built on a powerful set of functions. However, usually writing functions is considerably easier than creating a good user interface. The programmer should in fact design the appearance and behavior of the program prior to doing any programming, as the program's user interface bears the burden of allowing the users to solve their problems as efficiently as possible. This design stage lays the foundation for the rest of the program, and only after this step is done should any programming be attempted. For the remaining discussion of user interface and system design, it is assumed that the system has the capacity to solve any problem. The only questions being discussed herein will be those that relate to accessing and utilizing the power available in the system.

III. INTERFACE DESIGN PRINCIPLES

Examples of good engineering and user interface can be found everywhere: television sets, toasters, and cars are just a few examples. Each of these can be operated by relatively naive users with little or no training. Three factors greatly differentiate these devices from most software systems. The first factor is system complexity. Where a toaster only has two controls (an activation switch and a darkness control), an imaging program may have tens or hundreds of controls. Naturally, the more complex system will be more difficult to use. If you agree with this idea, then you have fallen into the fallacy of most software engineers—that ease of use must be traded off against system capability. With a well-designed user interface, even the most powerful imaging system can be used by a computer novice, though it may take some time for the user to learn how to optimally utilize the system.

The second differentiating factor is that televisions, toasters, and cars operate via physical, analogue controls. Controls such as the volume control and the steering wheel give the user an immediate physical feedback when manipulated. This makes a tremendous difference when learning how to use them. By our very nature, we require feedback to adjust actions and quickly achieve a desired result. By using such controls, we learn how our actions affect our surroundings. But what of imaging systems? Why do they often have such obtuse designs? The answer is usually laziness or ignorance on the part of the programmers. Later in this chapter, I describe how a poorly designed control can be modified to make it as easy to use as a volume control or a steering wheel.

Imaging systems also differ from the aforementioned devices in the maturity of their designs. Toasters, cars, and even television sets have existed as consumer items for decades, and have had the opportunity to evolve into their current elegant designs. Image processors, on the other hand, have only been generally available for less than a decade. Software development is a young science as well, making the task even more difficult. However, we can try to learn from the designs of other more mature devices, so that we do not take decades to develop an imaging system that is as easily used as a television set.

Earlier I claimed that it is possible to create imaging products that are easy to use. The original analogue enhancement systems had very simple interfaces, permitting interactive image enhancement. Their intuitive interface and simplicity made them feel familiar and natural, much like a common household appliance. Unfortunately, when the newer, digital systems were developed, the designers often ignored the elegant simplicity of the analogue processors in an attempt to bring more powerful operations to market in the shortest possible time. Basic operations performed on a computerized system often required great expertise to use. Just finding the appropriate command on a menu was difficult and the situation was exacerbated by the inadequacy of the controls.

For example, it is strangely amusing to examine the manner in which many imaging systems force you to adjust the contrast of a displayed image—by manipulating a graph of input versus output brightness, usually in a manner that even people familiar with programming and imaging do not find intuitive. Even though this is directly analogous to the method that the computer uses internally to modify the contrast, this is an extremely poor design, as it requires the user to know about the inner workings of the image processor. Why not use a simple contrast knob?

Unfortunately, few systems actually have physical knobs, so controls must be implemented using the computer screen. This in itself does not present any inherent problems. However, most programmers implement controls without much thought, so the controls operate in a nonintuitive manner. The programmer of a well-designed program should create the program's user interface with the knowledge that users expect to adjust on-screen controls in a manner analogous to physical controls. For the contrast control discussed above, the user's expectations dictate contrast modification via analogue controls such as dials or sliders, as illustrated in Figure 1.

Control A is simply a graphical representation of a contrast adjustment knob. It may be adjusted by dragging the indicator to the desired level or by indicating a desired level directly by pointing with a mouse. The problem with this control is that even though it is a direct visual analogue of a physical knob, it may be cumbersome to use since it requires motion along an arc. Control B is a slider, which maintains the advantages of an analogue control while remaining easy to manipulate on a computer, using only simple linear motion of the mouse. Control C shows the control used by many imaging systems. It has no physical analogue. Worse, it provides no intuitively obvious means by which to modify the contrast. Moreover, each program that uses this type of control may require one to use it in a different manner. In some cases, one has to draw the curve of the desired contrast function. In others, some arbitrary relationship exists between cursor motion and the slope and offset of the line. In all cases, method C is to be avoided as a primary contrast modification method as it has neither an intuitive meaning nor a physical analogue.

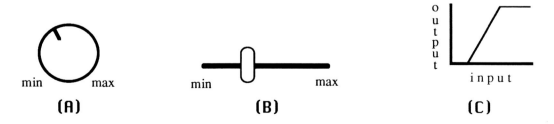

min   max   min   max   input

(A)   **(B)**   **(C)**

Fig. 1. An example of three possible controls for adjusting contrast.

However, using analogue controls does not guarantee a good user interface. The response of a control is as important to the overall user interface as is the method of control. Imagine how difficult it would be to drive a car if rotating the steering wheel did not immediately cause the car to change direction. One would have to steer by trial and error, until achieving the desired turn. In a physical system such as a car, it is clear that the operator needs continuous feedback, so why should a computer program be any different? A program that does not respond to changes interactively makes one feel as though one is "driving blind," and modifications become tedious at best. On the other hand, systems in which changes take immediate effect often have a natural responsiveness. This makes the system much easier to learn to use.

Not only does a system have to be interactive and responsive, it has to respond *appropriately*. Again using a steering wheel as an example, an appropriately scaled steering wheel would respond to both rough and fine changes. If it did not have the proper response and was too precise, one would have to rotate it through numerous rotations to change lanes or to make a U-turn. On the other hand, if it were not precise enough, the slightest change could induce a fatal change in direction. Fortunately one can almost always find a happy medium between these extremes of sensitivity.

Nonetheless, appropriate control responsiveness in an imaging system is not always as simple as adjusting the sensitivity of the steering wheel. Many of the controls encountered in imaging applications affect some aspect such as brightness and contrast that varies our perceptions of the image. Psychologists have long known that most sensory perceptions work on logarithmic scales as opposed to simple linear scales, allowing us to perceive a huge range of stimulus levels while maintaining appropriate sensitivity throughout the range. Our perceptual systems facilitate this by detecting differences in stimuli by their fractional variations.

For example, our eyes are sensitive to roughly a 5% change in brightness regardless of absolute brightness. This is as contrasted to discrimination based on absolute linear luminosity variations, which would be unnecessarily sensitive to fine variation at high stimulus levels and too rough to discriminate well at low levels. A simple linear scale is also terribly inefficient. To allow a 5% discrimination across four orders of magnitude of stimulus intensity, a linear scale must uniquely discriminate 200,000 different levels. On the other hand, a logarithmic scale requires only 189 unique levels to represent the same range with a 5% resolution throughout.

Unfortunately, programmers are used to dealing with numbers on a linear scale, as creating controls with a nonlinear scale is considerably more difficult. This has led many programmers to assume that linear scales are superior and should always be used. Hence, virtually all controls in use today are linear, and do not permit adjustments in a perceptually valid manner. This leads to controls that are awkward and frustrating to use.

These three items—selection of a proper control, interactive control response, and proper control scaling—provide the basis for developing an interactive system that is easily learned. Moreover, because it is easier to learn how to use such a system, one will be able to operate it effectively with less training. Yet there are other important aspects to be considered in the overall design of a complete system, including the quality and quantity of the functions, the menus, the interface to other laboratory devices, and the adaptability of the system.

IV. THE USER ENVIRONMENT

A key step in choosing or designing any system is to determine the environment in which it will be used. For a typical video microscopy system, the environment is well defined. A research laboratory can usually be characterized as a small, dark room in which numerous people with varied technical backgrounds run complex experiments on a tight schedule. The variety of experiments is virtually endless, though there are some commonly used pieces of equipment that are logical to control with the computer that runs the imaging application. This includes equipment such as shutters for the light source(s), filter changers for fluorescence microscopy, motorized stage and focus controls, as well as various image storage systems such as optical disk recorders and video printers.

Examining the information we have about the laboratory environment provides some constraints on the selection or design of a system. Knowing that labs may be dark implies that the system should be able to be controlled without having to use a keyboard, as keys may be difficult to locate in the dark. One popular option is to use an on-screen menu with a pointer driven by a mouse. However, it should also be noted that the system will undoubtedly be used extensively under normal lighting conditions, where a keyboard may be more efficient for experienced users, and therefore keyboard control should be available as well.

Next, because the system will be used by people with various levels of technical expertise, it must be designed so that it is self-explanatory. This requires a carefully worded set of menus, so that it is immediately obvious what a given function or option will do. In addition, in those cases where the exact meaning of an item is still unclear, it should be possible to access additional information pertinent to the current situation. Such context-sensitive on-line help can be extremely useful when learning how to use a system, as it can save tremendous amounts of time that might normally be spent searching through a system manual.

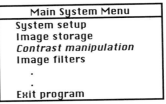

Fig. 2. The main menu of a typical image-processing system.

V. AN EXAMPLE OF IMPROVED DESIGN

The efficiency of using a system can be quite difficult to determine, as this depends greatly on personal preference. However, certain common-sense rules apply when designing software and user interfaces. Perhaps the most important of these rules dictates that a system should not hide frequently used operations in deeply layered submenus. This seemingly obvious rule is violated by numerous systems currently in use. The following example takes us through the steps necessary to enhance contrast using a typical poorly designed interface.

Upon entering the program, one is greeted by a main system menu with a number of choices, each pertaining to a different class of operations (Fig. 2). Finding the option which gives access to the contrast enhancement function is easy enough, so one selects the contrast manipulation option.

After making this selection, a submenu appears (Fig. 3). One is now faced with a more difficult decision: deciding which of

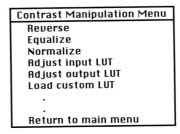

Fig. 3. A contrast-enhancement submenu.

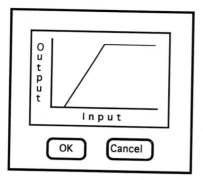

Fig. 4. An illustration of a poor control used for contrast enhancement in many systems.

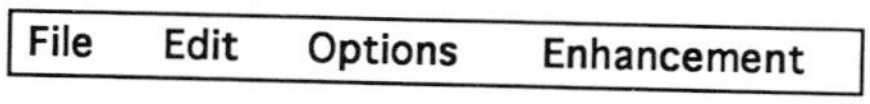

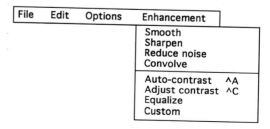

Fig. 5. A "menu bar" typical of modern user interfaces such as used by the Apple Macintosh and Microsoft Windows.

these operations lets one perform a simple contrast-enhancement operation.

After referring to the user's manual to discover which option actually lets one modify the contrast of the image on the screen, one selects *Adjust output LUT*. Now the program displays the dialogue box shown in Figure 4 on the image monitor, forcing one to change the locus of one's attention from the computer display screen to the image monitor.

After much experimentation, one learns how to modify the contrast (again, by referring to the manual), and finally is able to adjust the contrast.

This approach is known as a *layered* or *hierarchical* menu system, because menu items are arranged on a number of menus in a hierarchy. This method has the advantage that it presents one with all of the options available at a given time on a fairly small and simple menu. In addition, operations are usually grouped into logical classes, letting one locate functions easily. Unfortunately, this style of menu is cumbersome to use, as one has to move through numerous menus in order to get to the function one wants to use, and therefore it provides an unsuitable basis for an interactive system.

How can this approach be improved upon? First, an operation as commonly used as contrast manipulation should be readily accessible. An alternative to the above example might be to group the items in the menus

by the frequency of use. Unfortunately, this is usually done at the expense of the logical layout of functions in menus. A better approach would be to use a different style of menu altogether. One choice is a single menu containing all of the available functions. This format has been shown to be the most efficient style possible when the number of menu choices is relatively small (less than roughly 40 items). However, most applications quickly outgrow this number, rendering this format inadequate for large applications.

A good compromise between simple hierarchical menus and a single, efficient menu is the pull-down style menu, as exemplified by those used on the Apple Macintosh computer and in Microsoft Windows. Under this scheme, the main menu is displayed on a bar at the top of the screen (Fig. 5).

Using this format of menu, one selects a class of functions from a main menu bar in much the same way as in a standard hierarchical menu. On doing so, a small menu immediately "drops down" to reveal the functions available for that class. In the sample case where we wish to enhance the contrast, we would select *Enhancement* to reveal a submenu containing all of the available image enhancement options. Figure 6

Fig. 6. The enhancement menu of an imaginary image processor listing typical functions and "hot keys."

shows such a menu, typical for a simple system.

Next, one simply selects the desired option (*Adjust contrast*) to activate the function. At first glance, this menu scheme appears no more efficient than a normal hierarchical menu, as both styles require two steps to get to the contrast control. In practice, activating functions in a well-implemented drop-down menu feels like a single step and requires very little effort. In this case, for example, it takes only a single button press and release to activate the contrast adjustment control.

Another step has been added to make accessing functions more efficient. This is the addition of key shortcuts (CTRL + C in this example). These are simple keyboard commands that take one directly to the function in a single step, which can greatly improve efficiency in circumstances in which it is possible to use a keyboard.

If one has the opportunity to use such a system, one may still find that this procedure is less than ideal. It often requires a few significant hand and arm motions to activate a function. A more efficient system would allow one to keep one's most often used functions active and available at all times, while hiding less common functions on the pull-down menus.

Regardless of the menu format chosen, the functions should be designed with consideration of the previous discussion of controls. Continuing with our example, the modified contrast control would be designed to appear as shown in Figure 7.

Using this control, one simply moves the contrast and brightness sliders until the desired contrast is achieved. As described earlier, the contrast of the image should change dynamically as one varies the contrast controls. Note that an *Automatic adjustment* option has been added to simplify the operation of the system by automatically setting the contrast levels to an optimum value calculated by the system. This option should also be available on the original enhancement menu, saving an additional step if one

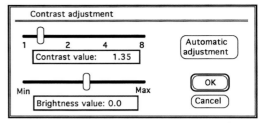

Fig. 7. A much-improved contrast-enhancement control. Note the use of intuitive sliders and direct readout of contrast values.

decides that automatic adjustment is sufficient for one's application. Such automatic functions can be extremely useful and save the user considerable amounts of time when using the system under normal conditions.

Is the method shown in this example the best one for adjusting contrast? Probably not. Even though it is acceptable, it also has some problems. First, if it were displayed on the same monitor that displays the image, it would take up and obscure an inordinate amount of space. This is a very common problem when designing graphical user interfaces of the type being described.

How can we reduce this control to an acceptable size while retaining its functionality? One method would be to remove some of the extraneous text and make the graphics smaller (Fig. 8). Using this modified control, we substantially reduce the screen area that it covers while retaining its full functionality. Note that the *Cancel* option has been removed, as has the title of this control box, as neither added to the function-

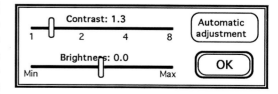

Fig. 8. A miniature contrast-enhancement control, useful in cases where the screen is not large enough to facilitate the larger control of Figure 7.

ality or clarity of the control. Canceling is achieved via an *Undo* command added to the main menu.

This control box could be made even smaller, if more information were removed and its size further reduced, though this could make it harder to use for a novice. Before doing so, it would be prudent to determine whether an average user could decipher the abbreviated control.

As can be seen from these simple examples, good design is at best a difficult task. Design improvements can be as trivial as rearranging the layout of a menu, or as far-reaching as choosing the proper system of menus to use. In all cases, a system should be designed with one goal in mind—to create a system that can be used in a minimum amount of time by a person with a minimum amount of experience.

VI. CONCLUSION

The evolution of image processing systems follows a course parallel to that of the digital computer. At first, systems were large and expensive, requiring teams of programmers to keep them working. Data were processed in batches, noninteractively, as computer time was both scarce and expensive as well as extremely difficult to use, and they had only the most rudimentary user interfaces. Over time, as costs dropped, smaller labs were able to afford powerful mini- and microcomputers that had the potential to solve many of their day-to-day tasks. However, most of the software was just as user-hostile as that found on the early mainframes.

In the early 1980s, public awareness of high-quality user interfaces grew, and developers made efforts to incorporate features such as interactive menus, control panels, and high-resolution graphical user interfaces. Unfortunately, few developers put in the research required to develop programs that were natural and intuitive to use. Most efforts resulted in systems that were only marginally easier to use than their ancestors.

Even today, the user interfaces of most modern imaging software fall far short of what is possible on personal computers. However, there is hope. Programmers developing imaging systems can learn a great deal by examining software and systems that behave in a user-friendly manner. Research and design, combined with today's powerful microcomputers, will yield systems with considerably more versatility and ease of use than in the past.

The future holds great promise for imaging and visualization in scientific research. Current development tools greatly simplify the once-onerous task of programming for graphical user interfaces, permitting the creation of new software with unprecedented ease. In turn, software can evolve far more quickly to user requirements, resulting in software which is much easier to use. We are finally entering a time when computers will fulfill the dreams of the researchers using them—the machines will become transparent, allowing their users to forget about them and return to the applications and research which demand their attention.

REFERENCES

Laurel B (ed) "The Art of Human-Computer Interface Design." Menlo Park, CA: Addison-Wesley.
"Common User Access: Advanced Interface Design Guide." International Business Machines Corp.
"Human Interface Guidelines: The Apple Desktop Interface." Menlo Park, CA: Addison-Wesley.

CHAPTER 5

Principles, Practice, and Applications of Video-Enhanced Contrast Microscopy

Dieter G. Weiss and Willi Maile

Electronic Light Microscopy, pages 105–140 © 1993 Wiley-Liss, Inc.

I. INTRODUCTION

The possibility of analogue and digital processing of microscope images in "real time," i.e., at video rates, has opened a variety of new ways to dramatically improve the quality of microscopic images and to create new applications of light microscopy. In video-enhanced contrast (VEC) microscopy, contrast is electronically increased in low-contrast or "flat" images. This process not only clarifies images containing details already visible to the eye, but enables one to increase contrast and magnification to an extent that objects can be visualized that are 5 to 20 times smaller than the limit of resolution (see Fig. 2 in Shotton, Chapter 1, this volume). This means that now positions and movements of biological objects as small as 15–20 nm, e.g., small membrane-bounded vesicles and microtubules, can be analyzed in the living state. Organelle morphology and motion can be quantitatively described by a number of parameters such as size, shape, contrast, velocity, straightness of path, reversals of direction, etc. With this technique, light microscopy gained access to a higher magnification range than was previously accessible exclusively by the electron microscopist. VEC microscopy yields the best image improvement with low-contrast objects and especially with weak phase objects. Since these are most common in the biosphere, it is there where there will be the great potential, while in materials science and mineralogy, where high-contrast objects prevail, this technique may stay restricted to some specialized applications, i.e., in colloid science. The new technique has already profoundly influenced our conception of the activity and dynamics of cytoplasm and cytoskeleton, and of cell motility. VEC microscopy also represents a very powerful means to monitor and quantify the effects of chemical agents or physical treatments on the living cell.

II. PRINCIPLES OF VIDEO-ENHANCED CONTRAST (VEC) MICROSCOPY

VEC microscopy was developed in the laboratories of S. Inoué [Inoué, 1981, 1986] and of R.D. Allen [Allen et al., 1981a,b; Allen and Allen, 1983]. Allen et al. [1981a,b] noted that the use of video microscopy with polarized light methods (for instance, Nomarski differential interference contrast) allows the introduction of additional bias retardation, which, after offset adjustment and analogue enhancement, permitted much better visualization of minute objects, and he termed his procedure Allen video-enhanced contrast (AVEC) microscopy.

Digital image processing may be performed once microscopic images picked up by a video camera have been converted into a digital signal. Image processors of many kinds may be used to manipulate this digital image in order to reduce noise in the image by digital filtering or averaging, to subtract undesired background patterns, to further enhance contrast digitally, or to perform measurements in the image (e.g., intensity, size, speed, or form of objects) (see Shotton, Chapter 2, this volume). It is only since procedures for noise reduction and contrast enhancement *in real time*, i.e., at video frequency, have become available that the microscopist has been able to generate electronically optimized pictures *while* working with the microscope.

For the following reasons, VEC microscopy is especially useful for the cell biologist, biochemist, and molecular biologist:

- Because it often enables even an inexperienced microscopist to generate relatively good microscopic images (although optimal results require a thorough knowledge of light microscopy!);
- Because it enables one to visualize small organelles and supramolecular aggregates in the living state;
- Because, under certain circumstances, it allows quantitative measurements of amounts, concentrations, transport, or metabolism of specific molecules;
- Because objects beyond the limit of resolution of conventional light microscopy can be visualized; and
- Because recording on videotape for the later assessment of dynamic processes is easy and cheap.

A. Analogue Contrast Enhancement

Understanding the image manipulations for analogue contrast enhancement is complicated, so that a short preliminary discussion of contrast, stray light, and optical resolution is required.

1. Stray light. Light distributed evenly over the image and contributing nothing to image detail is called stray light. It has many sources (Table I). In many cases, stray light prevents the use of otherwise optimal settings of the microscope. Often the theoretically achievable resolution has to be sacrificed by closing the condenser diaphragm, thus reducing the numerical aperture, in order to reduce stray light. When polarized light is used, there is usually an annoying contribution of unpolarized stray light, even at high extinction settings of the polarizers or prisms. Once an image has been captured

TABLE I. Stray Light, Which May Be Removed by Applying Offset, Arises From Various Sources

Bright-field Miroscopy
 Excessive condenser aperture
 Uncoated lens surfaces
 Reflected light from tube inner surfaces

Polarized Light and Interference Microscopy
 Optical rotation at lens surfaces
 Strain birefringence in lenses
 Light scatter due to dust, lens cement, etc.
 Surface imperfections in lenses
 Defects (holes) in polarizing filters
 Submaximal compensation

Fluorescence Microscopy
 Autofluorescence of any material in the light path
 Nonspecific localization of fluorochromes
 Incomplete removal of excitation light by
  fluorescence filters
 Bleed-through of one fluorescence emission
  while viewing the fluorescence of a second
  fluorochrome in a double-labeling application,
  due to imperfect or wrongly chosen filters

in analogue electronic form, such stray light can be removed from the image by applying a negative DC bias voltage to the video signal, called the offset or pedestal voltage. By applying suitable amplification or gain to the camera signal, the contrast is enhanced, and by using a variable offset the camera signal may be shifted to the appropriate region of gray levels for best visual contrast on the video screen (Figs. 1a,b, 2A–D).

2. Contrast. The intensity at each point of the optical microscope image is converted into a voltage signal by the television camera. Contrast can be controlled, with a factor of 100 or more, by the gain applied to the camera signal, provided the proper offset setting is used. Contrast (C) for the eye is perceived as the difference in brightness (or intensity) between the background (I_B) and the specimen (I_S) divided by the intensity of the background:

$$C = \frac{|I_S - I_B|}{I_B}. \tag{1}$$

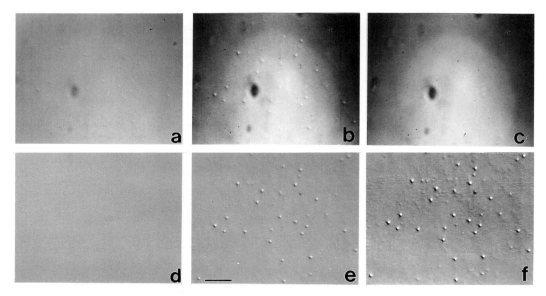

Fig. 1. VEC microscopy of a specimen with very weak contrast. The steps of image generation and improvement in VEC microscopy are demonstrated using subresolution-size polystyrene latex beads (diameter 60 nm) attached to the coverglass. **a–c** demonstrate analogue contrast enhancement and **d–f** digital background subtraction and digital contrast enhancement. **a:** In focus, not enhanced; **b:** in focus, analogue-enhanced; **c:** out of focus, with mottle; **d:** out of focus, mottle subtracted; **e:** in focus, mottle subtracted; **f:** digitally enhanced. The process of image improvement may be understood by reference to the corresponding computer simulation given in Figure 2. Figure 1a corresponds to Figure 2a, 1b to 2d, 1c to 2e, 1e to 2f. Microscope: Reichert-Leica Polyvar; processor: Hamamatsu C-1966 Photonic Microscope System. Bar, 5 μm.

Equation 1 shows the modulus of $I_S - I_B$, i.e., $|I_S - I_B|$, since contrast may be either positive or negative [Shotton, 1988].

3. Analogue contrast manipulation. The manipulation of contrast by gain and offset can be applied to any mode of optical microscopy. With techniques involving polarized light, considerable additional contrast may be achieved by adjusting the compensator to a higher bias retardation (AVEC microscopy, see below). In this case the resulting images are usually of inadequate visual contrast, because the denominator of Equation 1 is too high, due to excessive stray light (Fig. 3). However, in the electronic image, the negative offset voltage applied to the video signal acts in a manner analogous to a "negative brightness" (I_V), which is added to the denominator. Video contrast C_V is then

$$C_V = A \, \frac{|I_S - I_B|}{I_B + I_V}, \qquad (2)$$

where A is the electronic amplification or gain factor, and I_V is a negative DC bias voltage.

4. Contrast enhancement and visualization of objects. If very small objects or pinholes are imaged by some optical system, they will be seen as a blurred image or diffraction pattern known as the Airy pattern, consisting of a central Airy disk surrounded by the Airy rings (see Castleman, Chapter 3, this volume). Usually the Airy rings around larger objects are negligible. If, however, the size of the object becomes smaller than the wavelength of the light used, the diffraction pattern may be larger than the object. In the best lenses, where aberrations

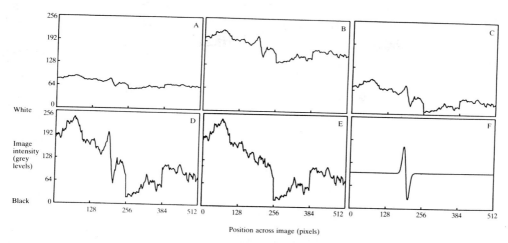

Fig. 2. A computer simulation of the contrast-enhancement process in VEC microscopy. Each graph represents the digitized gray-level intensity of a single horizontal line of 512 pixels from a video frame of a hypothetical Nomarski DIC microscope image, in which a low-contrast object is centered at pixel 200. The shear of the Nomarski image is such that this object appears brighter than background between pixels 180 and 200 and darker between pixels 200 and 220. **A:** Conventional Nomarski microscope image, optimized for direct visual observation. The overall illumination level is low, and the low-contrast object is barely visible within the noisy image (compare with Fig. 1a). **B:** Nomarski image after readjustment of the microscope's bias retardation to optimize $|I_S - I_B|$ for VEC microscopy (see text), increasing it by a factor of ×2.5 over A. The overall illumination level is now too bright for direct visual observation. **C:** Image resulting from B after adjustment of the video camera offset (black-level) control by an amount equivalent to subtracting 128 gray levels from the digitized image, to reduce the overall brightness. **D:** Final contrast-enhanced video image resulting from B after adjustment of the video camera offset control as in C *and* adjustment of its analogue gain control to amplify the resultant image by a further ×2.5. While the analogue image contrast-enhancement processes (B–D) have together amplified the original signal $|I_S - I_B|$ by ×6.25, so that the image now efficiently fills almost the entire dynamic video range from black to white, without saturation of the bright highlights and consequential loss of information, the signal-to-noise ratio is unaltered, remaining poor. Indeed the noise present but barely observable in the original microscopic image (A) has now been amplified to troublesome proportions, resulting in a background "mottle" that effectively obscures the image of the specimen (compare with Fig. 1b). **E:** An image of the background mottle due to imperfections in the optical and video system, contrast-enhanced as for D but in the absence of the specimen (compare with Fig. 1e). **F:** The resulting VEC Nomarski image of the idealized hypothetical specimen upon a featureless gray background, obtained by direct digital subtraction of the stored background mottle image (E) from the contrast-enhanced image of the specimen (D), and a subsequent digital black-level adjustment equivalent to adding 100 gray levels in order to display the resultant difference image within the dynamic range of the video signal from black to white (compare with Fig. 1e). (Reproduced with permission from Shotton, 1988.)

have been made negligible, image quality is only limited by the size of the spreading of each image point due to diffraction, i.e., by the size of the Airy pattern.

Objects smaller than the limit of resolution are imaged as Airy patterns whose amplitude (intensity) is small, but whose spatial dimensions are determined by the objective and cannot be reduced (Fig. 4). Applying video enhancement permits such normally invisibly weak Airy patterns to be visualized, but if two such objects are separated by a distance less than the limit of resolution, their diffraction images will merge into one. This means that such minute objects can be visualized although they

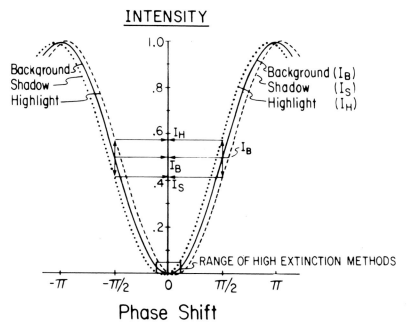

Fig. 3. Dependence of image brightness and contrast on phase retardation in Nomarski differential interference contrast (DIC) microscopy. Phase retardation is introduced and varied in DIC microscopy by laterally displacing a Wollaston prism or by setting the de Sénarmont compensator away from extinction. This converts positive and negative phase gradients of specimens to contrast, thus producing highlights (I_H) and shadows (I_S) relative to a neutral gray background (I_B). A phase shift of $\pi/2$ radians corresponds to $\lambda/4$, i.e., $\frac{1}{4}$ of a wavelength. It can be seen that theoretically the image contrast is highest at a phase retardation of about $\pi/2$ (vertical arrow). Due to stray light, the image background (I_B), as well as the shadows and highlights, are all of considerable intensity at this setting. This stray light has to be compensated for by the addition of a negative DC voltage (offset) in order to make use of the high image contrast. For reasons discussed in the text, smaller retardations ($\frac{1}{8}$ or $\frac{1}{10}$ of a wavelength) were recommended by Allen. (Reproduced with permission from Allen et al., 1981a.)

cannot be resolved (Fig. 4). Using Nomarski-DIC and VEC microscopy, biological structures of a size between 15 and 20 nm can be visualized, while other materials such as colloidal gold particles can be visualized down to diameters of 5 nm [theoretically even 1 or 2 nm: Mizushima, 1988; see also De Brabander et al., Chapter 6, this volume].

5. Contrast enhancement and resolution of objects. Figure 5 shows the intensity distribution across two closely spaced Airy patterns. One reason why resolution can be somewhat increased by contrast enhancement is that Rayleigh's criterion (Fig. 5a) of resolution (i.e., the depression of the combined intensity distribution between the peaks of two separate objects is 15% or more; see Castleman, Chapter 3, this volume for further details) which is appropriate for vision and photomicrography, is replaced by Sparrow's criterion [Hecht and Zajac, 1974] (Fig. 5c,d). The latter is applicable to electronic images because these can be readily enhanced by the above-mentioned procedures (transition from Fig. 5a to 5b), so that even a slight trough between the two objects (Fig. 5c) can be enhanced electronically to yield good separation (Fig. 5d). The increase in true resolution gained from this effect is a little less than twofold.

While video-based devices are instrumental in obtaining such unique spatial resolu-

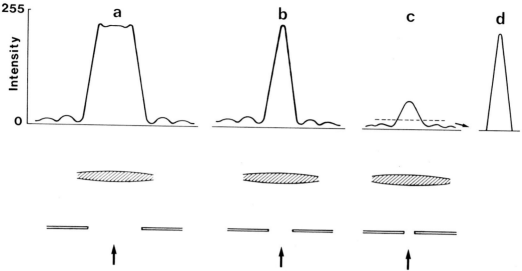

Fig. 4. Schematic of VEC microscopic imaging of subresolution-size objects. Positive objects (slits) (**a**) larger than, (**b**) equal in size, and (**c**) much smaller than the limit of resolution are imaged by transmitted light (arrows) using an ideal diffraction-limited optical system represented schematically by a single lens. The top panels show the resulting intensity (I) distributions across the images (diffraction disks, Airy disks). Before digitization, this corresponds to the voltage of the analogue video signal, i.e., brightness, along a video scan line. The subresolution-size object (c) yields a very low contrast "image" which cannot normally be distinguished from surrounding noise and therefore remains indiscernible by the eye. However, its contrast can be enhanced by applying offset and gain, i.e., applying a negative DC voltage of a magnitude indicated by the dashed line and subsequent electronic amplification. This results in the definition of a new black level (intensity zero) and a higher signal, as seen in **d**. As a result of such analogue contrast enhancement, objects much smaller than the limit of resolution (c) can be clearly visualized (d). However, their real size and shape cannot necessarily be inferred from the size or shape of their "images," image d of object c being inflated by diffraction effects to be equal in size to image b. (Reproduced from Weiss, 1987.)

tion and visualization, it can in practice be difficult to display and store this information because standard video devices, especially video recorders, have limited resolution (Section III B3, below). Good video camera–monitor combinations may reach 500–800 TV lines, which is very poor when compared with a good photograph of the same size whose spatial resolution can be several hundredfold better. There are two possibilities to circumvent this and capture the high-resolution images. One could use systems of a resolution of at least 1,000 TV lines throughout, i.e., camera, image processor (1,012 × 1,012 spatial resolution is possible with real-time capabilities), monitor, and recording equipment. This, although desirable and straightforward, would at present be too expensive to represent a practical solution. The other possibility is to image only a fraction of the field of view by magnifying it 2× or 4× more than was useful by conventional rules prior to picking it up by the video system (see discussion on transfer optics and sampling in Shotton, Chapter 1, and Castleman, Chapter 3, this volume).

6. Advantages of analogue contrast enhancement. The practical resolution of a VEC microscopy system is increased by a factor of about two over conventional microscopy. This is in part because the highest possible condenser numerical aperture can be used, since the resulting excessive image brightness due to stray light can be suppressed

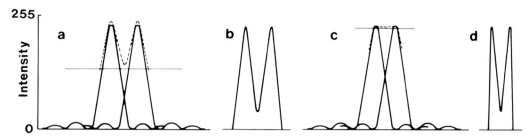

Fig. 5. Improvement of resolution by VEC microscopy. The image (Airy pattern) of a very small object is characterized by a central 0th order maximum and smaller maxima of 1st, 2nd, and higher orders. The overlapping images of two closely adjacent subresolution objects, with their summed intensity distribution (dashed), are shown in **a**. The two objects are resolved according to Rayleigh's criterion since the central depression is sufficiently deep to be clearly perceivable by the eye. If the contrast is manipulated by redefining the black level at the position indicated by the horizontal line in a, by either analogue (applying offset) or digital enhancement, and subsequently amplifying the signal (applying gain), a much better image results (**b**). In c the same objects are shown somewhat closer, so that they are not resolved according to Rayleigh's criterion. However, if contrast is enhanced as for b, even in this situation an image can be obtained which shows the two objects separated (**d**). Sparrow's limit of resolution is reached when there is no trough between the two peaks. (Reproduced with permission of Oxford University Press from Weiss et al., 1989.)

electronically with offset, and in part because Rayleigh's criterion of the limit of resolution is replaced by Sparrow's criterion. The gain in contrast is sufficient to visualize structures in living cells that are around one order of magnitude smaller than could be resolved or detected previously under the same specimen conditions. Additionally, the AVEC conditions reduce the diffraction anomalies caused by depolarization at lens surfaces and by residual strain birefrigence in the lenses that produce spurious detail and contrast in conventional polarized light-based interference techniques [Allen et al., 1981a,b; Hansen, 1988; Hansen et al., 1988].

7. Limitations of analogue contrast enhancement. Electronic noise is amplified along with the video signal in the enhancement process and may have to be reduced subsequently. If the optical system (including the slide, coverglass, and the camera target surface) contains dust, dirt, or manufacturing imperfections, these will create a fixed pattern of mottle that is enhanced along with the image. This can only be removed by digital processing. If the illuminating system is poorly designed or incorrectly aligned and focused (for Köhler illumination), the field may be unevenly illuminated. For optimal results in video-enhanced microscopy, the requirements for even illumination are much more stringent than with photomicroscopy. Within certain limits, however, uneven illumination can also be treated as fixed pattern "mottle" noise and removed by digital subtraction (Figs. 1, 2, 6).

Excess shading in the raw image, i.e., more than in Figure 6a, will limit the degree of analogue contrast enhancement that can be applied, because the shading that can be removed by subtraction may not be much more than that of Figure 6b. It may be useful in such cases to use the analogue shading correction function built into the camera control units of several manufacturers (e.g., Hamamatsu, Zeiss, Dage MTI). The use of a fiber-optic light scrambler which provides especially even illumination (see Section IIIB, below) would be even better.

B. Digital Image Processing

Many of the digital image processing routines used in VEC microscopy were available long before their value was recognized

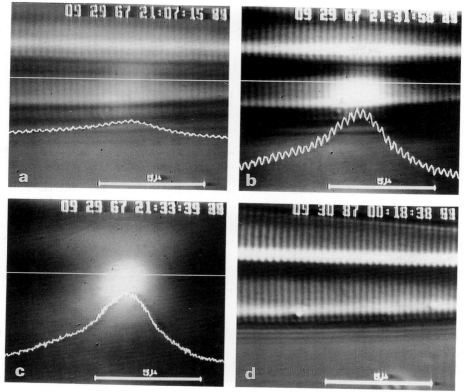

Fig. 6. Correction of uneven illumination (shading) in analogue-enhanced images by background subtraction. If, after fixing potential flaws in the optics and possibly correcting the shading by analogue adjustments of the video camera shading controls, an image appears as in **a**, this will show a very annoying "hot spot" after analogue enhancement (**b**). Subtraction of a specimen-free mottle image (**c**) results in an evenly illuminated image (**d**). This sequence of micrographs also demonstrates the usefulness of a calibration scale bar, of a time–date generator (month to $1/100$ sec), and of the display of the analogue intensity measurement along a video scan line (highlighted). The specimen is the test diatom *Amphipleura pellucida* with a known line spacing of 250 nm. Microscope: Axiophot; processor: Hamamatsu C 1966 Photonic Microscope System. (Reproduced with permission of Oxford University Press from Weiss et al., 1989.)

by microscopists [e.g., Castleman, 1979; Walter and Berns, 1981; Allen and Allen, 1983]. With the rapid development of faster and more powerful computers, many of these routines are now available at video frequency [Walter and Berns, 1981; Allen and Allen, 1983]. The principles of digital image processing and some of the procedures to be employed have been already described by Shotton, Chapter 2, this volume. In VEC microscopy, they are used for rapid preprocessing, i.e., the improvement of images prior to their storage on videotape or disk.

It should be noted, however, that individual frames, once stored, can also be subjected to further subsequent digital processing.

Following analogue contrast enhancement, the analogue TV signal (a temporal pattern of voltage changes) is digitized so that it can be stored in a frame store (i.e., RAM = random-access memory, which stores a full frame of the video picture signal; also called frame buffer) or manipulated by the arithmetric logic unit (ALU) of an image processor. For real-time subtraction (see Section IIB2, below) one frame store may be sufficient to strip the incoming video frames from its fixed pattern noise (mottle, background) and output them to a displaying or recording device. More complicated or composite operations may require two or three frame stores. In the most suitable processors, the instructions necessary to carry out a number of different arithmetic manipulations are controlled by firmware or by packages of macro software routines, so that an operator can easily learn to process images rapidly for his or her particular operation. The operations themselves may be specified manually in seconds, and are then carried out in "real time," i.e., repetitively during the intervals between each consecutive frame (i.e., 25 or 30 times per second, depending on the TV standard used).

During digitization, the image is subdivided into 512 or more lines, each of 512 or more picture elements (pixels). This spatial resolution is much poorer than that of a photograph of equivalent size. In principle, comparable resolution may be achieved by image analysis systems capable of handling images of a spatial resolution of $1,024 \times 1,024$ or better, but these are still very expensive. Usually, each pixel can assume one out of 256 (8 bit) gray levels, with zero being defined as black and 255 as white. If averaging or other multiframe procedures are to be used, the ALU must be capable of 16-bit operations, to avoid truncation errors.

1. Rolling average or jumping average. The rolling average function, also called recursive filtering, computes an exponentially weighted average of the last incoming image and the previously stored average image, such that the most recent frames dominate. In jumping average mode, a time average is computed from a predefined number of frames, and this is displayed for the duration of the accumulation of the next set of frames. Both averaging modes may be used to diminish electronic noise in the video signal, maximally by the square root of the number of frames averaged. The former smears and de-emphasizes any fast motion present, while the latter accentuates slow

motion. Averaging is generally used for intensified fluorescence microscopy, where photon fluctuations and electronic noise from the imaging device often present severe problems, and is advisable for VEC microscopy when high enhancement is used, i.e., when electronic noise becomes annoying. The Kalman filter [Erasmus, 1982] represents another way to produce a continuously displayed image with reduced noise (see Shotton, Chapter 2, this volume).

2. Mottle subtraction. Patterns of image imperfections (mottle) remain in the analogue image when the specimen is defocused or moved out of the field of view (Figs. 1c, 2E). Consequently, mottle can be averaged and stored in a video frame memory and then subtracted from each frame of the incoming video signal (Fig. 1d). This operation (mottle or background subtraction) results in a "clean" image lacking mottle (Figs. 1e, 2F). The same operation also eliminates inhomogeneities in background brightness (Fig. 6), if the contrast does not exceed the "window" of 256 gray levels which can be stored and displayed.

3. Digital contrast enhancement. The analogue-enhanced, mottle-subtracted image may not have sufficient contrast. In this case the image can be further enhanced digitally, for example, by stretching the histogram of gray levels (Fig. 1f). The procedure is analogous to analogue enhancement, but the selection is made digitally by choosing that restricted region of the gray levels containing the image information and expanding it to stretch the entire distance from black to white, i.e., to fill the entire 256 gray levels. This is done by assigning new gray levels to the original ones through the use of an output look-up table (LUT), as illustrated by Shotton in Chapter 2, this volume.

4. Study of motion by sequential or interval subtraction. The analogue-enhanced image can be subjected to *sequential subtraction* in order to observe and detect only moving elements. This is done by freezing a reference image without taking the specimen out of the field or out of focus, and then subtracting it from all incoming frames. Subtraction of an image (stored) from very similar subsequent images (i.e., the incoming live images) results in otherwise blank images in which moving elements cause image differences and so make their presence known (Fig. 7). This is an extremely sensitive means of motion detection, but works satisfactorily only with a very stable microscope stand under good temperature control. Any drift in focus or pressure applied to the stage or microscope body may result in a distorted image of the whole object. This mode gives both the position of the moving object at time zero (frozen and in negative contrast, i.e., a "missing object") and the live position of the moving object (Fig. 7). Distance measurements for velocity calculations are made very convenient by this technique. Sequential subtraction can

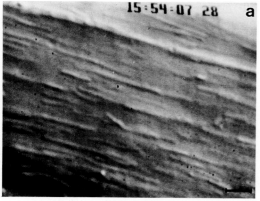

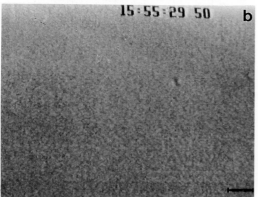

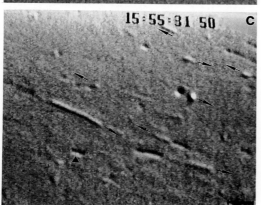

Fig. 7. Selective visualization of moving objects by sequential subtraction. a: VEC image of a bundle of 50–100 nerve fibers (axons) from fish olfactory nerve. The axons are C-fibers of 0.2 μm in diameter and therefore not resolvable. Due to their close packing they cannot be visualized by VEC-DIC microscopy, but elongated mitochondria and other organelles within them, mostly stationary with some undergoing rapid axonal transport, become visible [see Weiss and Buchner, 1988]. When a reference image taken at time 29.40 sec is subtracted from the incoming video image at time 29.50 sec, i.e., 0.1 sec or 5 frames later, almost no detail is visible in the resulting image (b) because all objects remained close to their original location. c: Two seconds later (at 31.50 sec) the moving organelles have become visible. Each moving organelle is depicted twice, once appearing in negative contrast (dark depression), marking the organelle's original location, i.e., the loci from which the organelles are now missing, and again in positive contrast (with arrows indicating direction of movement), marking their current locations. In this sequence, the movement of mitochondria (elongated) and of lysosomelike organelles (round) is observed, with most organelles, except one mitochondrion near the center, moving toward the lower right. The triangle denotes an organelle which disappeared by moving out of the focal plane. Microscope: Polyvar, Reichert/Leica; processor: Hamamatsu C 1966 Photonic Microscopy System. Scale bar, 2.7 μm.

also be used to visualize moving low-contrast objects in front of an immotile high-contrast structure or background. It was used to visualize thin actin filament bundles and tubular endoplasmic reticulum in plant cells although these were directly adjacent to a cell wall of very high contrast [Lichtscheidl and Weiss, 1988]. The low-contrast objects made themselves notable because they were agitated in Brownian motion.

Interval subtraction is an alternative mode of sequential subtraction that is programmed to refresh after predetermined intervals, i.e., a new "background" image to be subtracted from incoming video images is automatically chosen and stored after a certain, preselectable number of frames.

It is clear that a multitude of additional procedures to accentuate specific features can be devised and implemented with programmable digital image processors [e.g., Baxes, 1984; Anonymous, 1985; Walter and Berns, 1986; Häder, 1991; Jähne, 1991; Shotton, Chapter 2, this volume].

III. PRACTICE OF VIDEO-ENHANCED CONTRAST (VEC) MICROSCOPY

A. General Strategies for VEC Microscopy Image Improvement

The strategy when planning to use VEC microscopy should be as follows: If high image fidelity and detail are desired and the smallest objects of interest have very weak contrast, but are larger than the limit of resolution (approx. 200 nm), we need only the first steps of VEC microscopy image improvement, namely moderate analogue contrast enhancement (see Table II, steps 1–4). If shading occurs, analogue shading correction is usually sufficient. Digital processing for background correction or further enhancement may be dispensable in many cases. A fair amount of light is required, but can be achieved with differential interference contrast and all brighter techniques. Typical applications include studies of changes in cell form, movement of larger organelles,

or cell division (see Fig. 9 below as an example).

However, if one wishes to visualize the smallest objects possible, the full extent of VEC microscopy, including digital mottle subtraction, must be used (see Table II, steps 1–9), preferentially employing differential interference contrast (DIC) or anaxial illumination techniques. Visualization of microtubules (25 nm diameter) or of vesicles with diameters of 50 nm or less can be achieved. Here one needs the following functions of electronic image improvement: high analogue enhancement, high-performance polarized light microscopy (DIC or POL microscopy) according to Allen et al. [1981a,b] or Inoué [1981, 1986], and digital image processing including real-time background subtraction and digital contrast enhancement. If the resulting image is noisy, real-time averaging over two or four frames or real-time digital filtering should additionally be employed.

Video enhancement may not be achievable to its full extent if weak and/or small objects of interest are located next to objects of high contrast due to high absorption, birefringence, or other physical properties (e.g., cells rich in crystalline inclusions or dense and bulky organelles) because these would become too bright and too annoying when further contrast enhanced (see, however, Sections IIB4 and IIIF2).

When information on the third dimension is of interest or when relatively thick specimens (e.g., tissue slices) are used, contrast enhancement offers a number of ways to obtain remarkable results (see Sections IIIC, IIIF1, and IIIF2) and it may even be used in combination with confocal microscopy (see Section IIIF7), which is discussed in Chapters 9 to 14, this volume. Rapidly changing scenes and moving objects are, however, the true domain of straightforward VEC microscopy where confocal scanning microscopy is usually too slow, and where also motorized focusing and 3D image reconstruction are too slow to permit one to follow live objects on the microscope stage while they are visualized electronically on the monitor.

TABLE II. Steps for AVEC-DIC Microscopy

Step		Manipulation	Result
1.[a]	M[b]	Focus specimen and adjust microscope correctly for Köhler illumination	Image appears
2.	M	Open iris diaphragm fully	Optical image becomes too bright
3.	M	Set compensator up to 20° from extinction[c]	Optical image worsens
4.	C	Analogue enhance by manually adjusting gain and offset	High-contrast TV image with often disturbing mottle pattern appears
5.	M	Defocus or move specimen laterally out of field of view	Object disappears, mottle remains
6.	P	Average and store mottle image, then subtract mottle image from incoming video images	Absolutely homogeneous, light-gray ("empty") image appears
7.	M	Return specimen to focal plane	Clear image appears; if contrast is weak, go to Step 8
8.	P	Contrast enhance digitally (histogram stretching)	Contrast becomes optimal; if pixel noise is high, go to Step 9
9.	P	Use rolling or jumping averaging or digital filtering	Clear, low-noise and high-contrast image appears

[a]Before step 1, one should set the "brightness" and "contrast" controls of the monitor showing the processed image to their intermediate positions because the degree of enhancement will not be adequate in the recorded sequence if the monitor has been adjusted to an extreme setting. In order to utilize VEC microscopy to its full extent, make sure that the microscope objective and condenser front lenses are absolutely clean (check at least once daily) and that the lamp is always optimally adjusted and centered.
[b]The manipulations are performed at the microscope (M), camera control (C), or image processor (P), respectively.
[c]Particles need to appear in DIC microscopy images as if illuminated from above, i.e., with their bright part up, while vacuoles have the opposite shadows. If this is not the case, the camera has to be rotated 180° or the compensator or Wollaston prism has to be set to the opposite side with respect to the extinction position.

For many applications, intensified fluorescence microscopy (see Chapters 7 to 9, this volume) may be required to supplement VEC microscopy. The latter reveals the intracellular details, while the former may be employed to determine the identity of these objects using fluorescent antibodies or organelle-specific fluorescent dyes.

B. Equipment Considerations

Video-enhanced contrast microscopy is based on the combination of a versatile research microscope, a high-quality video camera with manual offset and gain controls, and a real-time digital image processor. The components required and the basic procedures have been extensively discussed elsewhere [see Inoué, 1986; Schnapp, 1986; Shotton, 1988; Weiss et al., 1989], so that a short summary may suffice here. It should be mentioned that the microscope and image processors required for VEC and video-intensified fluorescence microscopy are basically the same, the only fundamental difference being that different specialized cameras are needed for the two techniques.

1. Microscope. To utilize the technique to its full extent, a massive, stable high-quality research compound optical microscope equipped with transmitted light bright-field, dark-field, and differential interference contrast (DIC) is required (Fig. 8). Inverted microscopes are preferable in some cases, but

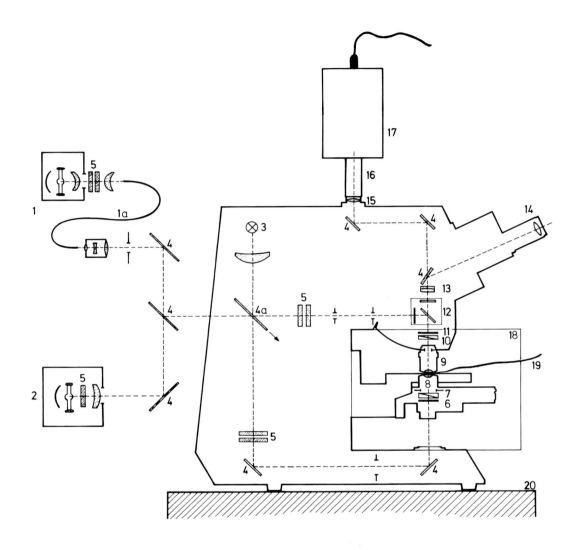

Fig. 8. The video microscope. Schematic representation of a microscope with the modifications required for both video-intensified microscopy (VIM) and video-enhanced contrast (VEC) microscopy. 1, Mercury arc lamp with fiber-optic light scrambler for high-magnification differential interference contrast (DIC) microscopy; 1a, quartz optical fiber; 2, alternatively: conventional mercury arc lamp illumination (for DIC or epifluorescence microscopy); 3, halogen lamp; 4, mirrors; 4a, removable mirror for switching lamps; 5, UV, IR (heat), and green filters; 6, polarizer; 7,10, Wollaston prisms for DIC; 8, high-NA oil-immersion condenser; 9, high-NA oil-immersion objective; 11, rotatable analyzer (preferably a rotatable graduated analyzer is combined with a λ/4 plate to form a de Sénarmont compensator); 12, epifluorescence filter cube; 13, additional magnification lens ($2\times$); 14, eyepiece; 15,16, projective lens ($1\times$ and extension tube ($2\times$ or $4\times$) or, alternatively, a high-magnification eyepiece ($16\times$ or $25\times$) and camera objective (50 or 63 mm); 17, video camera (Chalnicon, Newvicon, Pasecon, or CCD for VEC microscopy or SIT camera for VIM; for most purposes in the biomedical sciences, two video exits with one of each of the two classes of cameras permanently mounted is most desirable); 18, temperature-controlled chamber; 19, thermistor; 20, vibration isolation table. This setup is based on a Polyvar microscope (Reichert/Leica, Vienna). The parts of special importance for VEC microscopy are 1, 1a, 5, 8, 9, 13, 15, 16, 17, and often 20. (Reproduced from Weiss, 1987.)

usually upright ones are similarly well suited. Condenser and objective should be of the oil-immersion type when working with objective magnifications of 40 × and more. An exit port capable of projecting 100% (or at least 80%) of the light to the video camera is essential for high-magnification DIC work. To ensure that the optical image is sufficiently oversampled by the video camera as to remain diffraction-limited (see Shotton, Chapter 1, and Castleman, Chapter 3, this volume, for discussion) in order to visualize objects of sizes near and below the limit of resolution, additional transfer magnification of 2–6× between the primary image plane and the video camera target is required (Fig. 8, components 13, 15, 16). This is obtained, for example, by the use of a high-magnification ocular (16 × or 25 ×) and a video camera lens of focal length of 50 mm or 63 mm, or of additional magnification changers. This yields magnifications on the camera target of up to 500×, yielding an image on a monitor with a 25 cm screen width of 10,000 × or more. The use of at least one heat-absorbing and one heat-reflecting filter in the illumination path is essential to protect the specimen and the polarizers. A narrow-band green filter (546 ± 10 nm for mercury arc lamps) and a UV-absorbing filter are recommended for DIC work with most cell types, to prevent damage and loss of viability of living specimens (Fig. 8, component 5).

2. Illumination system. The illumination should be as bright and even as possible, to provide enough light for working at the highest magnifications. The high luminous density required makes the use of mercury arc lamps (usually 50 or 100 W) necessary for most high-magnification VEC microscopy applications. Intensity changes over time have to be minimized by the use of a stabilized DC lamp and power supply, because AC lamps usually start flickering to an extent intolerable for work at high-contrast enhancement so that they have to be replaced before reaching 100 working hours. The other requirement is that the lamp illuminates the aperture absolutely evenly. The highly concentrated light from arc lamps is in most microscopes distributed over the aperture with uniformity just sufficient for observation with the human eye or for microphotography. If, however, the contrast of a relatively flat microscope image is enhanced electronically to the extent used in VEC microscopy, then the central "hot spot" that stems from the bright spot of the arc becomes visible (Fig. 6) and will limit the extent to which analogue enhancement can be applied. In addition, with uneven aperture illumination, the diffraction pattern produced by each image point in the specimen departs from the ideal Airy disk, thus causing a loss of resolution [Inoué, 1986, p. 127].

It is now possible to make the illumination of the aperture uniform by inserting a fiber-optic light scrambler between lamp housing and the microscope body [Ellis, 1985; Inoué, 1986, p. 495f.]. The scrambler consists of a single-mode quartz optical fiber with light-collecting and -projecting optics. The fiber is curved in such a way that the light beam is scrambled by multiple internal reflection and forms a bright homogeneous patch of light when it exits the carefully polished end of the fiber. Despite its importance for VEC microscopy, microscope manufacturers usually do not provide microscopes with light scramblers, except for some Leitz microscopes, but most microscopes can be retrofitted later (Technical Video Ltd., Woods Hole, MA 02543, USA).

The light scrambler converts a three-dimensional light source (mercury arc) into a two-dimensional one (polished exit surface of the light fiber). This is probably a major reason why, with an optimal microscope and using VEC microscopy, Inoué [1989] reported true depth resolution of 0.1–0.2 μm for DIC and polarization microscopy. This is better than achieved with fluorescence and confocal microscopes, and better than the best lateral resolution obtainable. The light scrambler illumination, therefore, increases the degree of contrast enhancement that can be employed, helps one reach the theoretical lateral resolution, and increases depth resolution considerably.

3. Camera and image processor. The image is picked up by a high-quality video camera, usually a Chalnicon, Pasecon, or Newvicon, but today also some types of charge-coupled device (CCD) cameras are suitable [see Inoué, 1986, for technical details]. The analogue video signal is first subjected to analogue contrast enhancement by means of the manual gain and offset (also called "black level" or "pedestal") controls on the camera, and the resulting raw video image is then digitized by the digital image processor, permitting the various steps of digital image processing to be subsequently performed on each video frame as it is received. Cameras with built-in "automatic gain control" (AGC) are unsuitable unless this can be disabled and operated manually.

The most important requirement for the digital equipment used by the VEC microscopist is that it be fast, i.e, the image processor has to work in "real time," completing all the necessary processing of one video frame before the arrival of the next 33.3 msec or 40 msec later, depending upon the video standard employed (see Chapter 1, this volume). Similarly, for ease of interactive use, the entry of commands has to be restricted to the striking of a single or a few keys, or mouse manipulation of a single menu (see Inoué, Chapter 4, this volume).

A generalized VEC microscopy setup with its principal technical features and the proper connection of the ancillary equipment is given in Chapter 1, Fig. 4. In view of the rapid changes of the electronic devices used (processors, cameras, video recorders), it is necessary to inquire into the most recent performance data immediately prior to purchase. Generally, much care has to be spent to obtain the best possible recording device since this is the quality-limiting factor due to its limited spatial resolution. Images created by good cameras (600–800 TV lines) and processed by digital devices (512×512 pixel resolution) can, without loss in quality, be viewed and photographed from high-reso-lution monitors (600–1,200 TV lines). But even three-quarter inch (U-matic) or S-VHS recorders *reduce* the vertical resolution to 320–400 TV lines, and regular VHS and color systems reduce it further to around 250 TV lines. Optical memory disk recorders (OMDRs), although still high in price, can presently reach or somewhat exceed 450 TV lines and represent a more desirable solution. However, especially in this field, major improvements in technical performance are to be expected as time progresses. A very good introduction to and a wealth of technical details on video equipment can be found in Inoué [1986]. The group of companies offering specialized video equipment also undergoes rapid changes so that no definitive list can be given. The companies listed in Inoué [1986], Anonymous [1988], and Lacey [1989] may serve as a good basis.

C. Sample Preparation

In VEC microscopy, the same samples can be used as in conventional light microscopy. Live cells from tissue cultures should preferentially be grown on a coverglass. The specimen's region of interest should be close to the coverglass surface, where the best image is obtained. If the highest magnifications are intended, it may be that the optics can be adjusted for Köhler illumination only at this surface and a few tens of micrometers below (upright microscope), since high-magnification objectives are usually designed for optical imaging of objects at a distance of 170 μm from the front element. Thus for work with oil-immersion objectives it is recommended that a No. 0 coverglass (80–120 μm thick) be used instead of a normal No. 1 coverglass (approx. 170 μm thick), since this permits a greater working distance within the specimen. Note, however, that since the oil-immersion objective will now be focusing through water rather than glass, spherical aberration will be introduced. In Chapter 9, Shaw discusses the use of immersion oils of differing refractive indices to overcome this problem. Alternatively, for imaging deep within an aqueous specimen, a water-immersion objective may be employed to overcome the problem.

Aqueous samples have to be prevented from drying out by completely sealing the cover glass to the microscope slide. Nail polish may be used for this, or if a live specimen, such as microtubules, extruded cytoplasm, or cultured cells is being observed, VALAP is recommended. This consists of equal parts by weight of vaseline, lanolin, and paraffin (M.Pt. 51–53°C) and liquefies at around 65°C. It may be applied around the coverglass with a cotton-tip applicator or fine paintbrush, and rapidly hardens. If the specimen is in suspension, sample volumes of no more than 5–7 µl should be used with 22 mm × 22 mm coverglasses, in order to produce very thin specimens (ca. 10 µm thick) for best image quality.

If working with an inverted microscope, the specimen slide has to be inverted for fitting to the microscope with the coverglass underneath. With most microscope stages this will interfere with the VALAP sealant, and flat positioning of the slide will not be possible. It is recommended instead to use a metal frame the size of a regular slide and 0.8 to 1 mm thick to hold a sandwich of two coverglasses of dissimilar sizes [Schnapp, 1986; Weiss et al., 1989, 1990a].

If thicker specimen, such as tissue slices, vibratome sections, or nerve bundles are to be observed, only DIC or anaxial illumination [Kachar, 1985a] techniques are recommended. Only the first 10 or 20 µm closest to the objective will yield good images, since the image quality degrades quickly if one focuses deeper into the tissue. The opacity of living tissue can be greatly reduced when infrared or near-infrared (IR) light is used (see Section IIIF1 and Fig. 9.a).

D. Steps to Generate the Image

The steps required for image generation and improvement by VEC microscopy are summarized in Table II. They include procedures different from those used in conventional microscopy, which are required for the highest resolution and visualization of subresolution objects. The procedure described is that for AVEC-DIC (see below), but if DIC is not required, step 3 should simply be omitted. A more detailed description, including technical hints, is given elsewhere [Weiss et al., 1989].

Video contrast enhancement of microscopic images obtained using bright-field, dark-field, anaxial illumination, or fluorescence optics, rather than DIC, is very straightforward. It is performed by following the steps in Table II, with the exception of step 3. At step 2 it is important to make sure that the camera receives the proper amount of light to work near saturation. Some manufacturers have red and green control lamps built in to indicate this. One should see a moderately modulated image on the monitor, while a *very* flat image or no image indicates insufficient light. If this is the case, it may be necessary to readjust the illumination while observing the image on the screen, to remove any diffusers, or to go to a brighter lamp type. If there is more stray light than can be compensated by analogue offset, the intensity of illumination should be reduced using neutral density gray or other filters.

Much more striking image improvement is usually gained with DIC and polarization microscopy, but special attention and some explanation are required for step 3. Allen et al. [1981a,b] and Inoué [1981,1989] simultaneously described procedures of video contrast enhancement for polarized-light techniques which differed considerably in their approach but yielded very similar results. Although this is not the place to judge which one is more appropriate, we have to distinguish clearly between the two strategies in order to avoid confusion.

Allen named his techniques "Allen video-enhanced contrast" differential interference contrast and polarization (AVEC-DIC and AVEC-POL, respectively) microscopy. The AVEC techniques involve the introduction of additional bias retardation by setting polarizer and analyzer relatively far away

from extinction, in order to gain a high specimen signal, I_s. Allen suggested the use of
a de Sénarmont compensator setup [de
Sénarmont, 1840; Bennett, 1950] which consists of a quarter-wave plate (specific for the
wavelength used) in front of a rotatable analyzer. In DIC microscopy, alternatively but
less accurately, the desired bias retardation
can also be introduced by shifting the adjustable Wollaston prism. Allen recommended
a bias retardation of ¼ to ⅑ of a wavelength
away from extinction, with ⅑ as the best compromise between high signal (double arrows
in Fig. 3) and minimal diffraction anomaly
of the Airy pattern [Allen et al., 1981b; Hansen, 1988; Hansen et al., 1988]. The enormous amount of stray light introduced at
such settings (Fig. 3) is removed by an appropriately large setting of analogue and/or digital offset. In AVEC microscopy, the amount
of light required for saturation of the video
camera can be adjusted by setting the de
Sénarmont compensator or the Wollaston
prism further away from extinction, thereby
gaining a higher signal but also admitting
more stray light (Fig. 3). It is most desirable
to obtain saturation at a retardation of about
⅑ of a wave (20°) if the microscope–camera–
illumination combination does permit it,
since this setting provides the best resolution [Hansen, 1988; Hansen et al., 1988]. Further opening of the crossed polarizers beyond
20° will rarely improve the image further,
but it may introduce amounts of stray light
no longer manageable by offset.

The technique recommended by Inoué,
which in this chapter is called IVEC microscopy for distinction, aims to optically reduce
stray light and diffraction anomaly arising
from curved lens surfaces and other sources
(Table I) by employing extremely strain-free
objectives and the special rectifying lenses
developed by Inoué [Inoué, 1961]. The latter are commercially available only for a few
microscopes (some lines of Nikon) [e.g.,
Bajer et al., 1986] and are expensive. Inoué's
special optimized microscope [Inoué, 1961,
1986] is used at a polarizer setting very close
to extinction, which cannot be used for VEC

microscopy with many other instruments
because insufficient light is passed for near-
saturation of the video camera.

Thus in IVEC microscopy, stray light is
not admitted, since the polarizers stay close
to extinction and the special rectifying optics
further reduce the stray light. Consequently,
filters to reduce brightness are not required.
On the contrary, a very bright arc lamp, ideally with a fiber-optic illuminator, is necessary to saturate the camera. The AVEC
technique, on the other hand, electronically
improves primary optical images characterized by low contrast and relatively high stray
light content, arising from a "nonoptimal"
optical arrangement. In IVEC microscopy no
compromise is made regarding the optics,
and consequently less demanding electronic
steps are required to rescue the image. The
AVEC technique is, however, the one which
can be used with any good research microscope equipped with commercial film polarizers. The optimal compensator setting can
be experimentally evaluated between ¹⁄₁₀₀
and ⅑ of a wavelength (and with some loss
of resolution up to ¼), within the limits of the
ability of one's illuminating system to nearly
saturate the video camera and depending
on whether one needs optimal resolution
(at ⅑ of a wavelength [Hansen, 1988; Hansen et al., 1988]) or optimal visualization
(highest contrast was reported for ¹⁄₁₅ of a
wavelength in some microscopes [Schnapp,
1986]). But note that the dimensions of an
object imaged at the latter setting may be
different for different orientations, due to
the diffraction anomaly of the Airy pattern
[Allen et al., 1981b].

Steps 1–4 of the procedure given here yield
the final image, if digital processing is not
possible, or if only analogue enhancement
is required. In this situation, analogue
enhancement has to be stopped just before
the mottle or uneven illumination becomes
annoying. Analogue shading correction and
other types of analogue image improvement
may be applied, if one's camera control unit
offers them. The use of a light scrambler and
meticulous cleaning of the inner optical sur-

faces of the microscope and especially the lens surfaces in the projecting system to the camera (ocular and camera lens) usually helps considerably in permitting high levels of analogue enhancement.

For publication purposes, pictures may be photographed or filmed off the video screen by observing special procedures and hints which are discussed in detail by Inoué [1986, Chap. 12] or Weiss et al. [1989].

E. Interpretation of VEC Microscopy Images

Unlike EM images, in which the submicroscopic objects are truly resolved, the apparent sizes of objects seen in AVEC-DIC microscopy may not necessarily reflect their real size. Objects smaller than the limit of resolution, i.e., smaller than 100–250 nm, depending on the optics and the wavelength of light used, are inflated by diffraction to the size of the resolution limit, the Airy disk diameter (Fig. 4). While the size of the image does not enable one to decide whether one or several objects of a size smaller than the limit of resolution are present, the degree of contrast sometimes permits this judgment. A pair of adjacent microtubules would, for example, appear to have the same thickness as a single one, but the contrast would be about twice as high. Note, however, that if a large number of subresolution objects are crowded together, separated by distances less than 200 nm from one another, for instance, vesicles in very small cells or synaptic nerve endings, the confusing overlap of their Airy disk images will cause them to remain completely invisible, despite the fact that they could be clearly visualized if they were separated by more than the resolution limit (see Fig. 7a as an example of this effect).

F. Special Considerations for the Different Types of Contrasting Techniques

1. Differential interference contrast. This is the optical technique which in most appli-

cations is best suited for video contrast enhancement. The Nomarski DIC method gives in-focus, high-contrast, shadowcast images of phase details in which the direction of shadowing is opposite for phase-advancing and phase-retarding details [Allen et al., 1969]. The generation of contrast at high working aperture is limited to a very thin depth of field, with the result that this technique is unique in its ability to generate high-contrast optical sections only 200–300 nm or less in thickness, similar to those obtained by confocal microscopy (see Chapters 9 to 14). Since the depth resolution of DIC can be considerably better (see Section IIIB2) than in confocal fluorescence microscopy, it pays to use it for three-dimensional reconstruction [Inoué and Inoué, 1987; Inoué, 1990]. With the AVEC- or IVEC-DIC methods, the sensitivity of detection is increased to the level that transparent phase objects as small as 15 nm can be detected under optimal conditions (see examples in Section IVA below and in Figs. 1, 6, 7, 9, 10). Since living unstained tissue becomes more and more translucent when viewed with light of increasing wavelength, IR-DIC is recommended when there is a need to look deep into tissue samples, biopsies, or brain slices. Remarkable results have been reported with IR-VEC-DIC microscopy of brain slices, where this technique visualizes axons, dendritic spines, and electrode tips even at depths of 100 μm or more, thus opening new possibilities toward vision-guided electrophysiology [Dodt and Zieglgänsberger, 1990] (Fig. 9a).

2. Anaxial illumination. In anaxial (or oblique) illumination methods, the condenser aperture is unevenly illuminated to produce a differential (shadowcast) image of increased true resolution. In Abbe's original method, the iris was moved to one side of the front focal plane of the condenser. It is also possible, using a tungsten filament lamp with a hemispherical mirror, to displace the mirror to illuminate only half of the aperture [Kachar, 1985a]. In Hoffman's modulation contrast method [Hoffman,

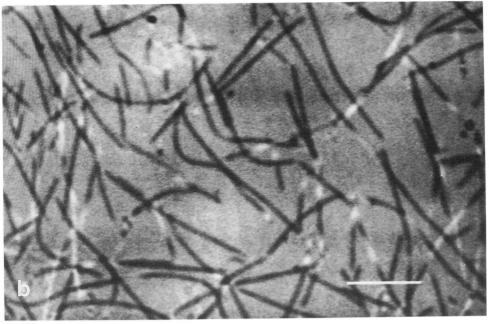

1977], undiffracted light is partially excluded by a trizonal plate in the condensor, which also serves to accentuate high spatial frequencies. The objective single sideband imaging technique developed by Ellis [1978] also yields very good results with video-enhanced contrast microscopy.

Video-enhanced contrast microscopy works very well for these anaxial techniques, because their main limitation of providing only low-contrast images can be overcome by electronic enhancement. High specimen birefringence does not interfere as much as it does in DIC microscopy, making these methods preferable for certain types of specimen. For example, vertebrate axons can be better seen with enhanced anaxial illumination than with DIC because their highly birefringent myelin sheaths cannot contribute excessive contrast. A further advantage is that the optical equipment is cheap: Only a bright-field microscope is required! It has

been reported that structure can still be satisfactorily observed inside tissue at depths at which the quality of DIC images would be much poorer [Kachar, 1985a].

3. Polarization microscopy. Sensitive visualization or photomicrography of the weakly birefringent biological structures was previously possible only by using a polarizing microscope with a rectifier (Nikon) that eliminated not only stray light due to depolarization at lens surfaces, but also the disturbing diffraction anomaly that can lead to spurious contrast and resolution [Allen et al., 1969; 1981b]. This rectifier consists of a zero-power meniscus lens and properly oriented half-wave plate [Inoué, 1961, 1986]. However, with video-enhanced contrast polarization microscopy (AVEC-POL), relatively sensitive observations may now be made with a normal polarizing microscope. At 1/9-wave bias retardation, the anomalous cloverleaf Airy disk diffraction pattern normally seen at extinction (zero retardation and crossed polars) is converted into a normal Airy disk, with the result that there is no longer spurious resolution or contrast. In addition, the diffraction anomaly due to weak strain birefringence in any of the lens components also disappears [Allen et al., 1981b]. The use of the highest quality optical components for polarized light microscopy *together* with video enhancement (IVEC-POL) results in the optimal visualization of extremely weak birefringent objects, such as individual microtubules, bacterial flagella, etc. [Inoué, 1986, 1989].

4. Phase contrast. Video enhancement can be used with great advantage in combination with almost any contrasting method of optical microscopy. However, in phase-contrast microscopy, phase objects exhibit minimal contrast in focus, and show opposite contrast above and below focus. It is, therefore, often very difficult to store an out-of-focus mottle image without significant contributions from the specimen. This limits the use of high analogue contrast enhancement in phase-contrast microscopy. However, if an empty "background" image

Fig. 9. a: Infrared VEC-DIC Microscopy. Live, unstained neurons in a 300 μm-thick brain slice from rat brain as visualized in infrared light (800 nm) by IR-VEC-DIC. When viewed with infrared light, tissue slices are much less opaque than with shorter wavelengths' light, and axons and dendrites can clearly be seen. This allows precise positioning of micropipettes for patch clamp recording, iontophoretic application of pharmacological agents, or microinjection of dyes or antibodies. Pyramidal cells and other neurons (center) from lamina II of neocortex at a depth of 50 μm from the surface. Zeiss Axiovert, 40× Plan Neofluar NA 0.75, 2.5× optovar, Hamamatsu C-2400 Newvicon camera with analogue contrast enhancement only. Bar, 10 μm. (Reproduced from Dodt and Zieglgänsberger, 1990.) **b:** Video-enhanced confocal laser scanning imaging of microtubules by epireflection microscopy. Taxol-stabilized microtubules (diameter 25 nm) were polymerized from pure tubulin free of associated proteins. Individual microtubules, by interference with reflection from the coverglass surface, appear black (close) or white (at a distance) in the image. Microscope: BioRad MRC 500 confocal laser scanning microscope. Framing rate 1 per second. Averaging over 50 scanning frames, analogue and digital computer enhancement and background subtraction. Bar, 4 μm. (Reproduced from Amos and Amos, 1991, with permission of Company of Biologists Ltd.)

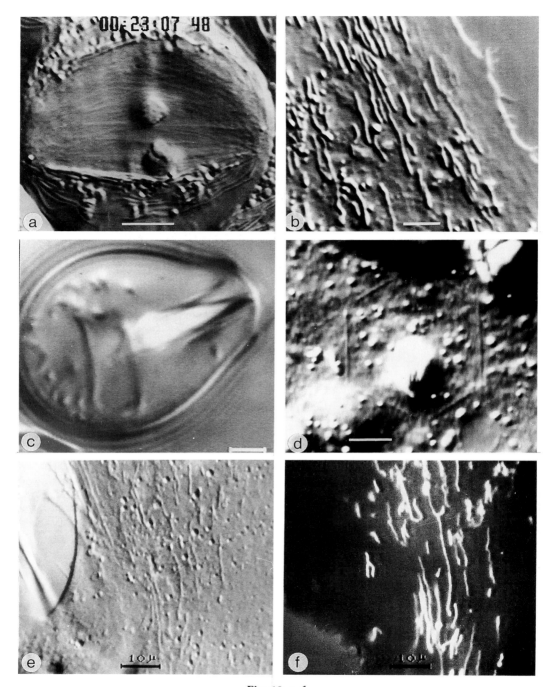

Fig. 10. a–f

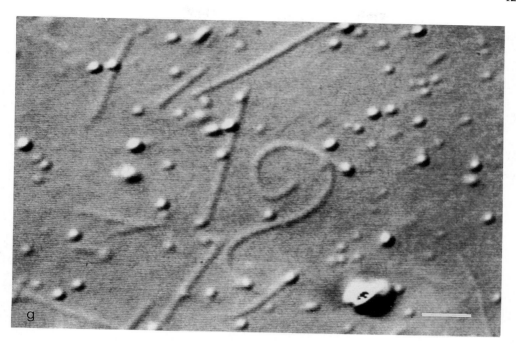

Fig. 10. Video-enhanced contrast microscopic images of cellular constituents in different cell types. **a:** Chromosome fibers in a living metaphase I spindle of *Pales ferruginea* (cranefly) spermatocytes. The fibers are small bundles and sometimes perhaps single microtubules whose "fir-tree"-like arrangement is visualized. AVEC-DIC. (Reproduced with permission from Bastmeyer and Fuge, 1987.) **b:** Mitochondria and smaller spherical organelles in the cytoplasm of an MDCK cell. AVEC-DIC. **c:** Isolated nematocyte of *Hydra* which can be stimulated in vitro to eject its tubular contents. From a collaboration of the authors with T. Holstein, Munich, FRG. AVEC-DIC. **d:** Hexagon consisting of bundles of actin filaments formed by filament gliding in droplets of extruded cytoplasm from the characean alga *Nitella*, which perform continuous planar rotation for over 1 hr. The large objects are loose chloroplasts, the small ones carbohydrate storage organelles. AVEC-DIC. From a collaboration of the authors with R. Jarosch, Salzburg, Austria. **e:** Mitochondria and other, spherical organelles in the cytoplasm of a neuroblastoma (N18) cell. **f:** Same region imaged by VIM after vital staining for mitochondria with Rhodamin 123. Note that all elongated organelles in e proved to be positive for the mitochondria-specific dye. This pair shows how the two techniques yield complementary information on fine details (e) and on the identity (f) of a subset of the multitude of organelles visualized by VEC microscopy. AVEC-DIC and VIM. (Reproduced from Weiss, 1987.) **g:** Microtubules, synaptic vesicles, and larger (prelysosomal) organelles from extruded squid axonal cytoplasm. AVEC-DIC. a–f: Polyvar microscope, Reichert/Leica, Vienna; g: Zeiss Axiomat. a–g: Hamamatsu C1966 Photonic Microscope System. Bars: (a,d) 5 μm, (b) 3 μm, (c,g) 2 μm, (e,f) 10 μm.

can be found laterally outside the specimen, phase contrast can be used with great advantage, especially with high-magnification objectives where the bright halos are less annoying [Allen, 1985; Lichtscheidl and Url, 1989].

5. Bright-field. Low intrinsic contrast, due to low concentrations of natural chromophores or especially vital stains, can be greatly enhanced by VEC microscopy. This can be of great advantage because dye toxicity is often less of a problem at low concentrations of dyes. Observation with light of the wavelength of maximum dye absorption further increases contrast. By using appropriate cameras with quartz lenses and windows, even UV microscopy, which yields high-resolution and high-contrast images, is feasible to some extent because the use of sensitive cameras allows one to reduce the intensity of UV illumination to a level at which live specimens survive the UV radiation for a significant period of time [Lichtscheidl and Url, 1987]. The problem of contributions from out-of-focus objects discussed for phase contrast may be similarly annoying in bright-field microscopy. The visualization with bright-field VEC microscopy of colloidal gold particles of 3–20 nm diameter is an especially powerful tool in cell biology (see Chapter 6 on *nanovid microscopy*).

6. Dark-field and fluorescence. With dark-field and fluorescence optics, images are produced which combine low light intensity with relatively high contrast. Additional contrast enhancement, therefore, is useful only in certain cases [e.g., Delgado et al., 1989]. Moreover, the low primary image intensity usually requires the use of special high-sensitivity cameras and digital noise reduction. Genuine stray light from nearby or out-of-focus features can be removed by offset to a limited extent, thus increasing contrast and visibility of weak objects close to big ones. However, most of the annoying contributions come from self-luminous out-of-focus objects that do not produce true stray light that can be removed in this way, because they are not evenly distributed over the image. Instead, special computational out-of-focus blur removal procedures may be employed, as discussed by Shaw in Chapter 9, this volume.

7. Reflection contrast and combination with confocal microscopy. Reflection microscopy visualizes zones of cellular attachment to glass in epi-illuminated specimens by surface reflection interference [Ploem, 1975; Izzard and Lochner, 1976; Bereiter-Hahn et al., 1979]. To reduce the considerable stray light caused by reflections from lens elements, "antiflex" optics were designed, comprising crossed polarizer and analyzer and a quarter-wave plate on the front surface of the objective [Ploem, 1975]. The possibility of removing stray light electronically by VEC microscopy now in many cases allows one to do simple epireflection microscopy with or without using polarizing elements (epibright-field or epipolarization) without antiflex lenses. For observing zones of attachment (black) of cells, or even microtubules [Amos and Amos, 1991] (Fig. 9b), it is essential to use a *high* illuminating aperture. On the other hand, information on the shape, thickness, and surface structure of cells can be obtained from the contour-mapping interference fringes that result when *low* numerical aperture illumination is used. Another very interesting variant using infrared reflection microscopy with image processing called Y-contrast was reported by Zand and Albrecht-Buehler [1989]. The use of epipolarization is especially useful for clean visualization of colloidal gold particles, since there are few if any endogenous cellular components that can be imaged by this technique.

Confocal scanning microscopes, when operated with techniques other than fluorescence, have certain advantages in image formation over conventional microscopes but may create low-contrast images. It may be advisable, therefore, to enhance the contrast of such images. For those real-time confocal microscopes that generate an apparently continuous optical image which is subsequently captured using a video camera (see Chapters 13 and 14, this volume), this may be achieved by video contrast enhancement, as described above. Figure 9b shows an example of individual microtubules imaged by laser scanning confocal microscopy with a combination of analogue and digital contrast enhancement.

G. Spatial Measurements and Motion Analysis

Video microscopy has the further advantage that images in video format, either live or previously recorded onto tape or optical disk, can be analyzed by analogue devices such as a video analyzer or a video micrometer, to determine, for instance, the intensity distribution along a given line (Fig. 6) or the distance between two features. Such analogue devices are, however, far less versatile than digital image analyzers.

Using a digital image analyzer, parameters such as the size, length, width, area, perimeter, and the center of gravity may quickly be determined automatically for one or more objects which can be differentiated from the rest of the image according to their intensity by setting an upper and lower gray-level threshold (binarization of the image). When applied to a series of frames, this makes accessible the analysis of motion, such as the movement of individual organelles in intact cells or along free microtubules, or the growth of objects (nuclei, cells, microtubules) [De Brabander et al., 1986; Weiss et al., 1986, 1988, 1990b; Geerts et al., 1987; Weiss, 1987]. The position of objects can be determined with extremely high precision (down to 1 nm) using video microscopy and either suitable optical equipment [Kamimura, 1987] or computational cross-correlation [Gelles et al., 1988; Inoué, 1989, for discussion].

Digital techniques may be used to improve the imaging of moving objects: The "trace" operation adds frames at predetermined intervals to the frame memory, thereby generating images showing multiple positions of moving objects. Jumping averaging over several seconds makes visible processes too slow to be detected otherwise, such as cell growth, chromosome movements, and cell locomotion. Rolling averaging can be used as a filter to *remove* from the image all objects moving faster than a certain velocity, depending on the number of frames averaged. Conversely, subtraction of sequential in-focus images can be used to *image only* moving objects, while stationary ones are absent from the resulting image (Fig. 7). This can also be used as a velocity filter: As the interval between sequential images in an interval subtraction mode is made longer and longer, slower and slower movements, which are otherwise excluded, will become part of the resulting image.

Quantitative motion analysis can be conveniently performed by extracting the x- and y-coordinates of moving objects from a series of video images. The resulting data may then be quantitatively evaluated [Weiss et al. 1986, 1990b; Cohn et al., 1987; Kamimura, 1987; Gelles et al., 1988; Smith, 1988; Soll, 1988] by classical time-series analysis [Chatfield, 1975]. Scenes derived from the real-time image processor are first stored on videotape or on optical disk video recorder, since storage of sequences of images in a computer memory would be possible only for a limited number of frames (one 512×512 image of 256 gray levels [8 bit] takes up ¼ Mbyte of memory). The sequences are then played back through an X,Y-tracker which detects, by thresholding, a bright object in a small user-defined and -positioned frame that can be moved as a cursor within the image on the monitor. The frame is then automatically positioned over the center of the particle, whose coordinates are read out through a serial or parallel interface to a personal computer. This can either be done at video rate or more slowly, as desired. The positional data may then be further analyzed in the personal computer [Weiss et al., 1986, 1988; Smith 1988]. Plotting the x-coordinate against the y-coordinate yields a trace of the moving object, while plotting position against time yields the velocity function. Similarly, acceleration behavior and directional changes can be displayed as functions of location or time. The movement can be analyzed for regular features by the standard techniques of time-series analysis, such as autocorrelation and fast Fourier transform (FFT) analysis for detecting frequential components (rhythmic or regular motion, oscillations), or cross-correlation for similarity of the motion of different organelles [Koles et al., 1982a,b; Weiss et al., 1990b]. A typical sequence showing several steps of a time-series analysis is shown in Figure 11.

If the object is not sufficiently distinguishable by contrast to permit automatic tracking, but is nevertheless visible by eye, an interactive frame-by-frame analysis has to

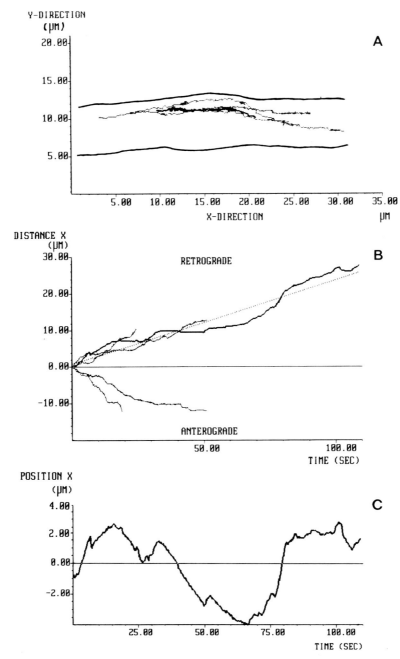

Fig. 11. Organelle motion analysis. A series of graphs demonstrating steps in the motion analysis of lysosomelike organelles from a human neuron (fresh nerve biopsy material from the sural nerve). **A:** Path of several organelles moving toward the right-hand side on the video screen. The outline of a myelinated axon is indicated by the heavy lines.

B: Positions of the organelles in the main direction of movement plotted versus time. The slope of the curves (between a given pair of points) indicates the instantaneous velocity. The average velocity is given by the regression line as indicated by a dotted line for one organelle. The movement of this organelle is further analyzed in C to E. **C:** Positional

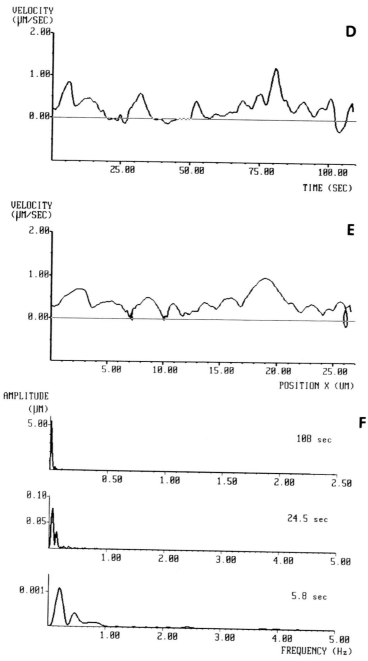

deviations from the hypothetical constant movement or trend (regression line in E). This step is called trend removal. **D:** Instantaneous velocity function plotted versus time. **E:** Instantaneous velocity versus position in the main direction of movement. **F:** power spectra obtained by fast Fourier transform (FFT) of positional deviations from the hypotheti-cal constant movement (regression line in B) for three particles. The absence of frequency information, except for very low frequencies, indicates that the velocity fluctuations are mainly random. Most of these parameters characterizing this movement can also be averaged for a group of particles. For more information, see Weiss et al. [1990b].

be performed. For this, a tape recorder with "single frame advance" feature is used, and from each consecutive, or nth, frame the coordinates are extracted manually by moving the cursor of the tracker to overlie the position of the feature of interest and read the coordinates to the computer [Weiss et al., 1988]. Simultaneous analysis of a multitude of moving objects such as organelles, swimming microorganisms, or sperm cells may be achieved at or near video rates by sophisticated dedicated equipment for motion analysis [Weiss et al., 1990b].

H. Intensity Measurements

The intensity of each digital picture element (pixel) is represented by a number, in images digitized to 8-bit accuracy, between 0 and 255. Intensity in a microscope image means absorption, phase retardation, fluorescence intensity, or birefringence, depending on the technique used. If we ensure that only one of these contributes to the image, we can quantify these properties in the different objects comprising the image. This is easily done by using specific fluorochromes in fluorescence, and by applying monochromatic light in bright-field absorption imaging, but the other parameters may be more difficult to isolate.

It should be made clear that microscope photometers, which use photomultipliers to measure the integrated light intensities over a given area of the image defined by a slit or diaphragm, yield highly accurate intensity measurements, typically with a dynamic range of several thousand, which contain, however, no spatial information. Measuring intensities in the TV image, on the other hand, although giving a photometric resolution of only 256 or less gray levels, has the advantage of providing this information spatially resolved over 512 × 512 pixels. Intensity in a final video image is the result of many steps, and cannot necessarily be assumed to still be linear or to follow Lambert-Beer's law. It is, therefore, essential, if absolute quantitation is required, to calibrate

the system with known standard samples observed using the same optical and electronic settings. This, and the relatively low photometric resolution, are the two grave drawbacks of such intensity measurements using video cameras and 8-bit digital images. In Chapter 9, Shaw discusses the advantages of cooled CCD array cameras for accurate photometric quantitation over a wider dynamic range.

IV. TYPICAL APPLICATIONS OF VIDEO-ENHANCED CONTRAST (VEC) MICROSCOPY

A. General Applications

VEC microscopy is certainly not appropriate for stained material or other specimens which already have high contrast. On the contrary, specimens which are extremely weak in contrast or are even invisible by conventional microscopy are best suited. An overview of what can be seen by this method in the living cell has been given by Allen [1985] and Inoué [1986]. Some typical examples are also presented in Figure 10. With VEC microscopy, one can see for the first time *all* classes of membrane-bounded organelles and even certain classes of cytoskeletal filaments [Travis et al., 1983; Hayden and Allen, 1984; Allen et al., 1985; Koonce and Schliwa, 1985; Schnapp et al., 1985; Schnapp, 1986; Weiss, 1986] (Fig. 10a,b,e). It should be stressed that these details become visible only after image processing, but remain invisible when one looks down the microscope or at enlarged photomicrographs.

Well-suited specimens for VEC microscopy are micelles, colloids, liposomes, and single- or double-layer membranous material [Kachar et al., 1984], bacterial flagella [Block et al., 1991], the live DNA double helix with actively transcribing genes for rRNA [Trendelenburg et al., 1986; Spring and Trendelenburg, 1990], synaptic and other small cytoplasmic vesicles [Allen et al., 1982; Weiss, 1986] (Fig. 10e,g), artificial latex particles of 60 nm and smaller (Fig. 1), and

TABLE III. Some Parameters Which Can Be Measured by VEC Microscopy in Living Cells or Tissue Slices[a]

1. Number, position, area, size, and shape of free cells
2. Number, position, area, size, and shape of organelles (mitochondria, endoplasmic reticulum, Golgi apparatus, lysosomes, endosomes, nucleoli, nucleus) (identification may require specific labeling and intensified fluorescence microscopy)
3. Motion analysis of organelles and microtubules (velocity, fluctuation of velocity, straightness of path, rhythmicity, directionality, pauses, contribution of Brownian motion)
4. Cytoplasmic viscosity (Brownian motion) (requires microinjection of calibrated beads or magnetic particles)
5. Elastic properties of organelles, microtubules, membranes, and cells
6. Local motile activity in cells (variance of the pixel intensity over time)
7. Determination of dry matter in free cells by video microinterferometry
8. Distribution and amount of internal gold-labeled antibodies (requires microinjection) and receptor and antigen transport studies; similar studies on surface receptors
9. In situ hybridization with gold-labeled probes (at the cellular and chromosomal level)

[a]The source images are usually recorded on videotape, with moderate spatial and high temporal resolution. They may be analyzed at low magnification to yield values for many cells in a field or at higher magnifications to study the regional distribution of a given parameter in various parts of one cell.

cytoskeletal elements such as microtubules [Allen, 1985; Allen et al., 1985; Vale et al., 1985a; Schnapp, 1986; Weiss, 1986] (Fig. 10a,g) or actin bundles [Allen and Allen, 1981; Kachar, 1985b] (Fig. 10d).

A number of interesting special applications are based on video contrast enhancement in combination with other contrasting techniques (Table III). Double-beam interference techniques such as that designed by Jamin-Lebedeff yield quantitative data on the flux of dry mass within single, living cells when combined with VEC microscopy [Brown and Dunn, 1989].

Colloidal gold coupled to antibodies is an important tool in immunocytochemistry at the EM level [De Mey, 1984]. Colloidal gold in sizes down to 5–10 nm can also be detected and localized using bright-field or epipolarization *nanovid* VEC microscopy [Geerts et al., 1987; De Brabander et al., 1988]. Different sizes of colloidal gold will have nearly the same apparent size (0.1–0.2 μm, depending on focus), but will occupy different gray levels. Consequently, they can be distinguished by intensity, especially after pseudocolor conversion, where they can be presented in different colors. Thus, double labeling with different sizes of colloidal gold is feasible, although it remains impossible to distinguish between one large gold particle and two smaller ones of equal total mass lying closer than 200 nm apart. The *nanovid* technique has been used to observe microtubule-based motility, surface receptor density, and internalization, as well as the further fate of endosomes, as discussed in Chapter 6 of this volume by De Brabander et al.

Besides observing the behavior and redistribution of the antigens tagged with gold particles in living cells, VEC microscopy can be used to quickly screen gold-labeled EM specimens in the light microscope. The same is true for semithin and sometimes thin, unstained, plastic-embedded EM sections [see example in Weiss et al., 1989]. In high-magnification light microscopic autoradiography, the specimen plane and the film emulsion plane containing the silver grains are too distant to be seen simultaneously in the same focal level. However, Montag et al. [1988] pointed out that a contrast-enhanced image of the specimen can be combined with the silver grain pattern by addition into one frame store of the image processor.

Intensified fluorescence microscopy (see Shotton, Chapter 1, and Hung Leo et al., Chapter 7, this volume) is especially well suited for the quantitative evaluation of morphological parameters (size, shape, number) of subcellular objects (Table III), because different classes of organelles can be visualized selectively with specific vital dyes or antibodies. It should be noted that such analysis is usually *not* undertaken in real time and thus requires only an *image analysis system,* whereas for the initial recording of video-intensified fluorescence microscopy sequences, as for VEC microscopy discussed here, a *real-time digital image processor* is required. It should be emphasized once more that in many applications intensified fluorescence and VEC microscopy are complementary and that a combination of the two is most useful and highly desirable (Fig. 10e,f).

B. Cytoskeleton, Cell Motility, and Motion Analysis

Since VEC microscopy yields the best image improvement with unstained specimens, the cytoplasm, its structural components, and their dynamics become amenable to study. In this field the new technology brought so much insight that our former static, electron microscopy-based understanding of the cytoskeleton and of the cytoplasmic structures has been changed profoundly to a much more lively, highly dynamic view [Weiss, 1986; Shotton, 1988; Weiss and Galfe, 1991]. When viewing cytoplasm of living cells with VEC microscopy at 10,000 × magnification, the lively motion of organelles gives a most striking impression [see Allen, 1987].

Not only have the fine structure and the distribution of cytoskeletal elements and organelles been studied but much attention has also been focused on intracellular motility. Although in cells some organelles may be in Brownian motion, the majority moves steadily along visible or invisible cytoskeletal elements [Weiss et al., 1990b]. This motion is highly nonrandom and is a consequence of the active, ATP-dependent cytoplasmic transport systems characteristic of and essential for all eukaryotic cells [Schliwa, 1984; Weiss, 1986]. These transport systems are known to depend on microtubules in animal cells and on actin filaments in plant cells [Schliwa, 1984; Kuznetsov et al., 1992]. Cytoskeletal elements and cytoplasmic transport along them can be visualized and analyzed quantitatively by VEC microscopy in neu-

rons and dissociated axoplasm (fast axonal transport) [Allen et al., 1982, 1985; Brady et al., 1982, 1985; Vale et al., 1985a,b,c; Weiss, 1986; Weiss et al., 1988, 1990a, b; Goldberg and Burmeister, 1989; Kuznetsov et al., 1992], in plant cells (protoplasmic streaming) [Allen and Allen, 1981; Kachar, 1985b; Allen and Brown, 1988; Lichtscheidl and Weiss, 1988], in protists [Travis et al., 1983; Koonce and Schliwa, 1985, 1986], and in tissue culture cells [Hayden et al., 1983; Hayden and Allen, 1984; De Brabander et al., 1985; Weiss et al., 1986; Maile et al., 1987]. There are examples of such movements visualized by VEC microscopy available on film and videotape [Weiss and Lichtscheidl, 1987; see also the video tape supplements of *Cell Motility and the Cytoskeleton,* starting with Vol. 11, 1990, especially Sanger and Sanger, 1991].

Several groups have developed or applied specific software to quantitatively analyze organelle motion in neurons [Lynn et al., 1986; Weiss et al., 1986, 1988, 1990b; Cohn et al., 1987; Gelles et al., 1988; Smith, 1988], in human nerve biopsies from patients with toxic and hereditary neuropathies [Breuer and Atkinson, 1988; Gulden et al., 1988; Weiss et al., 1990b], or in cultured cells [Maile et al., 1987; Sheetz et al., 1989] (Fig. 11). Some of the quantitative parameters which can be obtained by motion analysis are given in Table III. These have already been analyzed for several cell types and a preliminary classification of various types of cell organelle motion has been suggested [Weiss et al., 1986, 1988].

The visualization of moving organelles by VEC microscopy, the quantitative analysis of their motion, and the determination of positional information with nanometer accuracy (Section IIIG) led to a breakthrough in the analysis of the molecular mechanisms of force generation. First, the process of microtubule gliding was discovered by AVEC-DIC microscopy [Allen et al., 1985; Allen and Weiss, 1985]. This motility was then used as an assay to measure the enzyme activity and to purify the microtubule-dependent force-generating enzymes ("motors") kinesin [Vale et al., 1985c] and cytoplasmic dynein [Paschal et al., 1987]. Similarly, the actin-based organelle motor myosin-I was detected [Adams and Pollard, 1989] and the enzymes' characteristics and pharmacology were determined [Brady et al., 1985; Vale et al., 1985c; Cohn et al., 1987; Paschal and Vallee, 1987; Euteneuer et al., 1988; Gilbert and Sloboda, 1989; Weiss et al., 1991]. Analogous studies of actin filament gliding were performed in parallel, but required video-intensified fluorescence techniques due to their smaller diameter of ~6 nm [Yanagida et al., 1984; Kron and Spudich, 1986; Kuznetsov et al., 1992]. Another process extensively studied by VEC microscopy has been that of microtubule subunit assembly and disassembly (treadmilling and dynamic instability) [Cassimeris et al., 1988; Hotani and Horio, 1988; Seitz-Tutter et al., 1988; Walker et al., 1988; Weiss et al., 1988].

C. Toxicology, Pharmacology, and Experimental Pathology

Morphometric parameters suitable to describe cellular and subcellular changes due to toxic influences include form factors of cells and organelles [e.g., Dunn and Brown, 1987; Maile et al., 1987], and directional, meshwork, and length parameters, as well as the size, position, motility, and numbers of organelles, all of which can be derived from single video frames or sequences by classical image analysis [see Castleman, 1979; Baxes, 1984; Anonymous, 1985; Loats et al., 1988; Bradbury, 1989; Jähne, 1991; and Shotton, Chapter 2, this volume].

By determining a large number of the above-mentioned parameters by VEC microscopy (Table III), perhaps augmented by those measurable with intensified fluorescence microscopy, a whole battery of new tests based on advanced microscopic techniques arises [Weiss, 1987]. If the parameters are determined at selected time intervals in a given set of cultured cells, multiple endpoints for toxicology can be measured simultaneously. In addition, since quantitative

data for these parameters can be calculated from the live images in real time and encoded in the form of gray-shaded or pseudocolor images, the events can also be *continuously recorded* to generate video films of the complete sequence of intracellular events during and after exposure to toxic or pharmacological agents. Video microscopy thus allows multiparametric studies to be performed with cultured cells, which yield a wealth of information that could formerly be obtained only by animal experimentation. In addition, video microscopy enables one to directly observe and quantify certain metabolic, physiological, and morphological parameters which were *not* accessible by animal experimentation.

When toxicologists study the toxicity of chemicals, they would like to know whether these agents have adverse effects on living cells, at what concentration the cells begin to deviate from their normal behavior, their biochemistry and morphology, and finally at what concentration and time point they die. The latter question is the easiest to study, and for many compounds is the only one that could formerly be answered. Therefore, in many of the in vitro tests developed, cells were studied at late time points when irreversible or lethal damage has occurred. For this reason, little was learned about the mechanism of the toxic compound's action or its site of attack. However, many other parameters can now be accurately quantified by video microscopy [Giuliano et al., 1990; Lemasters et al., 1990] (Table III) and yield information difficult to obtain by other techniques on the cell's response at sublethal concentrations of toxic agents [e.g., Maile et al., 1987].

Intracellular motion is not only a multifaceted event that can be described quantitatively by many parameters, it also appears to be especially suited for toxicity studies, since normal motility requires that most cellular components and physiological parameters are undisturbed. This can be seen from the fact that a large variety of different kinds of toxic agents can all severely impair cytoplasmic transport [Weiss, 1987]. Thus monitoring intracellular motion is likely to be indicative of toxic influences of a multitude of substances, interfering, for example, with the cytoskeleton, the organelles, the plasma membrane and endomembrane systems, the ionic milieu in the cytoplasm, the energy metabolism, and other processes. Cytoplasmic motion analysis is thus proposed as a means of obtaining a family of very sensitive parameters for toxicity monitoring. Cultured cells may be studied for acute toxicity on glass coverslips [Maile et al., 1987] or they can be observed for several hours in one of a variety of perfusion chambers [Dvorak and Stotler, 1971; McKenna and Wang, 1989; Ince et al., 1990].

The application of video microscopy to in vitro toxicology and pharmacology is still in its infancy, but seems to be of great potential. It is therefore to be expected that video microscopy in the near future will catalyze a major shift from animal to in vitro experimentation in the fields of toxicology, pharmacology, and experimental pathology.

V. CONCLUSION

VEC microscopy represents one of a group of relatively new techniques described in this volume, in which the microscope has to be seen as a valuable quantitative measuring device rather than merely a producer of "pictures." In addition to its primary application in Nomarski DIC imaging, VEC microscopy has lead to increasing attention being paid to a variety of other high-resolution microscopic techniques suitable to image phase objects, that due to their low intrinsic contrast or low image brightness have not hitherto become popular [Hoffman, 1977; Ellis, 1978; Kachar, 1985a].

There are only a few drawbacks to this kind of video microscopy. The present moderately high price of video microscopy setups is determined by the fact that a high-quality microscope is required and that additional optical parts have to be added to convert it into a true video microscope. It is to be

hoped that the construction of straightforward dedicated *video microscopes* in the future will lead to considerable improvement in quality and cost reduction. PC-based or stand-alone digital image processors are now offered for about a quarter of the price of the first-generation devices of five or six years ago. The potential initial difficulty in mastering the details and theory of this video technique can for the most part be ascribed to the fact that many of our universities do not train the present generation of biologists well enough in optical microscopy. However, it can be expected that this will change very soon, since the rumor is spreading quickly that the new electronic technologies have made light microscopy once more one of the most powerful and versatile tools for research in the life sciences.

ACKNOWLEDGMENTS

D.G.W. expresses his gratitude to his collaborators Günther Galfe, Josef Gulden, Franz Keller, Willi Maile, Elisabeth Möncke-Buchner, and Dieter Seitz-Tutter, who helped to establish and develop the video microscopic techniques in our Munich laboratory and contributed figures from their current work. The initiation of this research was made possible by support from the Wilhelm-Sander-Stiftung. The present research of the authors is supported by grants from Bayerisches Staatsministerium für Wissenschaft und Kunst and Bayerisches Staatsministerium für Landesentwicklung und Umweltfragen, and by Deutsche Forschungsgemeinschaft. We thank L.A. and W.B. Amos, M. Bastmeyer, H.U. Dodt, and T. Holstein for their permission to reproduce videomicrographs from their work.

REFERENCES

Adams RJ, Pollard TD (1989): Binding of myosin I to membrane lipids. Nature 340:566–568.
Allen NS, Brown DT (1988): Dynamics of the endoplasmic reticulum in living onion epidermal cells in relation to microtubules, microfilaments, and

intracellular particle movement. Cell Motil Cytoskel 10:153–163.

Allen RD (1985): New observations on cell architecture and dynamics by video-enhanced contrast optical microscopy. Ann Rev Biophys Biophys Chem 14:265–290.

Allen RD (1987): The microtubule as an intracellular engine. Sci Am 255(2):42–49.

Allen RD, Allen NS (1981): Videomicroscopy in the study of protoplasmic streaming and cell movement. Protoplasma 109:209–216.

Allen RD, Allen NS (1983): Video-enhanced microscopy with a computer frame memory. J Microsc 129:3–17.

Allen RD, Allen NS, Travis JL (1981a): Video-enhanced contrast, differential interference contrast (AVEC-DIC) microscopy: A new method capable of analyzing microtubule-related motility in the reticulopodial network of Allogromia laticollaris. Cell Motil 1:291–302.

Allen RD, David GB, Nomarski G (1969): The Zeiss-Nomarski differential interference equipment for transmitted-light microscopy. Z Wiss Mikrosk Mikro Tech 69:193–221.

Allen RD, Metuzals J, Tasaki I, Brady ST, Gilbert SP (1982): Fast axonal transport in squid giant axon. Science 218:1127–1129.

Allen RD, Travis JL, Allen NS, Yilmaz H (1981b): Video-enhanced contrast polarization (AVEC-POL) microscopy: A new method applied to the detection of birefringence in the motile reticulopodial network of Allogromia laticollaris. Cell Motil 1:275–288.

Allen RD, Weiss DG (1985): An experimental analysis of the mechanisms of fast axonal transport in the squid giant axon. In Ishikawa H, Hatano S, Sato H (eds): "Cell Motility: Mechanism and Regulation." Tokyo: University of Tokyo Press, pp 327–333.

Allen RD, Weiss DG, Hayden JH, Brown DT, Fujiwake H, Simpson M (1985): Gliding movement of and bidirectional transport along native microtubules from squid axoplasm: Evidence for an active role of microtubules in cytoplasmic transport. J Cell Biol 100:1736–1752.

Amos LA, Amos WB (1991): The bending of sliding microtubules imaged by confocal light microscopy and negative stain electron microscopy. J Cell Sci (Suppl) 14:95–101.

Anonymous (1985): "Image Analysis. Principles and Practice." Published by Joyce Loebl Ltd., distributed by IRL Press, London.

Anonymous (1988): Buyers Guide to Biotechnology Products and Instruments. Science 239:G73, G174–G180.

Bajer AS, Sato H, Mole-Bajer J (1986): Video microscopy of colloidal gold particles and immuno-gold labelled microtubules in improved

rectified DIC and epi-illumination. Cell Struct Funct 11:317–330.

Bastmeyer M, Fuge H (1987): Arrangement and dynamics of spindle structures in crane fly spermatocytes seen with video-enhanced contrast differential interference contrast microscopy. Chromosoma 95:51–56.

Baxes G (1984): "Digital Image Processing." Englewood Cliffs, NJ: Prentice Hall.

Bennett HS (1950): Methods applicable to the study of both fresh and fixed materials. The microscopical investigation of biological materials with polarized light. In McClung CE (ed): "Handbook of Microscopical Technique." New York: Harper & Row (Hoeber), pp 591–677.

Bereiter-Hahn J, Fox CH, Thorell B (1979): Quantitative reflection contrast microscopy of living cells. J Cell Biol 82:767–779.

Block M, Fahrner KA, Berg HC (1991): Visualization of bacterial flagella by video-enhanced light microscopy. J Bacteriol 173:943–946.

Bradbury SJ (1989): Micrometry and image analysis. In Lacey AJ (ed): "Light Microscopy in Biology. A Practical Approach." Oxford: IRL Press, ch 7, pp 187–220.

Brady ST, Lasek RJ, Allen RD (1982): Fast axonal transport in extruded axoplasm from squid giant axon. Science 218:1129–1131.

Brady ST, Lasek RJ, Allen RD (1985): Video microscopy of fast axonal transport in extruded axoplasm: A new model for study of molecular mechanisms. Cell Motil 5:81–101.

Breuer AC, Atkinson MB (1988): Fast axonal transport alterations in amyotrophic lateral sclerosis (ALS) and in parathyroid hormone (PTH)-treated axons. Cell Motil Cytoskel 10:321–330.

Brown AF, Dunn GA (1989): Microinterferometry of the movement of dry matter in fibroblasts. J Cell Sci 92:379–389.

Cassimeris L, Pryer NK, Salmon ED (1988): Real-time observations of microtubule dynamic instability in living cells. J Cell Biol 107:2223–2231.

Castleman KR (1979): "Digital Image Processing." Englewood Cliffs, NJ: Prentice Hall.

Chatfield C (1975): "The Analysis of Time Series." London: Chapman & Hall.

Cohn SA, Ingold AL, Scholey JM (1987): Correlation between the ATPase and microtubule translocating activities of sea urchin egg kinesin. Nature 328:160–163.

De Brabander M, Geuens G, Nuydens R, Moeremans M, DeMey J (1985): Probing microtubule-dependent intracellular motility with nanometer particle video ultramicroscopy (nanovid ultramicroscopy). Cytobios 43:273–283.

De Brabander M, Nuydens R, Geerts H, Hopkins CR (1988): Dynamic behavior of the transferrin receptor followed in living epidermoid carcinoma

(A431) cells with nanovid microscopy. Cell Motil Cytoskel 9:30–47.

De Brabander M, Nuydens R, Geuens G, Moeremans M, De Mey J (1986): The use of submicroscopic gold particles combined with video contrast enhancement as a simple molecular probe for the living cell. Cell Motil Cytoskel 6:105–113.

De Mey J (1984): Colloidal gold probes in immuno-cytochemistry. In Pollak JM, van Noorden S (eds): "Immunocytochemistry." Bristol, London, Boston: Wright PSG, pp 82–112.

de Sénarmont H (1840): Sur les modifications que la réflexion spéculaire à la surface des corps métalliques imprime à un rayon de lumière polarisée. Ann Chim Phys (2nd ser.) 73:337–353.

Delgado RM, Fink MJ, Brown Jr M (1989): Imaging of submicron objects with the light microscope. J Microsc 154:129–141.

Dodt HU, Zieglgänsberger W (1990): Visualizing unstained neurons in living brain slices by infra-red DIC-video microscopy. Brain Res 537:333–336.

Dunn GA, Brown AF (1987): A unified approach to analysing cell motility. J Cell Sci (Suppl) 8:81–102.

Dvorak JA, Stotler WF (1971): A controlled-environment culture system for high resolution light microscopy. Exp Cell Res 68:144–148.

Ellis GW (1978): Advances in visualization of mitosis in vivo. In ER Dirksen, DM Prescott, CF Fox (eds): "Cell Reproduction: In Honor of Daniel Mazia." New York: Academic Press, pp 465–476.

Ellis GW (1985): Microscope illuminator with fiber optic source integrator. J Cell Biol 101:83a.

Erasmus SJ (1982): Reduction of noise in TV rate electron microscope images by digital filtering. J Microsc 127:29–37.

Euteneuer U, Koonce MP, Pfister KK, Schliwa M (1988): An ATPase with properties expected for the organelle motor of the giant amoeba, *Reticulomyxa*. Nature 332:176–178.

Geerts H, De Brabander M, Nuydens R, Geuens S, Moeremans M, De Mey J, Hollenbeck P (1987) Nanovid tracking: A new automatic method for the study of mobility in living cells based on colloidal gold and video microscopy. Biophys J 52:775–782.

Gelles J, Schnapp BJ, Sheetz MP (1988): Tracking kinesin-driven movements with nanometre-scale precision. Nature 331:450–453.

Gilbert SP, Sloboda RD (1989): A squid dynein isoform promotes axoplasmic vesicle translocation. J Cell Biol 109:2379–2394.

Giuliano KA, Nederlof MA, DeBiasio R, Lanni F (1990): Multi-mode light microscopy. In Herman B, Jacobson K (eds): "Optical Microscopy for Biology." New York: Wiley-Liss, pp 543–557.

Goldberg DJ, Burmeister DW (1989): Looking into growth cones. Trends Neurosci 12:503–506.

Gulden J, Weiss DG, Clasen B (1988): The velocity fluctuations of organelles transported in crustacean and human axons are random. Eur J Cell Biol (Suppl 46) 22:60.

Häder D-P (1991): "Image Analysis in Biology." Boca Raton, FL: CRC Press, p 363.

Hansen EW (1988): Overcoming polarization aberrations in microscopy. In Chipman RA (ed): "Polarization Considerations for Optical Systems." Proc SPIE 891:190–197.

Hansen EW, Conchello JA, Allen RD (1988): Restoring image quality in the polarizing microscope: Analysis of the Allen video-enhanced contrast method. J Opt Soc Am A5:1836–1847.

Hayden JH, Allen RD (1984): Detection of single microtubules in living cells: Particle transport can occur in both directions along the same microtubule. J Cell Biol 99:1785–1793.

Hayden JH, Allen RD, Goldman RD (1983): Cytoplasmic transport in keratocytes: Direct visualization of particle translocation along microtubules. Cell Motil 3:1–19.

Hecht E, Zajac A (1974): "Optics." Reading, MA: Addison-Wesley, p 564.

Hoffman R (1977): The modulation contrast microscope: Principles and performance. J Microsc 110:205–222.

Hotani H, Horio T (1988): Dynamics of microtubules visualized by darkfield microscopy: Treadmilling and dynamic instability. Cell Motil Cytoskel 10:229–236.

Ince C, Beekman RE, Verschragen G (1990): A microperfusion chamber for single-cell fluorescence measurements. J Immun Meth 128:227–234.

Inoué S (1961): Polarizing microscope. In Clark GL (ed): "The Encyclopedia of Microscopy." New York: Reinhold, pp 480–485.

Inoué S (1981): Video image processing greatly enhances contrast, quality, and speed in polarization-based microscopy. J Cell Biol 89:346–356.

Inoué S (1986): "Video Microscopy." New York: Plenum Press, p 582.

Inoué S (1989): Imaging of unresolved objects, superresolution, and precision of distance measurement with video microscopy. In Taylor DL, Wang Y-L (eds): "Methods in Cell Biology." New York: Academic Press, vol 30, ch 3, pp 85–112.

Inoué S (1990): Whither video microscopy—towards 4-D imaging at the highest resolution of the light microscope. In Herman B, Jacobson K (eds): "Optical Microscopy for Biology." New York: Wiley-Liss, pp 497–512.

Inoué S, Inoué TD (1987): Computer-aided stereoscopic video reconstruction and serial display from high-resolution light-microscope optical sections. Ann NY Acad Sci 483:392–404.

Izzard CS, Lochner LR (1976): Cell-to-substrate contacts in living fibroblasts: An interference reflexion study with an evaluation of the technique. J Cell Sci 21:129–145.

Jähne B (1991): "Digital Image Processing." Berlin: Springer-Verlag, p 337.

Kachar B (1985a): Asymmetric illumination contrast: A method of image formation for video light microscopy. Science 277:766–768.

Kachar B (1985b): Direct visualization of organelle movement along actin filaments dissociated from Characean algae. Science 227:1355–1357.

Kachar B, Evans DF, Ninham BW (1984): Video-enhanced differential interference contrast microscopy: A new tool for the study of association colloids and prebiotic assemblies. J Colloid Interface Sci 100:287–301.

Kamimura S (1987): Direct measurement of nanometric displacement under an optical microscope. Appl Opt 26:3425–3427.

Koles ZJ, McLeod KD, Smith RS (1982a): The determination of the instantaneous velocity of axonally transported organelles from filmed records of their motion. Can J Physiol Pharmacol 60:670–679.

Koles ZJ, McLeod KD, Smith RS (1982b): A study of the motion of organelles which undergo retrograde and anterograde rapid axonal transport in Xenopus. J Physiol (Lond) 328:469–484.

Koonce MP, Schliwa M (1985): Bidirectional organelle transport can occur in cell processes that contain single microtubules. J Cell Biol 100:322–326.

Koonce MP, Schliwa M (1986): Reactivation of organelle movements along the cytoskeletal framework of a giant freshwater ameba. J Cell Biol 103: 605–612.

Kron SJ, Spudich JA (1986): Fluorescent actin filaments move on myosin fixed to a glass surface. Proc Natl Acad Sci USA 83:6272–6276.

Kuznetsov SA, Langord GM, Weiss DG (1992): Actin-dependent organelle movement in squid axoplasm. Nature 356:722–725.

Lacey AJ (1989): "Light Microscopy in Biology—A Practical Approach." Oxford: IRL Press, p 329.

Lemasters JJ, Nieminen AL, Gores GJ, Dawson TL, Wray BE, Kawanishi T, Tanaka Y, Florinecateel K, Bond JM, Herman B (1990): Multiparameter digitized video microscopy (MDVM) of hypoxic cell injury. In Herman B, Jacobson K (eds): "Optical Microscopy for Biology." New York: Wiley-Liss, pp 523–542.

Lichtscheidl IK, Url WG (1987): Investigation of the protoplasm of Allium cepa inner epidermal cells using ultraviolet microscopy. Eur J Cell Biol 43:93–97.

Lichtscheidl IK, Url WG (1989): Video-enhanced phase contrast microscopy: A tool for the study of endoplasmic reticulum in Allium cepa inner epidermal cells. Inst Phys Conf Ser 89:675–678.

Lichtscheidl IK, Weiss DG (1988): Visualization of submicroscopic structures in the cytoplasm of Allium cepa inner epidermal cells by video-enhanced contrast light microscopy. Eur J Cell Biol 46:376–382.

Loats HL, Lloyd DG, Pittenger M, Tucker RW, Unnerstall JR (1988): Biomedical image analysis applications. In Swedenburg CE, Conklin JJ (eds): "Imaging Techniques in Biology and Medicine." San Diego: Academic Press, pp 1–75.

Lynn MP, Atkinson MB, Breuer AC (1986): Influence of translocation track on the motion of intra-axonally transported organelles in human nerve. Cell Motil Cytoskel 6:339–346.

Maile W, Lindl T, Weiss DG (1987): New methods for cytotoxicity testing: Quantitative video microscopy of intracellular motion and mitochondria-specific fluorescence. Mol Toxicol 1:427–437.

McKenna NM, Wang Y-L (1989): Culturing cells on the microscope stage. In Wang Y-L, Taylor DL (eds): "Methods in Cell Biology." New York: Academic Press, Vol 29, ch 12, pp 195–205.

Mizushima Y (1988): Detectivity limit of very small objects by video-enhanced microscopy. Appl Opt 27:2587–2594.

Montag M, Trendelenburg MF, Spring H (1988): Video microscopic image processing facilitates the evaluation of light microscopic autoradiography at high magnification. J Microsc 150:245–249.

Paschal BM, Vallee RB (1987): Retrograde transport by the microtubule-associated protein MAP 1C. Nature 330:181–183.

Paschal BM, Shpetner HS, Vallee RB (1987): MAP 1C is a microtubule-activated ATPase which translocates microtubules in vitro and has dynein-like properties. J Cell Biol 105:1273–1283.

Ploem JS (1975): Reflection-contrast microscopy as a tool for investigation of the attachment of living cells to a glass surface. In von Furth R (ed): "Mononuclear Phagocytes in Immunity, Infection and Pathology." Melbourne, London: Blackwell, pp 405–421.

Sanger JM, Sanger JW (1991): "Cellular Motile Processes: Molecules and Mechanisms. Microtubule-Based Motility." Cell Motil Cytoskel, Video Suppl 2. New York: Wiley-Liss [31 contributions by different authors, 1 hour VHS format video tape; ISBN 0-471-56170-3 (NTSC standard) or ISBN 0-471-56171-1 (PAL standard)].

Schliwa M (1984): Mechanisms of intracellular organelle transport. In Shay JW (ed): "Cell and Muscle Motility," Vol 5. New York: Plenum, pp 1–84.

Schnapp BJ (1986): Viewing single microtubules by video light microscopy. Meth Enzymol 134: 561–573.

Schnapp BJ, Vale RD, Sheetz MP, Reese TS (1985): Single microtubules from squid axoplasm support bidirectional movement of organelles. Cell 40:455–462.

Seitz-Tutter D, Langford GM, Weiss DG (1988): Dynamic instability of native microtubules from squid axons is rare and independent of gliding and vesicle transport. Exp Cell Res 178:504–512.

Sheetz MP, Turney S, Qian H, Elson EL (1989): Nanometre-level analysis demonstrates that lipid flow does not drive membrane glycoprotein movements. Nature 340:284–288.

Shotton DM (1988): Review: Video-enhanced light microscopy and its applications in cell biology. J Cell Sci 89:129–150.

Smith RS (1988): Studies on the mechanism of the reversal of rapid organelle transport in myelinated axons of Xenopus laevis. Cell Motil Cytoskel 10:296–308.

Soll DR (1988): "DMS," a computer-assisted system for quantitating motility, the dynamics of cytoplasmic flow, and pseudopod formation: Its application to Dictyostelium chemotaxis. Cell Motil Cytoskel 10:91–106.

Spring H, Trendelenburg MF (1990): Towards light microscopic imaging of hydrated 'native' ribosomal RNA genes. J Microsc 158:323–333.

Travis JL, Kenealy JFX, Allen RD (1983): Studies on the motility of Foraminifera. II. The dynamic microtubular cytoskeleton of the reticulopodial network of Allogromia reticularis. J Cell Biol 97:1668–1676.

Trendelenburg M, Allen RD, Gundlach H, Meissner B, Tröster H, Spring H (1986): Recent improvements in microscopy towards analysis of transcriptionally active genes and translocation of RNP-complexes. In Peters R, Trendelenburg M (eds): "Nucleocytoplasmic Transport." Berlin: Springer-Verlag, pp 95–112.

Vale RD, Reese TS, Sheetz MP (1985c): Identification of a novel, force-generating protein, kinesin, involved in microtubule-based motility. Cell 42:39–50.

Vale RD, Schnapp BJ, Reese TS, Sheetz MP (1985a): Movement of organelles along filaments dissociated from the axoplasm of the squid giant axon. Cell 40:449–454.

Vale RD, Schnapp BJ, Reese TS, Sheetz MP (1985b): Organelle, bead and microtubule translocations promoted by soluble factors from the squid giant axon. Cell 40:559–569.

Walker RA, O'Brian ET, Pryer NK, Soboeiro MF, Voter WA, Erickson HP, Salmon ED (1988): Dynamic instability of individual microtubules analyzed by video light microscopy: Rate constants and transition frequencies. J Cell Biol 107:1437–1448.

Walter RJ, Berns MW (1981): Computer-enhanced video microscopy: Digitally processed microscope images can be produced in real time. Proc Natl Acad Sci USA 78:6927–6931.

Walter RJ Jr, Berns MW (1986): Digital image processing and analysis. In Inoué S (ed): "Video Microscopy." New York: Plenum Press, pp 327–392.

Weiss DG (1986): Visualization of the living cytoskeleton by video-enhanced microscopy and digital image processing. J Cell Sci (Suppl) 5:1–15.

Weiss DG (1987): Videomicroscopic measurements in living cells: Dynamic determination of multiple end points for in vitro toxicology. Mol Toxicol 1:465–488.

Weiss DG, Galfe G (1991): Video-microscopic techniques to study the living cytoplasm. In Häder D-P (ed): "Image Analysis in Biology." Boca Raton, FL: CRC Press, ch 8, pp 135–158.

Weiss DG, Galfe G, Gulden J, Seitz-Tutter D, Langford GM (1990b): Motion analysis of intracellular objects: Trajectories with and without visible tracks. In Alt W, Hoffmann G (eds): "Biological Motion." Lecture Notes in Biomathematics, Vol 89. Berlin: Springer Verlag, pp 95–116.

Weiss DG, Keller F, Gulden J, Maile W (1986): Towards a new classification of intracellular particle movements based on quantitative analyses. Cell Motil Cytoskel 6:128–135.

Weiss DG, Langford GM, Seitz-Tutter D, Keller F (1988): Dynamic instability and motile events of native microtubules from squid axoplasm. Cell Motil Cytoskel 10:285–295.

Weiss DG, Lichtscheidl IK (1987): The cytoplasm of Allium cepa inner epidermal cells observed by AVEC-DIC microscopy. Scientific Film, Österr. Inst. f. d. wiss. Film ÖWF, Vienna, Film P2117, 16 mm, 12 min, soundtrack.

Weiss DG, Maile W, Wick RA (1989): Video microscopy. In Lacey AJ (ed): "Light Microscopy in Biology. A Practical Approach." Oxford: IRL Press, ch 8, pp 221–278.

Weiss DG, Meyer M, Langford GM (1990a): Studying axoplasmic transport by video microscopy and using the squid giant axon as a model system. In Gilbert DL, Adelman WJ Jr, Arnold JM (eds): "Squid as Experimental Animals." New York: Plenum Press, pp 303–321.

Weiss DG, Seitz-Tutter D, Langford GM (1991): Characteristics of the motor responsible for the gliding of native microtubules from squid axoplasm. J Cell Sci (Suppl) 14:157–161.

Yanagida T, Nakase M, Nishiyama K, Oosawa F (1984): Direct observation of motion of single F-actin filaments in the presence of myosin. Nature 307:58–60.

Zand MS, Albrecht-Buehler G (1989): Long-term observation of cultured cells by interference-reflection microscopy: Near-infrared illumination and Y-contrast image processing. Cell Motil Cytoskel 13:94–103.

CHAPTER 6

Nanovid Microscopy: Imaging and Quantification of Colloidal Gold Labels in Living Cells

M. De Brabander, H. Geerts, R. Nuyens, R. Nuydens, and F. Cornelissen

I. INTRODUCTION

Nanovid microscopy is an acronym for nonometer particle video microscopy. This relatively new approach uses submicroscopic gold particles as markers that are visualized by video-enhanced contrast microscopy. It is particularly useful for investigations of dynamic phenomena in living cells. A recent review covers the basic aspects of the ap-

Electronic Light Microscopy, pages 141–155 ©1993 Wiley-Liss, Inc.

proach and its potential applications [De Brabander et al., 1989]. In this chapter, emphasis is put on the detailed technical aspects.

II. THE PRINCIPLE OF NANOVID MICROSCOPY

A. Basic Aspects of Nanovid Microscopy

In order for a marker to be visible and unambiguously discernible within or on living cells in the light microscope, it must provide some optical characteristic which is not shared by other subcellular constituents. The organic molecules (proteins and lipids) that build cellular organelles are essentially translucent. They do not adsorb or scatter sufficient light for them to be visible in ordinary light microscopy. They interact with light mainly by producing phase shifts. The latter phenomenon is exploited by optical techniques such as phase-contrast and differential interference contrast microscopy (DIC) to generate a visible contrast.

Colloidal gold particles, on the other hand, are very strong light scatterers. They can thus generate contrast (difference in intensity in the image plane against the background) simply by scattering light out of the objective lens aperture. Because of their small size, this contrast is essentially invisible to the unaided eye. However, when the optical signal is transformed into an analogue or digital electronic signal, it can be manipulated so as to enhance the contrast and to visualize clearly gold particles as small as 20 nm in diameter. According to the Rayleigh limit (size of the scatterer << wavelength of the light) the scattered light intensity is proportional to the sixth power of the radius (Fig. 1). This means that doubling the size of a gold probe increases the scattering intensity by a factor of 64. Consequently the detection limit decreases sharply with the size. On the other hand, the scattering intensity increases with the fourth power of the wave number (the inverse of the wavelength). Therefore, using blue or UV illumination enhances the detection by increasing the contrast. However, the sensitivity of the camera tube and absorption of the optical system limits the useful wavelength. In addition, when observing living cells, the eventual phototoxic effect of UV illumination must be taken into account.

For nanovid microscopy, the following basic steps are essential:

1. The preparation is observed in bright field, taking care to obtain the highest optical resolution possible. Bright-field illumination maximizes the contrast generated by the gold particles (scattering) and minimizes the contrast produced by cellular constituents (phase effects).

2. The image is captured by a high-resolution black-and-white video or charge-coupled device (CCD) camera and displayed on a monitor.

3. The contrast is enhanced by putting an offset to the analogue signal, i.e., subtracting an equal amount of background intensity from all pixels, and by subsequently amplifying the remaining peaks of the video signal. This basic approach to analogue video contrast enhancement has first been advocated by Inoué [1981] and Allen et al. [1981] to amplify resolution and detection by polarization microscopy and DIC.

4. The extreme contrast enhancement which is necessary to visualize the gold particles also amplifies the contrast produced by disturbing background elements (dust, lens imperfections, etc.). An image of this background "mottle," obtained with the preparation slightly out of focus, is stored in a digital frame memory. It is subtracted in real time from the images produced with the preparation in focus.

5. The digital image can then be treated in various ways by using look-up tables, filtering procedures, etc., to further enhance the sensitivity of the detection, or for quantitative purposes (see below).

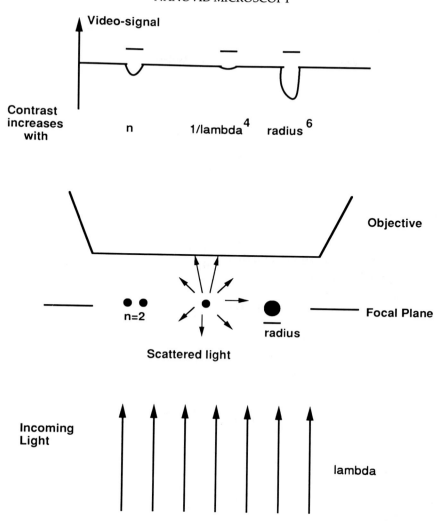

Fig. 1. The principle of image formation in nanovid microscopy is illustrated. In transmission mode, a fraction of the incoming light is scattered out of the objective aperture. The amount of scattered light is proportional to the sixth power of the radius, to the number of scatterers, and to the inverse fourth power of the wavelength.

These basic principles are used in all kinds of video-enhanced contrast microscopy. More detailed information can be found in the excellent monograph on video microscopy by Inoué [1986], in a review by Shotton [1988], and in Chapters 1 and 5 of this volume. In the next sections we deal primarily with the specific requirements for the optimal use of nanovid microscopy.

B. Essential Features of Nanovid Microscopy

1. Single probe detection. The basic outstanding feature that distinguishes nanovid microscopy from other *in vivo* labeling techniques is that it detects single markers. In practice we find that, with the present technology, single gold particles with a diameter of about 20 nm can be followed in or on living cells. The resolution is thus well below the theoretical resolution limit of light microscopy (200 nm). However, when two or more gold particles are less than 200 nm apart, they are seen as one particle. Because of their greater scattering power they can be discerned as distinct from a single particle (Fig. 1). If we assume there is no multiple scattering of light between gold probes themselves (a safe assumption with regard to the low gold concentration normally used in nanovid microscopy), the resulting contrast of a single cluster is proportional to the number of gold particles, observed within the resolution of the light microscope. This feature is exploited in the absolute quantification of the gold label (see below).

2. **Continuous observation.** Unlike fluorescent probes [Taylor and Wang, 1980], the gold marker does not fade and can be followed indefinitely in real time. In addition, no phototoxic effect from fluorescent bleaching is induced.

3. **High-resolution localization.** Switching from bright field to DIC allows a very accurate localization of gold markers within or on cells in relation to cell organelles (Fig. 2).

4. **Optical sectioning.** Using high numerical aperture optics, the contrast generated by out-of-focus label is negligible (see below). In practice, effective optical sections of less than 0.5 μm can be made with a 1.3 NA objective (Fig. 3). Indeed, the small size of the observed colloidal gold (20–40 nm), in principle, permits it to be regarded as an ideal point. Therefore, the digital image formed by a single colloidal gold particle could be regarded as the point spread function of the whole optical system, including the microscope, camera system, and subsequent electronics. The property of colloidal gold as an ideal point function in the light microscope could in principle be exploited for measuring the 3D point spread function for use in optical sectioning and reconstruction experiments [Agard, 1984]. The gold itself is ideally suited as a probe in these experiments, although to our knowledge no application in this direction has yet been reported.

5. **Variable contrast.** Because the contrast can be increased or decreased at will, optimal information can be obtained both from regions where the label is very dense and from sparsely labeled areas [De Brabander et al., 1989].

6. **Easy combination with other markers.** Gold labeling can be combined with fluorescent markers to extend the possibilities of double and triple labeling [De Brabander et al., 1988].

7. **Correlated light and electron microscopy.** Because the gold label is electron-dense, it is an ideal marker for subsequent electron microscopy. This allows one to observe the dynamic behavior of labeled components in a living cell, to make a record of the distribution in the entire cell, and to determine accurately the localization at the ultrastructural level in thin sections through the same cell [Geuens et al., 1989].

8. **Size.** The size of the single small gold label (20 nm plus the layer of adsorbed antibodies) is within the range of large proteins. It can enter the smallest vesicles such as coated pits and can diffuse throughout the cytoplasm. In contrast, a colloidal gold probe of 30 nm has a diameter five times larger than a conventional protein, with a proportionally reduced mobility as far as free diffusion is concerned. Indeed, the diffusion coefficient is not related to the mass, but to the size of the complex. In two-dimensional lipid-mediated diffusion, the diffusion is even less dependent on the size of the complex [Saffmann and Delbruck, 1975]. We have shown previously that the two-dimensional mobility of a 40 nm gold-labeled anti-Thy-1 antibody on the surface of 3T3 cells ($D = 0.1–0.2$ μm^2/sec) is only two times less than the mobility of the same, but fluorescently labeled, anti-Thy-1 antibody measured with photobleaching [Ishihara et al., 1988; De Brabander et al., 1991]. However, the access of 20–40 nm gold probes to compartments consisting of very dense protein meshworks is most certainly limited. Further improvements in detection of even smaller colloidal gold probes are

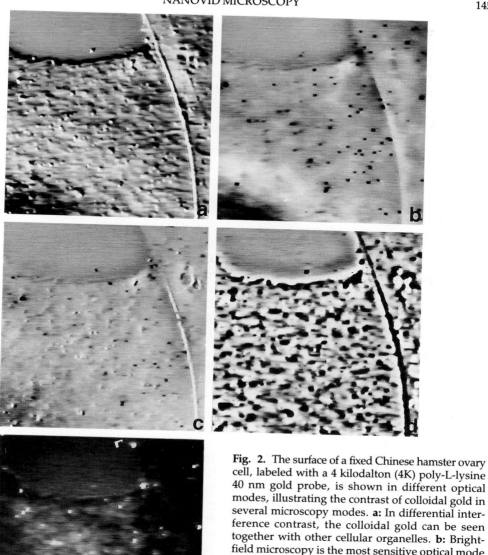

Fig. 2. The surface of a fixed Chinese hamster ovary cell, labeled with a 4 kilodalton (4K) poly-L-lysine 40 nm gold probe, is shown in different optical modes, illustrating the contrast of colloidal gold in several microscopy modes. **a:** In differential interference contrast, the colloidal gold can be seen together with other cellular organelles. **b:** Bright-field microscopy is the most sensitive optical mode for a clear detection of colloidal gold probes. **c:** Oblique illumination is a quick way to visualize colloidal gold probes together with cellular elements. **d:** Closing the condenser aperture gives a "phase-contrast" appearance. The contrast generated by the colloidal gold is buried in the general contrast enhancement of the other organelles. **e:** Epipolarization efficiently detects colloidal gold by the reflection of depolarized light through a crossed polarizer–analyzer combination.

expected to yield important information on the structure and sizes of such protein meshworks [Geerts et al., 1987; Luby-Phelps et al., 1987].

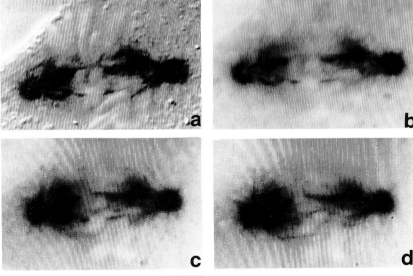

Fig. 3. A set of bright-field images at different focal planes in a late bipolar prometaphase PTK2 cell, fixed 1 min after microinjection with Paramecium tubulin and visualized with an indirect 10 nm colloidal gold immunocytochemical stain. The power of optical sectioning is illustrated qualitatively. (From Geuens et al., 1989, by copyright permission of the Rockefeller University Press.)

9. Multivalency. Gold probes made according to standard procedures are covered with multiple copies of the ligand. This can be expected to induce crosslinking. The effective valency of antibody-labeled gold probes is not well known. Preliminary studies show that about six antibodies stick to the surface of a 15 nm gold probe, the vast majority of which are not active in binding the epitope. In addition, our experience is that the multivalency as well as the specificity of the gold probe depends very much on the kind of stabilizing agents. In some cases, this multivalency induces crosslinking between epitopes, which contributes, among other things, to an increased proportion of aggregated clusters.

It may thus be useful for certain applications to construct univalent probes by mixing the ligand (e.g., F$_{ab}$ fragments) with a sufficient amount of an inactive protein.

III. PRACTICAL GUIDE TO NANOVID MICROSCOPY

A. Making the Preparation

Individual gold particles can only be discerned with high-resolution oil-immersion objectives. The preparation must therefore be sufficiently thin not to exceed the short focal distances of objective and condenser. The simplest way to achieve this is to make thin chambers consisting of a microscopic slide, spacers made of strips of Parafilm® melted onto the slide by heating at 60°C until they become translucent, and a coverslip. The chamber is sealed with VALAP (a 1:1:1 mixture of vaseline, lanolin, and paraffin]. Ports can easily be made for perfusion by capillary action. Several other types of more sophisticated microperfusion chambers have been described [Salmon and Ellis, 1985].

B. Gold Probes

Details on the preparation of colloidal gold sols and procedures to adsorb antibodies or other ligands onto the particles can be found in some recent reviews [De Mey, 1984]. The concentrated gold sols should be diluted in the physiological buffer or culture medium immediately before use. Working dilutions are dependent on the nature of the probe, the number of binding sites, etc. Usually, starting from a gold sol that has an optical density of 1.0 at 530 nm, a dilution of $\frac{1}{20}$ in culture medium is a good starting point. Aggregates of gold particles can be removed by centrifugation at 1,000g for 15 sec.

C. The Microscope

The major functions of nanovid microscopy are to detect gold particles at high resolution and to distinguish them unambiguously from cellular components. This can be achieved optimally using bright-field illumination with high numerical aperture optics. Because the detection of the gold particles relies most heavily on the extreme electronic contrast enhancement, it is imperative to minimize any contrast produced by cell constituents through aberrations and incorrect adjustment of the microscope. The following requirements are found to be essential:

- Monochromatic light using a narrow-band green interference filter
- High numerical aperture condenser and objective
- Oil immersion on both condenser and objective
- Correct Köhler illumination
- Availability of differential interference contrast and preferably also of phase contrast and epifluorescence

These requirements are available on all modern research microscopes. We feel it is also important to consider the ease with which one can switch from one optical mode to another, as well as the availability of additional ports to introduce filters, polarizers, etc.

For most purposes, illumination can be done with a simple halogen lamp. Using a mercury of xenon arc will provide crisper images but it is less easy to adjust the light intensity. We have no experience with the light scrambler described in Chapter 5 of this volume [Ellis, 1985]. Because it provides an even illumination of the entire field, it should make subsequent contrast enhancement much easier.

Besides the usual optical techniques (DIC, phase), a few simple procedures allow one to bring out cellular detail quickly while working in the bright-field mode. One is to close the condenser diaphragm, which produces a phase-contrast-like effect (Fig. 2d). Another is to block approximately one-half of the illuminating light path. This produces asymmetric oblique illumination and gives a DIC-like contrast (see Fig. 2c).

D. Analogue Contrast Enhancement

There are now an increasing number of suitable video and CCD cameras and dedicated video microscopy systems available on the market. The choice will depend largely on the available budget. Very much can be achieved with simple high-resolution monochrome cameras.

The specifications given by the manufacturers are not usually very helpful. It is our experience that the best way to proceed is to try different systems, preferably on one's own microscope, with a critical preparation, e.g., cells labeled with 20 nm gold particles coupled to polylysine [De Brabander et al., 1991].

Analogue contrast enhancement is done simply by gradually increasing the gain and the offset (or pedestal or black level) until an optimal image of the gold particles is obtained. A few characteristics that must be checked are the following:

- Manual control of gain and offset
- The sensitivity of the camera must be such that it can be saturated using a high-magnification image.

- The camera resolution must ideally be somewhat better than that of subsequent display and recording devices. For most purposes, 500–700 lines in the horizontal direction are sufficient.
- The dynamic range of the camera must be sufficiently large to cope with relatively large contrast differences.
- The resolution and in particular the dynamic range of the video monitor are at least as important as those of the camera. Fortunately many inexpensive monochrome monitors are available that perform very well. Most color monitors do not provide sufficient contrast.
- If one wants to do quantitative measurements, the linearity of the gain is important and geometric distortion of the image must be minimal. This can be checked by rotating a preparation containing fixed gold particles and measuring a fixed specimen distance in various directions on the screen.
- For the study of dynamic phenomena, a broadcast-quality videotape recorder (for instance, 1-inch, ¾-inch U-matic, or super VHS) usually is sufficient. For slow phenomena, a time-lapse feature is sometimes very useful.

E. Digital Image Processing

This topic is dealt with in detail elsewhere. A few basic practical guidelines:

- The system must be able to store a background image and to subtract this pixel by pixel in real time from subsequent images.
- On this corrected image, further digital processing in real time must be available.
- Averaging the images over two to four frames will produce a greatly improved image of the gold particles without diminishing the time resolution by a critical level. Look-up tables that allow one to enhance the contrast by changing the gain function and to segment the image

contrast will further increase the visibility of the gold label and are required to produce binary images for subsequent quantification (see below).
- Here, too, it is important to consider the ease with which the various procedures can be implemented (see Chapter 4 by Inoué, in this volume). While doing experiments with living cells, one does not have the time for lengthy procedures.

IV. ASPECTS OF QUANTITATIVE NANOVID MICROSCOPY

In this section, some crucial aspects of digital image processing and quantitative microscopic analysis are described.

A. Detection of Colloidal Gold Particles in a Video Image

As the gray value of pixels representing gold in nanovid microscopy have a lower value than the background (i.e., they look black), an obvious segmentation (discrimination) between gold and background can be made by simple thresholding, provided that background shading of the image field has previously been removed during "mottle subtraction." This operation classifies all pixels with a gray value below a certain threshold value as belonging to colloidal gold particles. It can be performed in real time by use of look-up tables. However, even with the best mottle subtraction, subtle illumination variations can yield a varying background, the gray value of which can drop below this threshold and give rise to false-positives. Especially in neurite extensions of neuronal cells, the boundary of the neurite can have a gray value within the same range as the colloidal gold itself. Furthermore, because the number of colloidal gold pixels is very low, they contribute only marginally to the image histogram. As a consequence, classical automatic calculations of this threshold are unable to provide a reasonable value.

Therefore, a method was developed which exploits the relationship between 1 pixel and its neighbors. The attentive reader will have noticed that the image of a single colloidal gold particle is blurred to a spot of 200–400 nm diameter (2–3 pixels at the magnification used). Detailed analysis of the gray values reveals a kind of two-dimensional Gaussian-shaped valley. A gray-scale morphology technique [Sternberg, 1986] is applied using a convolution with a profile resembling this Gaussian form (Fig. 4). Gray-scale dilation is different from classical binary dilation, in that the dilated pixel is calculated from

$$d(x,y) = \text{Max}(i,j) \{a(x + i, y + j) + b(i,j)\} \quad (1)$$

where $a(k,l)$ and $d(k,l)$ are the original and dilated image, respectively, and $b(i,j)$ is the kernel used. Conversely, erosion is defined as

$$e(x,y) = \text{Min}(i,j) \{a(x + i, y + j) - b(i,j)\}. \quad (2)$$

The closing operation is then a sequence of a dilation followed by an erosion. This operation can be compared to the movement of a ball, slightly bigger than the greatest signal of a gold cluster, rolling over the two-dimensional array of gray values (Fig. 4). This resultant image is then subtracted from the original image, yielding a much clearer contrast for the colloidal gold probes. An additional advantage of this method is that the threshold in this processed image is only dependent on the contrast difference between the gold and its local background (a property which is related intimately to the size of the colloidal gold) and relatively independent of global illumination. This obviates the need for sophisticated threshold algorithms. But the algorithm itself (involving for a 512 × 512 image some 8 million linear and nonlinear operations) takes some 70 sec on a conventional MicroVax II machine. On dedicated image processors, this is reduced to only a few seconds.

Once the segmentation of the image into objects (= gold probes) and background has been accomplished, a labeling procedure marks the size and position of each of the colloidal gold particles. For dynamic studies, this operation has to be repeated on all images. The result is a coordinate map of all particles over time.

B. Absolute Quantitation of Colloidal Gold in a Video Image

In certain applications of cytochemistry and histochemistry, the experimenter is interested in the absolute quantity of a specific class of protein in a certain region of a tissue. Suppose the problem of multivalency is solved, and the experimenter wants to know the absolute amount of colloidal gold present in a given region of the image. This problem is approached in the following way. As the amount of scattered light is proportional to the number of colloidal gold particles, the absolute contrast of any black dot representing gold is proportional to the number of particles. Indeed, it turns out that if the contrast is calculated as the difference between the local background intensity and the mean gray value of the gold cluster, all clusters have values which are multiples of a fundamental scattering power [De Brabander et al., 1989]. This fundamental value is then the scattering power of a single colloidal gold particle. This has been verified in detail for a test preparation of gold sedimented on a coverslip, the gold density of which was calculated from both electron microscopy and the video image in nanovid microscopy.

An important caveat has to be made here. This approach assumes that all gold particles lie within a single plane of focus. A gold particle lying out of focus will have lower contrast. This restricts the use of this method to surface labeling, or those applications where the area of interest at least can be safely assumed to lie within a single plane of focus.

In principle, if one is willing to apply 3D optical reconstruction algorithms (see Shaw, Chapter 9, this volume), the contrast difference related to out-of-focus contributions

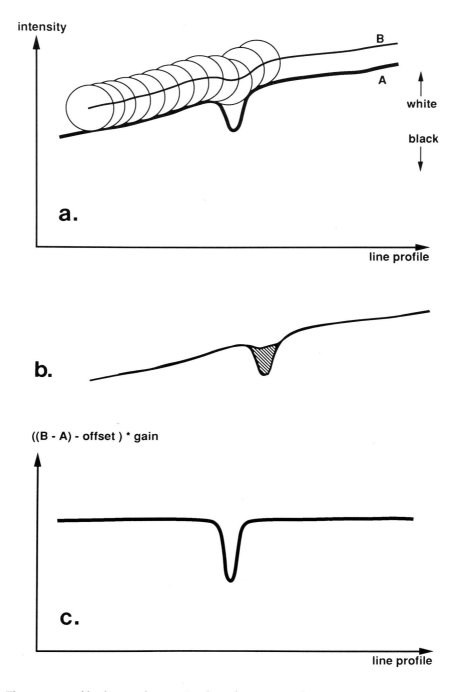

Fig. 4. The concept of background correction by gray-scale morphology. **a:** A kernel element is used to scan the original two-dimensional gray-level surface in the gray-scale morphology operation. For illustration, the effect is shown on a one-dimensional line. **b:** The closing is a gray-scale dilation, followed by a gray-scale erosion (see text). **c:** The final difference between original image (line A) and closed image (line B) essentially eliminates variations in the background and allows a simple thresholding to be performed in order to detect the colloidal gold particles.

could be integrated. To our knowledge, this has never been tried. An interesting question relates to the use of double labeling in videomicroscopy. In electron microscopy, due to their clear signals and their unimodal size distributions, colloidal gold particles have been used extensively in double-labeling experiments. In light microscopy, however, the image of a single gold probe, be it a 20 or 40 nm probe, is blurred to the resolution limit of the optical system (200–300 nm). As discussed above, only the contrast is dramatically dependent on the size. In conventional video camera systems with a relatively low dynamic range, this difference in signal is so large that the two particles can never be visualized within the same image. In addition, the analogue enhancement somewhat arbitrarily enhances the contrast to the optimal limit for visual detection. However, in modern large dynamic range slow-scan CCD camera systems, there is no need for analogue enhancement and the difference in scattering power could in principle be exploited to perform double-labeling experiments on a single image.

C. Measurement of Aggregation

Within the limits imposed in the above paragraphs on the potential multivalency of the gold probe and the optical resolution of the light microscope, the detection of individual gold probes can be correlated with the spatial organization of membrane components. This is of interest in the study of protein–protein interactions on cell surfaces.

A purely descriptive technique is the box method (see Fig. 5). The area is divided in 25 square boxes and the number of individual gold probes (n_k) is calculated in each box. If $<n>$ is the mean value of n_k, then the value

$$I = \Sigma (n_k - <n<)^2/n_k \qquad (3)$$

can be shown to follow a χ^2 distribution [Diggle, 1979]. The higher the value, the more aggregation and phase separation. For instance, this parameter allows the quanti-

tation of dynamic membrane organization, such as the patching and capping of cell surface proteins on mobile cells or the onset of phase separations on cell surfaces.

D. Dynamic Studies

The most exciting applications of nanovid microscopy use its ability to serve as a specific label for molecules in living cells. From the particulate character of the colloidal gold probes, mobility parameters can be easily calculated from the instantaneous coordinates of individual particles. This distinguishes this technique from relaxation approaches, such as fluorescence recovery after photobleaching (FRAP) [Axelrod et al., 1976], where the integrated recovery of a whole population of particles is followed after a concentration gradient has been imposed abruptly.

This technique is used now routinely in our laboratory to study axonal transport in living neuronal cells. By describing these experiments in some detail we will illustrate some crucial aspects of dynamic and quantitative nanovid microscopy.

N4 neuroblastoma cells were seeded on ethanol-sterilized coverslips and cultured at 37°C. Gold particles of 40 nm were stabilized with polyethyleneglycol (PEG) and bovine serum albumin (BSA) as described by De Mey [1984]. These particles have a negative surface charge. The probes were added to the living cell preparations with 50 µl of a 1/20 dilution in phosphate buffer saline (PBS) with 1 g/l glucose some 16 hr before the actual experiment. The neuroblastoma cells apparently took up the colloidal gold by endocytosis.

The procedures for actual visualization of the colloidal gold have been presented in detail above and will not be described again. The temperature of the microscope stage was kept at 37°C by an airstream incubator. Because the depth of field was extremely small at this resolution, it was absolutely mandatory to keep the plane of focus permanently controlled. The image was pro-

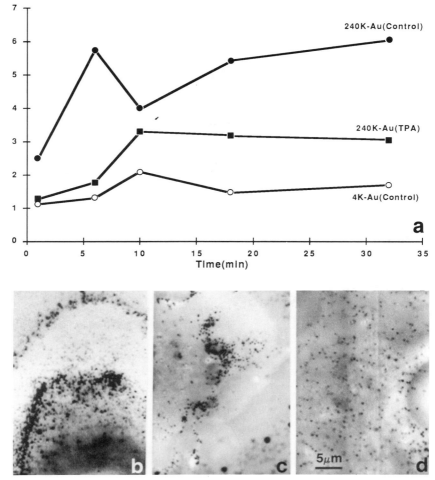

Fig. 5. The index of aggregation I, calculated as Equation 1, for three situations and different time points in a dynamic study of membrane reorganization [De Brabander et al., 1990]. **b:** Bright-field image of 240 kilodalton (240K) poly-L-lysine colloidal gold probe on PTK$_2$-cells after accumulation has occurred in control cells. This is reflected by high values for the aggregation index. **c:** Bright-field image of the same 240K poly-L-lysine colloidal gold probe in TPA-treated PTK$_2$-cells. Patches are clearly distinguishable, which yield intermediate values for the aggregation index. **d:** Bright-field image of a 4 kilodalton (4K) poly-L-lysine colloidal gold probe on PTK$_2$-cells showing a random distribution. The calculated value for aggregation remains low.

jected onto the camera of a video microscope system used in the normal mottle subtraction mode. To exploit the full temporal dynamics of the video signal (25 frames/sec), the rolling averaging mode was not used. The video signal was fed into a Sony V05850P U-matic videorecorder. Recordings were made in real time for periods of 10 min.

The automatic tracking of the particles in the nanovid setup (Fig. 6) proceeds as follows. Briefly, a rectangular area of interest around an axon is defined interactively with the aid of a mouse-driven cursor on the image. As the number of all subsequent digital operations is proportional to the size of this area, it is mandatory to try to keep the area of interest as small as possible. This can

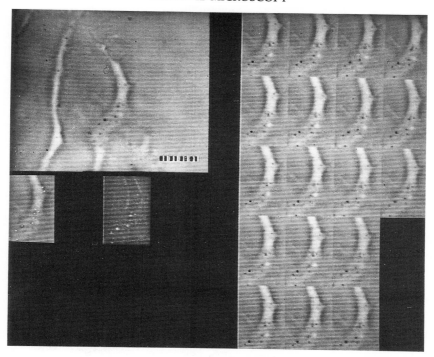

Fig. 6. Dynamic acquisition of a nanovid video sequence in the S200 Image Processor System. At the upper left is the actual last digitized image and the help area from which the area of interest is copied into the rest of the frame buffer. The copied areas are aligned, as shown in the right part of the image, so as to maximally cover the whole 2K × 2K frame buffer. In the lower left part of the image, the result of gray-scale morphology and segmentation on the small area of interest is shown. The detected colloidal gold particles are shown as white blobs.

be done, for instance, by aligning the axon with one of the two axes of the optical system. The videotape is then started and the replayed images are digitized by the Image Processor System 200. This system has a contiguous frame memory of 2,000 by 2,000 pixels, each pixel being 16 bits deep. This means that small areas of interest can be accumulated so as to cover the whole frame buffer. Depending on the orientation and the area of the axon in study, we can put between 200 and 500 images of this area of interest in this frame buffer. After a video frame is digitized into the working area, the area of interest is copied immediately to another part of the frame memory, so as to fill up the whole buffer. This sequence of activities allows us to digitize at a rate of 5 frames/sec and the movements to be continuously tracked for some 40 to 100 sec. This activity can be repeated several times to digitize the full 10 min of observation. The presence of two parallel 40 Mbytes/sec transfer buses on the image processor board allows us, in principle, to perform the digitalization and the copying simultaneously, reaching a digitization speed of 25 frames/sec. This can be of interest for fast phenomena, such as Brownian motion.

The digital information obtained by digitizing full frames contained in 10 min of videotape is almost 4 Gigabyte. By carefully choosing the area of interest, we can reduce this information to a mere 80 Mbyte. The calculation of particle coordinates further reduces this information to 500–2,000 Kbyte. The final data reduction factor is thus on the order of 2,000 to 10,000.

After digitization, the coordinates of each gold particle are extracted from the consecutive images by applying the gray-scale morphology operation described above to each of the areas of interest in the complete frame buffer. This yields good segmentation of the colloidal gold objects. The resultant set of coordinates is saved on disk for later analysis.

In this coordinate map, we can then proceed to detect saltations, Brownian motion, and stop sequences. A saltation is defined as a motion which is continuous for up to 1.25 sec and has a minimal velocity of 1 μm/sec. The complete coordinate map is searched for the appearance of such movements, after which a complete histogram of jump lengths and velocities is created [Geerts et al., 1987]. Brownian motion is detected by constructing $<x^2>$ versus elapsed time T relations and looking for linear relationships [Geerts et al., 1987; Sheetz et al., 1989]. The slope of the linear regression line is proportional to $2nD$, where n is the dimension of the Brownian motion (which is either 1 or 2) and D the diffusion coefficient. If the relationship between $<x^2>$ and T is not linear but shows upward deflections, we have evidence for uniform flow. If, on the other hand, the deflection is downward, hindered diffusion and the existence of microdomains can be detected [De Brabander et al., 1991].

As the segmentation of gold particles with the gray-scale morphology technique takes some 60–80 min for the full 80 Mbyte of image information, we routinely acquire the data first and dump the whole frame buffer temporarily to Winchester disk. This also alleviates the workload of the image processor during the daytime, clearing it for other real-time activities. The analysis can then proceed off-line in batch mode overnight. In this way, we can acquire some 20 min of videotape per hour, which then can be analyzed at night to yield a complete set of coordinates.

V. CONCLUSION

In this chapter, we have illustrated the practical features of nanovid microscopy.

The technique has grown from a most enjoyable form of microscopic observation of the living cells into a very powerful quantitative microscopic tool. It permits mobility to be observed nearly at the molecular level with a high sensitivity and in great detail. Therefore, our dream is that nanovid microscopy will stimulate and excite many biologists on their way to unravelling the living state of the cell.

ACKNOWLEDGMENTS

The skillful and patient photographic work of L. Leijssen, G. Verheyen, and G. Jacobs is greatly appreciated. Part of this work has been supported by the IWONL, Brussels.

REFERENCES

Agard D (1984): Optical sectioning microscopy: Cellular architecture in three dimensions. Ann Rev Biophys Bioeng 13:191–217.

Allen R, Allen N, Travis J (1981): Video-enhanced contrast differential interference contrast (AVEC-DIC) microscopy: A new method capable of analyzing microtubule-related motility in the reticulopodial network of Allogromia laticollaris. Cell Motility 1:291–302.

Axelrod D, Koppel D, Schlessinger J, Elson E, Webb W (1976): Mobility measurements by analysis of fluorescence recovery after photobleaching. Biophys J 16:1055–1069.

De Brabander M, Geerts H, Nuydens R, Nuyens R (1989a): Detection of gold probes with video-enhanced contrast microscopy: Nanovid microscopy. Am J Anat 185:282–295.

De Brabander M, Nuydens R, Geerts H, Hopkins C (1988): Dynamic behavior of the transferrin receptor followed in living epidermoid carcinoma cells (A431) with nanovid microscopy. Cell Motil Cytoskel 9:30–47.

De Brabander M, Nuydens R, Geerts H, Nuyens R (1989b): Detection and use of gold probes with video-enhanced contrast light microscopy. In Verkleij A, Leunissen J (eds): "Immunogold Staining in Cell Biology." Boca Raton: CRC Press, pp 217–232.

De Brabander M, Nuydens R, Ishihara A, Holifield B, Jacobson K, Geerts H (1991): Lateral diffusion and retrograde movements of cell surface components on single motile cells, observed with nanovid microscopy. J Cell Biol 112:111–124.

De Mey J (1984): In Polak J, van Noorden S (eds):

"Immunocytochemistry." London: Wright PSG, pp 92–110.

Diggle JP (1979): On parameter estimation and goodness-of-fit testing for spatial point patterns. Biometrics 35:87–101.

Ellis G (1985): Microscope illuminator with fiber optic source integrator. J Cell Biol 101:83a.

Geerts H, De Brabander M, Nuydens R, Geuens S, Moeremans M, De Mey J, Hollenbeck P (1987): Nanovid tracking: A new automatic method for the study of mobility in living cells based on colloidal gold and video microscopy. Biophys J 52:775–782.

Geuens S, Hill A, Levilliers N, Adoutte A, De Brabander M (1989): Microtubule dynamics investigated by microinjection of Paramecium axonemal tubulin: Lack of nucleation but proximal assembly of microtubules at the kinetochore during prometaphase. J Cell Biol 108:939–953.

He N, Hui S (1985): Electron microscopic observation of domain movement in reconstituted erythrocyte membranes. Proc Nat Acad Sci USA 82:7304–7308.

Inoué S (1981): Video image processing greatly enhances contrast, quality and speed in polarization based microscopy. J Cell Biol 89:346–356.

Inoué S (1986): "Video Microscopy." New York: Plenum Press.

Ishihara A, Hou Y, Jacobson K (1987): The Thy-1 antigen exhibits rapid lateral diffusion in the plasma membrane of rodent lymphoid cells and fibroblasts. Proc Nat Acad Sci USA 84:1290–1293.

Ishihara A, Holifield B, Jacobson K (1988): Analysis of lateral redistribution of a monoclonal antibody complex plasma membrane glycoprotein which occurs during cell locomotion. J Cell Biol 106:329–343.

Kucik D, Elson E, Sheetz M (1989): Forward transport of glycoproteins on leading lamellipodia in locomoting cells. Nature 340:315–317.

Luby-Phelps K, Castle P, Taylor L, Lanni F (1987): Hindered diffusion of inert tracer particles in the cytoplasm of mouse 3T3 cells. Proc Natl Acad Sci USA 84:4910–4913.

Saffmann P, Delbruck M (1975): Brownian motion in biological membranes. Proc Natl Acad Sci USA 72:3111–3113.

Salmon T, Ellis G (1975): A new miniature hydrostatic pressure chamber for light microscopy: Strain-free optical glass windows facilitate phase contrast and polarized light microscopy of living cells. Optional fixture permits simultaneous control of pressure and temperature. J Cell Biol 65:587–602.

Sheetz M, Turney S, Qian H, Elson E (1989): Nanometre-level analysis demonstrate that lipid flow does not drive membrane glycoprotein movements. Nature 340:284–288.

Shotton DM (1988): Video enhanced light microscopy and its applications in cell biology. J Cell Sci 89:129–150.

Sternberg S (1986): Grayscale morphology. Comp Vis Graph Image Process 35:335.

Taylor D, Wang Y (1980): Fluorescently labelled molecules as probes of the structure and function of living cells. Nature 284:405–410.

Yechiel E, Edidin M (1987): Micrometer-scale domains in fibroblast plasma membranes. J Cell Biol 105:755–760.

CHAPTER 7

Vital Staining and Time-Lapse Video-Intensified Fluorescence Microscopy of Organelles in Living Cells

A. Hung Leo, K. Lahteenmaki, and T.E. Kreis

Electronic Light Microscopy, pages 157–179 © 1993 Wiley-Liss, Inc.

I. INTRODUCTION

To fully understand dynamic processes within cells, they need to be visualized in vivo. By microscopic analysis of single, living cells, information about various forms of movements and interactions of a number of cytoplasmic structures can be obtained. Furthermore, an increasing number of reagents allows visualization of such dynamic processes by specific perturbations or modifications. Cytoplasmic structures or organelles can be specifically visualized or identified in vivo by fluorescent marker dyes, microinjected fluorescently tagged tracer proteins or antibodies. Fluorescent probes are very useful reporter molecules for this purpose, mainly for the following reasons: (1) Their application is experimentally straightforward and relatively easy; (2) fluorescent probes may be both very sensitive and specific; (3) sensitive microscopy systems are available which allow continuous visualization of fluorescence signals in living cells over considerable periods of time, without causing damage; and (4) software for acquisition, processing, and quantitative analysis of recorded data is currently being developed.

In this chapter we focus on the application of sensitive video-intensified fluorescence microscopy and time-lapse recording systems for direct visualization and characterization of dynamic properties of fluorescently labeled cytoplasmic organelles in living cells. A good overview of fluorescence microscopy of living cells in culture has recently been published [Taylor and Wang, 1989; Wang and Taylor, 1989]; therefore, we illustrate here only selected examples of vital staining and time-lapse video-intensified fluorescence microscopy and emphasize also the practical and technical aspects.

II. FLUORESCENCE VITAL STAINING OF CELLULAR ORGANELLES

In principle, each structure (or protein) on the cell surface or in the cytoplasm may be visualized by one of a variety of microscopical techniques (including differential interference contrast, dark-field, nanovid microscopy, etc., which will not be discussed here), provided the appropriate specific probes are available. We focus in this section on approaches to label cytoplasmic membrane-bounded organelles with fluorescent probes in living cells. The list of probes which may be used for these purposes is not intended to be complete, but it comprises a selection of examples to illustrate and summarize the basic strategies for their application and their usefulness [for a review, see also Wang and Taylor, 1989]. Probes which will specifically stain cellular organelles may be delivered into cells either by addition to the cell culture medium or by direct introduction into the cytoplasm (e.g., by microinjection).

A. Small Fluorescent Molecules Which Accumulate in Cytoplasmic Organelles When Added to the Cell Culture Medium

Probes which have proven useful for fluorescent labeling of living cells include the dicarbocyanine dye $DiOC_6(3)$, which can be used for labeling of the endoplasmic reticulum [Terasaki et al., 1984], the fluorescent ceramide analogue C_6-NBD-ceramide, which may be applied to give specific labeling of elements of the Golgi apparatus [Lipsky and Pagano, 1985], acridine orange, which accumulates in acidic organelles such as endosomes and secretory vesicles [Robbins and

Marcus, 1963; Kreis et al., 1989], and rhodamine 123, a specific marker for mitochondria [Johnson et al., 1980]. These four fluorescent probes all give intense and organelle-specific fluorescent signals, and their application is rather easy [for review, see Wang and Taylor, 1989]. Here, we discuss only the application of acridine orange and the fluorescent lipid analogue (C_6-NBD-ceramide) in more detail, since the organelles visualized with these two probes will serve as examples for time-lapse video-intensified fluorescence microscopy in living cells.

1. Acridine orange. The weak base acridine orange is a convenient vital marker for acidic compartments such as endosomes, lysosomes, or secretory vesicles, and may be used in a variety of different cell types [e.g., Robbins et al., 1964; Herman and Albertini, 1984; Matteoni and Kreis, 1987; Kreis et al., 1989]. It is relatively nontoxic and rapidly partitions into the cytoplasm when added to the cell culture medium [Robbins and Marcus, 1963; Hart and Young, 1975; Gluck et al., 1982]. Acridine orange becomes protonated in an acidic environment. The net positive charge on the molecule then prevents its free diffusion through membranes, thereby leading to an accumulation and aggregation in the lumen of these acidic organelles. The aggregation of acridine orange results in a red-shift of its emission spectrum. This allows the specific detection (with appropriate filter sets, see below) of acidic organelles in living cells.

We usually label cells for 3–5 min with 20 μM acridine orange (from a fresh 2 mM stock solution [6 mg/10 ml] in H_2O; acridine orange is from Calbiochem, D-6232 Bad Soden, Germany) in Hanks' balanced salt solution [Hanks and Wallace, 1949] at room temperature. Labeled cells are washed with four sequential rinses in Hanks' buffer of 1 min each, and then visualized by video-intensified fluorescence microscopy (rhodamine channel). Staining of acidic organelles is bright, and living cells may be visualized continuously for up to 30–60 min without significant damage of their motile properties [Matteoni and Kreis, 1987].

2. C_6-NBD-ceramide. The fluorescent lipid analogue N-{6-[(7-nitrobenzo-2-oxa-1,3-diazol-4-yl)amino]caproyl}-sphingosine (C_6-NBD-ceramide) has been developed for studying the synthesis, molecular sorting, and intracellular transport of lipids in animal cells [Pagano and Sleight, 1985]. Addition of this fluorescent lipid probe to cells in culture leads initially to an intense fluorescent staining of the Golgi apparatus and subsequent labeling of the cell surface [Lipsky and Pagano, 1983, 1985]. C_6-NBD-ceramide is metabolized in the lumen of the Golgi apparatus to fluorescent sphingomyelin and glucosylceramide, and these membrane-impermeable products remain entrapped in the lumen of the Golgi apparatus and Golgi-derived organelles before they appear on the cell surface. Cooling of the cells to 20°C inhibits transport from the trans-Golgi network to the cell periphery and leads to accumulation of the C_6-NBD-ceramide metabolites in the Golgi apparatus, and thus to the specific fluorescent staining of this organelle [van Meer et al., 1987; Ho et al., 1989].

Unilamellar liposomes containing NBD-ceramide (Molecular Probes, Eugene, OR 97402) are prepared with egg yolk phosphatidylcholine (Sigma) and octylglucoside (Sigma) as described [Ho et al., 1989]. We usually dilute NBD-ceramide-containing liposomes 5–100-fold into Hanks' buffer containing 0.2 mg/ml bovine serum albumin (BSA) and incubate cells for 30 min at 0–10°C [for alternative labeling protocols, see also Lipsky and Pagano, 1983, 1985]. To accumulate the fluorescent ceramide metabolites in the Golgi apparatus, cells are incubated at 20°C for 30–60 min. BSA is kept in the incubation medium in this and all subsequent steps to remove NBD-ceramide derivatives from the cell surface [van Meer et al., 1987]. Bright fluorescence (fluorescein channel) may be visualized in the Golgi apparatus after incubation of the cells at 20°C, and dynamic processes of Golgi elements may then be monitored by video-intensified fluorescence microscopy for up to 90 min [Ho et al., 1989, 1990].

B. Uptake of Fluorescent Fluid Phase Markers for Visualizing Cytoplasmic Organelles

A large variety of other fluorescently labeled probes has been used for labeling of the cellular endocytic compartments [for a review, see Wang and Taylor, 1989]. These probes have either been ligands for specific receptors, for instance α_2-macroglobulin [Maxfield et al., 1978; Willingham et al., 1979], insulin and epidermal growth factor [Schlessinger et al., 1978], and transferrin [van Renswoude et al., 1982], or fluid phase markers which are pinocytosed by the cells, such as Lucifer Yellow [Swanson et al., 1985] or fluorescein isothiocyanate-labeled dextran [Ohkuma and Poole, 1978; van Deurs et al., 1984]. This approach of fluorescently labeling endocytic organelles is quite easy, since the probes can simply be added to the cell culture medium. Subsequent incubation of the cells at appropriate temperatures results in the accumulation of the fluorescent probes in the intracellular compartments.

C. Microinjection of Fluorescently Tagged Proteins into Cells for Visualizing Cytoplasmic Organelles

This approach is potentially the most versatile, since any given cytoplasmic organelle may be specifically labeled. However, it requires the availability of the appropriate fluorescently tagged tracer proteins, or antibodies specific for epitopes present on the cytoplasmic surface of the organelle of choice. Lack of useful specific antibodies and proteins, and perhaps difficulties with introducing proteins into some cells, may have prevented this approach from being a more popular one. Yet, one example of a successful approach for visualizing dynamic properties of cytoplasmic organelles had been described, involving labeling of the biosynthetic pathway of viral transmembrane glycoproteins. Fluorescently labeled antibodies raised against synthetic peptides derived from sequences of the cytoplasmic domain of these transmembrane proteins were microinjected into virus-infected cells and movement of organelles containing the viral transmembrane glycoproteins could be visualized and recorded in vivo [Arnheiter et al., 1984].

III. VISUALIZATION OF FLUORESCENTLY LABELED ORGANELLES

Visualization of fluorescent signals in vivo requires a sensitive detection system [see, for example, Willingham and Pastan, 1978]. Furthermore, cells need to be cultured on the microscope stage during periods of recording. We discuss the basic equipment requirements for video-intensified microscopy in this section [see also Taylor and Wang, 1989].

A. Cell Culture

Cells have to be cultured for various periods of time on the stage of the microscope to visualize dynamic properties by fluorescence. Usually strict temperature control is necessary. Various forms of cell culture chambers, with or without thermostat facilities, are commercially available, or they may be constructed according to specific requirements in a mechanical workshop. Accessibility of cells for microinjection during microscopical analysis may be advantageous. A thermostatic chamber which enables easy removal and replacement of the glass coverslip containing the cell culture is preferable, since it allows manipulations of cells prior or subsequent to observation and makes analysis of parallel experiments possible. We are using a "homemade" stainless steel culture chamber with an external jacket connected to a thermostat, allowing continuous circulation of a thermostatic fluid at constant temperature within the range of $-20-60°C$ [Matteoni and Kreis, 1987]. Cells grown on round coverslips with a diameter of 22 mm are placed into the chamber and fixed and sealed using a screw ring. Labeling of cells can usually be done in a

Video-Enhanced Microscopy System

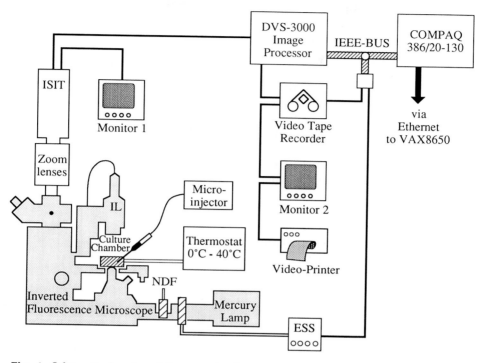

Fig. 1. Schematic drawing of the setup for the video-intensified microscopy system. For details, see text. ESS: electronic shutter system; IL: illuminator; ISIT: intensified silicon intensifier target camera; NDF: neutral density filters.

separate dish and several coverslips may be processed together.

We use either of two standard media for cell culture in the chamber on the microscope stage, dependent on the length of time of analysis. For relatively short periods of time (up to about 30 min), cells are kept in Hanks' buffer supplemented with 0.2 mg/ml BSA. However, when cells are monitored for more than 30 min, we recommend complete culture medium lacking phenol red (this red pH indicator may interfere with the detection of low rhodamine signals) which is adjusted to atmospheric conditions (lower concentration of carbonate buffer and supplemented with 10–50 mM Hepes, pH 7.4) [for further experimental details, see Ho et al., 1989]. Cells may usually be cultured on

the microscope stage for several hours without problems with bacterial contamination.

B. Setup of a Video-Intensified Microscopy System

An example for a basic video-intensified fluorescence microscopy setup which contains the essential instruments is depicted schematically in Figure 1.

We prefer an inverted microscope (for example, Leitz Diavert) equipped for transmitted and fluorescence light microscopy, because cells are more easily manipulated on the stage during visualization. An inverted microscope allows more flexibility with the design of a thermostatic culture chamber and media changes can be performed more easily during the experiments. Furthermore, cells can be microinjected using the same microscope.

A 50 W mercury lamp is usually sufficient for fluorescence excitation, since only low fluorescent light levels are monitored. A combination of neutral density filters (1.4–250×) can be inserted in the excitation light path to minimize the exposure of the cells to light. Furthermore, the periods of excitation may be controlled via an electronic shutter system (iris attached next to the mercury lamp). Timing and editing of excitation, image processing, and recording may be done by computer (see below). We use either of the following filter sets for detection of fluorescein or rhodamine signals: BP450-490, RKP510, LP515 (Leitz I2/3; fluorescein single stain), BP450-490, RKP510, BP525/20 (Leitz L2; fluorescein double stain with rhodamine), and BP515-560, RKP580, LP580 (Leitz N2.1; rhodamine).

An intensifier camera (ISIT camera, DAGE MTI) is connected via zoom lenses to the microscope exit. Other cameras like silicon intensifier target (SIT) or charge-coupled device (CCD) cameras, for example, may be used instead, dependent on the requirements and the budget available. Advantages of the ISIT camera are that it is one of the most sensitive cameras and that it works in real time (in contrast to most CCD cameras). The resolution of the ISIT camera is significantly below the resolution of the optical system of the fluorescence microscope, but in practical terms is usually sufficient to resolve the majority of the labeled organelles. In fact, endosomes or secretory vesicles labeled with specific antibodies after fixation of cells could be previously visualized and resolved in vivo in the same cells by acridine orange- or fluorescein-labeled transferrin gold [Matteoni and Kreis, 1987; Kreis et al., 1989]. Since the resolution of ISIT cameras is limited, it is advisable to insert a zoom-lens system in front of the camera to magnify areas within cells. The zoom system which we are using consists of an extension tubus (0.1 or 0.5×)

and the zoom lenses (5–12.5×, Leitz Vario-Okular), which project an image plane onto the video camera. Unfortunately, however, this increase in magnification is traded off against loss in light intensity.

The very weak signals detected by the intensifier camera usually require image processing to enhance the signal-to-noise ratio. A variety of image processing systems are commercially available. Since flexibility and accessibility to modifications of the programs for image processing may be desirable, we have chosen a simple version of an image processor (Hamamatsu DVS-3000), which we then connect to another (personal) computer (PC; Compaq 386/20-130). This image processor has the basic processing functions (including on-line background subtraction of averaged or recursive filtered images), which are often sufficient for initial acquisition of images. It digitizes analogue images so that they may be transferred subsequently into the memory of a PC.

For monitoring of an experiment, at least two video screens are necessary, one displaying the initial unprocessed image and the other displaying the final processed image.

C. Acquisition of Images of Fluorescent Signals

Perhaps the most important aspect to consider in visualization and image capturing of fluorescent signals from living cells is that every care should be taken to expose the labeled cells to the absolute minimum levels of excitation light necessary.

Images may be recorded by different instruments, depending on the quality of the images required.

1. Video printer. Single frames (directly from the camera, or from the image processor, etc.) may be printed onto paper. These prints are not of very high quality, but are certainly good enough for routine documentation and/or for recording of experiments. Color video printers are useful for pseudocolor images of double-labeled specimens.

2. Polaroid camera. Photographs of good quality may be taken on Polaroid negative film (Polaroid 665). Prints may be obtained by magnification of the negative. Examples of images obtained from fluorescently labeled cells taken with a Polaroid camera from the video screen are shown in Figures 2 and 3.

3. Videotape recording. Until very recently, this was the only convenient way (in Europe) of recording time series of images. Images can be recorded on VHS tape (½"; more recently also on super-VHS, with improved quality) or on U-matic tape (¾"). Furthermore, time-lapse recording can be performed with appropriate instruments (see below). Recording of time series on videotape is easy and cheap. Prints and further analysis of acquired data may be done from a video monitor by replaying the recorded tape, albeit with some loss of image resolution.

4. Optical disk recording. More recently, machines for recording onto optical disks have become commercially available (e.g., Sony, LVR 6000 Analog Draw System). They are relatively expensive, but the images stored retain high quality, they allow time-lapse recording, and the location, display, and replay of stored frames can be easily and rapidly edited.

5. PC memory. Images may also be stored in the memory of a PC connected to the image processor (see also below). Negatives and prints of recorded data may be obtained through a printer and/or negative writer which may be connected as accessories to the computer. Due to the high amount of digital information contained within frames of images of fluorescently labeled cells, large memory capacities are required when time-lapse series need to be recorded into the memory of a PC. The quality of the stored frames, however, is excellent, since it equals the digital image obtained by the image processor.

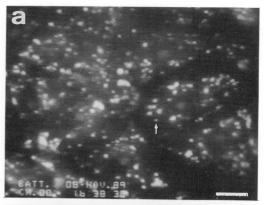
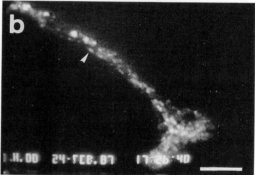
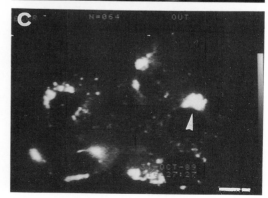

Fig. 2. Vital staining of cytoplasmic organelles. Endocytic organelles and secretory organelles and endosomes were labeled with acridine orange in HeLa **(a)** and AtT20 **(b)** cells, respectively. **c:** A group of Vero cells with the Golgi apparatus labeled with NBD-ceramide metabolites at 20°C, as described in the text. Living cells were visualized with the video-intensified microscopy system and Polaroid photographs were taken from the video monitor. The arrow indicates an endocytic organelle (a), the small arrowhead a secretory vesicle or endosome (b), and the larger arrowhead a Golgi apparatus (c). Bars: (a,c) 25 μm, (b) 10 μm.

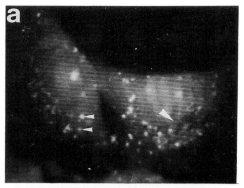
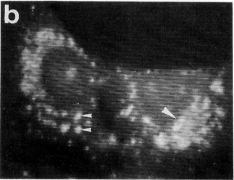
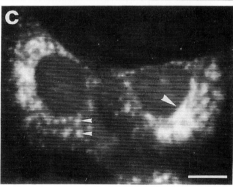

Fig. 3. Staining of endosomes and lysosomes in living NRK cells. NRK cells were incubated for 30 min at 20°C with fluorescently labeled transferrin-gold (fl-TFG) and for a further 15 min in medium lacking the fluorescent ligand. The ligand was then visualized by video-intensified fluorescence microscopy of the living cells, which were maintained at 20°C **(a)**. Photographs are taken by a Polaroid camera from the video monitor after play-back of the videotape. Immediately after recording of the distribution of the fl-TFG, the cells were labeled with acridine orange and the pattern of acridine orange organelles was also recorded by the video-intensified microscopy system **(b)**. Subsequently, the same cells were fixed, labeled with antibodies specific for lyso-

IV. TIME-LAPSE RECORDING OF MOVEMENT OF FLUORESCENTLY LABELED ORGANELLES

It should be emphasized again that every care should be taken to avoid excessive illumination of the fluorescently labeled cells to prevent unnecessary photodamage.

Most of the features implemented in the video-intensified microscopy system used for time-lapse recording of fluorescently labeled cells are logical consequences of trying to avoid cell damage during visualization and recording. Care should also be taken that the fluorescent probes delivered to the cells for monitoring of dynamic properties in the cytoplasm do not have any toxic effects during the time period of analysis. It is often very helpful to fix and immunolabel cells with antibodies directed against specific relevant cytoplasmic structures subsequent to the recording of dynamic processes.

A. Optimizing the Video-Intensified Microscopy System for Time-Lapse Recording

Dynamic events in cells may occur over relatively short (up to 30–60 min) or longer (several hours) periods of time. Fast processes require continuous recording (time-lapse recording may not always be necessary), whereas sufficient information to document cellular processes which occur more slowly can be obtained only by discontinuous (and/ or "slow" time-lapse) recording.

1. Continuous recording. Since we always attempt to monitor and record dynamic processes in fluorescently labeled cells at the lowest possible excitation light levels, image processing prior to recording is necessary. This initial image processing is performed,

somes, and visualized by the video-intensified microscopy system **(c)**. Small arrowheads point to identical organelles labeled with fl-TFG, acridine orange, and antibodies against lysosomes. Large arrowheads indicate the area of clustered lysosomes. Bar, 10 μm. (Reprinted from Matteoni and Kreis, 1987.)

for example, with the DVS 3000 (see Fig. 1), and it includes direct application, *on-line*, of the following functions implemented in the processor: (1) adjustment of contrast and brightness, (2) contrast enhancement, and (3) recursive filtering.

The intensity of the excitation light, sufficient to obtain fluorescence signals for monitoring organelles within labeled cells, is adjusted by neutral density filters (see Fig. 1). Signals are usually adjusted so that they are barely visible on the monitor showing the unprocessed image. We routinely reduce light for visualization of NBD-ceramide derivatives in Golgi elements, or acridine orange in acidic organelles 1,000–2,000-fold [Matteoni and Kreis, 1987; Ho et al., 1989; see also Figs. 2 and 3]. Next, the number of frames for averaging with the average addition algorithm ("recursive filter") is chosen. The formula for average addition is

$$V_n = V_{n-1} + (V_{in} - V_{n-1}) / N,$$

where V_n is the output image, V_{n-1} is the output image for the previous frame, V_{in} is the input image, and N equals 2, 4, 8, . . . , 128. For this function, a larger value of N improves the quality of the image, but decreases resolution of moving objects (due to the averaging procedure, moving organelles may appear as streaks rather than as defined spots; see Fig. 5c,f for an example). Thus, for optimal viewing of moving organelles, both the intensity of the signal and the resolution in time must be taken into account. We usually set N to 32, 64, or 128 for analyzing the movement of acridine orange- or NBD-ceramide-labeled organelles in vivo.

Recording is routinely done, in parallel, in real time on U-matic tape and speeded up (10-fold) on VHS tape. This allows both replay of the dynamic processes visualized at faster speed to emphasize and locate more easily the dynamic events, and provides at the same time a separate, good-quality recording of these processes, relevant frames of which may be transferred later on to the PC (see below).

Readjustment of focus may be necessary during the course of recording (losing focus may be due to instability of the microscope, temperature changes, etc.). This may be performed relatively easily at any time during recording, either manually or via a motor-driven focusing device.

2. Discontinuous recording. Discontinuous recording with discontinuous illumination may be preferable when cellular processes to be visualized proceed slowly (within hours). Thus, instead of continuous recording of cells, which requires continuous excitation, frames are taken only at certain time intervals. These intervals may be chosen, and illumination of cells will occur only during the acquisition of these images. Instead of using the recursive filtering function of the image processor, the signals of a given number of frames may be integrated once, and this image is then recorded on videotape (preferably U-matic) or directly transferred to the PC memory (see below). Software for editing the timing of intervals, opening of the shutter system and excitation of the cells, acquisition and processing of the image, and transfer of this image to videotape and PC memory has been developed (Lahteenmaki and Kreis, unpublished).

It is also important that the fluorescent marker remains associated with the cytoplasmic structures of interest throughout the visualization period.

B. Example 1: Movement of Acridine Orange-Labeled Endosomes and Secretory Vesicles in Living Cells

Immunolocalization in fixed cells gives a snapshot picture of the precise distribution of organelles (or other cellular components) at one point in time (the moment of fixation of the cells; Figs. 3c, 5h, 7b); yet very little information can be obtained about their dynamic properties. For a number of studies, however, it is essential to know the "history" of movements of a given organelle within a defined interval of time. This knowledge about features of movements of specifically labeled organelles may provide the basis for characterizing interactions of the organelles with other cytoplasmic components, such as cytoskeletal structures.

In this first example of the application of time-lapse video-intensified fluorescence microscopy, secretory granules and endosomes (both are acidic classes of organelles) are labeled with acridine orange in AtT20 cells grown on glass coverslips (Figs. 2b, 5) [for experimental details, see Kreis et al., 1989]. The coverslips are then mounted into a culture chamber thermostatically maintained at 37°C, the labeled cells are visualized, and their movements are recorded (speeded up 10-fold by time lapse) for various periods of time using the video-intensified microscopy system (see Fig. 1) [Kreis et al., 1989]. Recorded sequences can now be replayed and analyzed (Figs. 4, 5). Two aspects of the information which can be obtained by this approach are emphasized here. First, the overall movement of the labeled organelles can be described (Fig. 4), and second, it can also be determined whether endosomes and secretory vesicles exhibit different patterns of movement (Fig. 5).

Several parameters can be determined from time-lapse recordings to describe the overall movement of cytoplasmic organelles, including the direction of movement, maximal velocities, and the fraction of moving organelles. Basically, two approaches may be taken to determine these parameters; either it is done manually by copying onto a transparency each displacement of an organelle, displayed on the video screen by replaying the time-lapse recording (Fig. 4), or the analysis is done by computer (Table I; Figs. 9–12). The first approach is tedious and time-consuming, not fully unbiased, and not very precise, but it can be done without further equipment and software; the latter is more sophisticated and statistically more reliable, but it requires development of the appropriate software (discussed in more detail below). The "manual" analysis of the movements of acridine orange-labeled organelles in AtT20 cells led to several con-

clusions. Labeled organelles move by saltations according to the definition of Rebhun [1972]. Velocities of saltatory movement (i.e., distance of uninterrupted movement over time) of 3–5 μm/sec could be determined, but it appeared that velocities of movement of individual organelles were not constant. Moreover, periods of movement lasting up to 20 sec were followed by pauses ranging from 1 sec to over 30 sec (see Fig. 4). Interestingly, it was found that individual fluorescent organelles saltated in either the antrograde or retrograde direction, but not both [Kreis et al., 1989]. Quantitative analysis of the time-lapse recordings revealed that over 95% of acridine orange-labeled organelles did not reverse the direction of their long saltations between the cell body and the tip, or vice versa. Furthermore, about 60% of the labeled organelles moved anterogradely and 40% retrogradely [see Kreis et al., 1989].

This aspect of analysis of the overall movement in AtT20 cells revealed that acidic organelles labeled with acridine orange fall into two distinct classes, those that saltate anterogradely and those that saltate retrogradely [Kreis et al., 1989]. The second aspect discussed here concerns the further characterization of organelles after having visualized their movement; in this example a further characterization of the two classes of moving acidic organelles is necessary. Since cells can be fixed at the end of the period of visualization and recording, they can be immunolabeled with antibodies against appropriate organelle markers and the nature of the moving organelles may be determined (Fig. 5). Cells are fixed most rapidly (using standard fixation reagents) when still mounted in the culture chamber. Coverslips with the attached cells may now be removed from the culture chamber for immunofluorescence labeling. It may be helpful to label the areas containing the cells which have been visualized with a special mark, to allow easy relocation for comparison. To avoid potential problems such as damaging of the glass coverslips during removing them from the chamber, or refind-

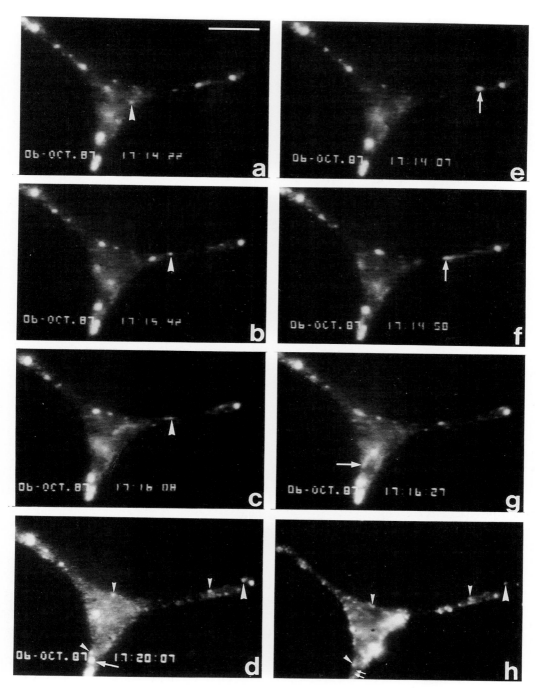

Fig. 4. Movement of acridine orange-labeled endosomes and secretory vesicles in processes of AtT20 cells. Labeling of cells with acridine orange and visualization and recording of the movement of fluorescently labeled organelles by video-intensified fluorescence microscopy is described in the text. The larger arrowheads in **a–d** and the arrows in **e–g** depict the positions of organelles, at the time points indicated on the video frames, during movement in the anterograde (arrowheads) and retrograde (arrows) directions, respectively. The cell body is downward (not shown), and the process has a branch point. At the end of the recording, the cell was fixed and a last frame of the acridine orange-labeled organelles was taken **(d)**. Cells were then labeled with antibodies against ACTH **(h)**. Small arrowheads in d and h indicate examples of acridine orange-labeled organelles that contain ACTH. Note that the retrogradely moving organelle (arrow in d) does not contain ACTH (small double arrows in h). Bar, 10 μm. (Reprinted from Kreis et al., 1989.)

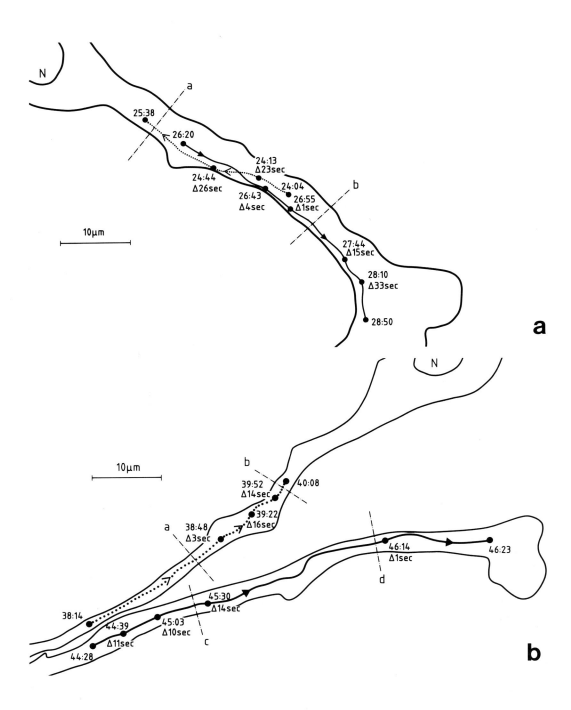

Fig. 5. Analysis of movement of acridine orange-labeled endosomes and secretory vesicles in processes of AtT20 cells. Tracks of anterogradely (——>——) and retrogradely (·····<·····) moving organelles are shown (**a**, same cell as shown in Fig. 2b). Breaks during the saltatory movements are indicated (●) and the time points (min:sec), as well as the periods of resting (Δ in sec) are given. Organelles passing the lines (—·—) indicated were counted [for details, see Kreis et al., 1989]. The cell body of the lower cell in **b** is toward the left side. N: nucleus.

TABLE I. Statistical Analysis of Movement of Organelles[1]

Name	Track Length (μm)	Distance (μm)	[M]oving [S]tationary [D]irected [R]andom	Mean Velocity (μm/sec)	Maximal Velocity (μm/sec)	Direction	Elapsed Time (sec)
0	17.0	11.7	M/D	0.09	0.30	74	190.0
107	12.0	11.9	M/D	0.17	0.27	73	70.0
1	12.8	8.4	M/D	0.07	0.24	39	190.0
2	8.8	1.2	M/R	0.03	0.11	276	320.0
200	2.1	1.5	S	0.10	0.11	49	20.0
177	2.3	1.5	S	0.11	0.16	13	20.0
43	5.1	3.9	M/D	0.03	0.05	130	180.0
3	9.9	5.6	M/D	0.07	0.16	69	140.0
261	10.8	1.0	M/R	0.07	0.13	186	150.0
160	6.2	2.9	M/R	0.06	0.24	222	110.0
188	1.4	0.5	S	0.07	0.08	263	20.0
5	16.0	2.1	M/R	0.09	0.25	58	180.0
151	5.9	4.2	M/D	0.20	0.25	172	30.0
152	8.0	4.9	M/D	0.05	0.21	52	160.0
201	6.3	1.9	M/R	0.08	0.14	254	80.0
165	2.2	0.1	S	0.04	0.06	285	50.0
156	2.2	2.2	S	0.22	0.22	44	10.0
144	1.5	1.5	S	0.07	0.08	51	20.0
299	4.1	3.1	S	0.14	0.18	258	30.0
274	15.9	10.7	M/D	0.11	0.22	174	140.0

Frame interval:		(0, 59)	Movement threshold:		5.00
Total number of organelles analyzed:		313	Direction threshold:		0.50
Total number of tracks analyzed:		68	Distance threshold:		0.50
Objective:	63	Moving:	47%	Random:	50%
Zoom factor:	1.3	Stationary:	53%	Directed:	50%

[1]Data obtained from computational analysis of time-lapse recordings of organelle movement in acridine orange-stained AtT20 cells. This example shows the first 20 records from a data set of 68 organelle tracks analyzed.

ing the relevant cells after immunolabeling, we suggest that all steps involved in the labeling of the cells may also be performed directly within the culture chamber. Using antibodies against adrenocorticotrophic hormone (ACTH), it could be shown here that secretory granules (containing ACTH) showed saltatory movement biased to the anterograde direction, whereas endosomes saltated predominantly in the retrograde direction (see Fig. 5) [Kreis et al., 1989].

Many of the probes used for vital staining of cellular organelles cannot be fixed with the standard methods. They will thus be washed away during the immunolabeling procedure. Despite some disadvantages associated with this behavior of the probes for vital staining (the distribution of the probe cannot be compared directly within the fixed and immunolabeled cells with any subsequent marker), it also offers advantages. Acridine orange, for example, exhibits strong fluorescence in both the rhodamine and the fluorescein channels, making parallel fluorescence labeling of other structures very difficult. Since it can be quantitatively removed from the cells by washing, however, subsequent double immunofluorescence labeling may be performed easily.

C. Example 2: Movement of NBD-Ceramide-Labeled Golgi Elements in Living Cells

This example illustrates how the tracks of moving organelles visualized in living cells can be compared to the cytoskeletal structures (Fig. 8). Information of this kind will help to identify and assay for components involved in the cytoplasmic machinery supporting organelle motility. Also, long-term analysis may be performed to visualize reorientation of Golgi apparati according to specific stimuli.

The Golgi apparatus in interphase animal cells appears usually as a rather immobile organelle. Two experimental model systems can be used, however, to characterize movement of Golgi elements in living cells. The first experimental model system takes advantage of the fact that the Golgi apparatus fragments and scatters randomly throughout the cytoplasm in cells with depolymerized microtubules (Fig. 7a); depolymerization of interphase microtubules may be induced by addition of the drug nocodazole to cells (or will occur naturally at the onset of mitosis). Removal of the drug will result in repolymerization of the microtubule network and subsequent reformation of a compact Golgi apparatus around the microtubule organizing center (Fig. 7b,c) [for a review, see Thyberg and Moskalewski, 1985]. The initially scattered Golgi elements congregate within 30–60 min, and the movements during this relocation process may be visualized in vivo using C_6-NBD-ceramide as a vital stain for the Golgi elements (the experimental protocol is schematically summarized in Fig. 6) [see also Ho et al., 1989]. This process is relatively rapid and has defined starting and ending points. The other experimental model system takes advantage of virus-mediated cell fusion and subsequent syncitia formation. The Golgi apparati congregate to one final central location in syncitia. This relocation of "intact" Golgi apparati may also be monitored using the fluorescent lipid as a marker [Ho et al., 1990]. This latter process, however, is rather slow (2–5 hr) and discontinuous recording is more appropriate to avoid bleaching of the probe and concomitant cell damage.

To determine whether movement of Golgi elements during reclustering after nocodazole washout occurs along individual microtubules, drug-treated cells are labeled with

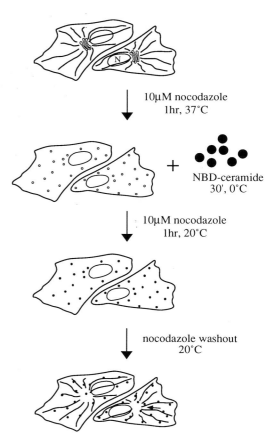

10µM nocodazole
1hr, 37°C

+ NBD-ceramide
30', 0°C

10µM nocodazole
1hr, 20°C

nocodazole washout
20°C

Fig. 6. Protocol for visualization of reclustering Golgi elements labeled with NBD-ceramide in living cells. Interphase microtubules are depolymerized and the Golgi apparatus is induced to break up into Golgi elements scattered throughout the cytoplasm by incubating fibroblasts for 1 hr at 37°C with 10 µM nocodazole in cell culture medium. The cells are then labeled by incubation in the presence of nocodazole with liposomes containing NBD-ceramide during 30 min on ice and an additional 1 hr at 20°C to accumulate the fluorescent lipid metabolites in the Golgi elements. Subsequent removal of nocodazole from the cell culture medium leads to repolymerization of microtubules and congregation of Golgi elements in a perinuclear region; this reclustering process of the fluorescently labeled cytoplasmic organelles can be visualized by video-intensified fluorescence microscopy.

C_6-NBD-ceramide, mounted in the culture chamber maintained at 20°C, and movement of fluorescent organelles is recorded by video-intensified microscopy continuously during 30–60 min, following removal of nocodazole from the culture medium (Fig.

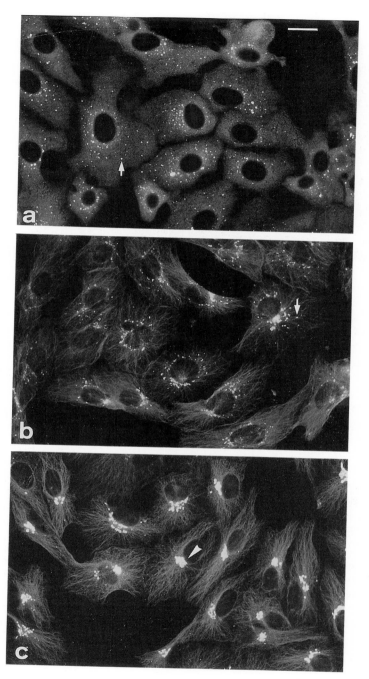

Fig. 7. Reclustering of Golgi elements in the region of the microtubule organizing center. The experimental protocol described in the legend to Figure 6 was applied; however, Golgi elements and microtubules were labeled in fixed and permeabilized Vero cells with specific antibodies at 0 min, when the microtubules are completely depolymerized **(a)**, and 15 min **(b)** after removal of nocodazole. **c:** The normal distribution of microtubules (green label, murine monoclonal antibodies against tubulin visualized with fluorescein-labeled secondary antibodies against mouse IgG) [Kreis, 1987] and the Golgi apparatus (orange label, antibodies against galactosyltransferase visualized with rhodamine-labeled secondary antibodies against rabbit IgG) [Roth and Berger, 1982] in untreated cells. An intact Golgi apparatus is indicated with an arrowhead in c, and scattered or reclustering Golgi elements are indicated with arrows in a and b, respectively. Bar, 25 μm. Figure appears in color in Color Figure Section.

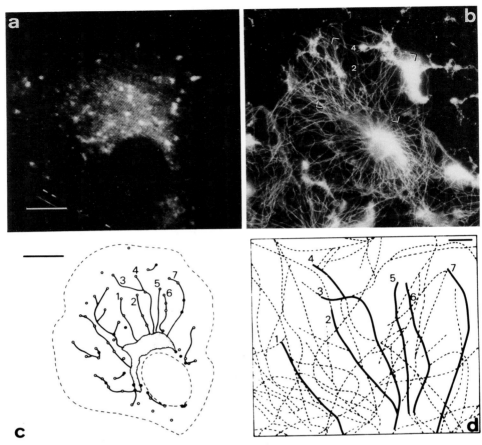

Fig. 8. Movement of Golgi elements along microtubules in Vero cells. Nocodazole-treated cells were labeled with NBD-ceramide as described in the text **(a)**. The reclustering NBD-labeled Golgi elements were recorded by video-intensified fluorescence microscopy for 30 min after the nocodazole washout at 20°C, and the tracks of the moving Golgi elements were traced from play-back of this recording **(c)**. Cells were then pre-extracted, fixed, and immunolabeled for microtubules immediately after recording **(b)**. **d** is an enlarged schematic drawing of the microtubules (dashed and solid lines) within the window in b where the microtubules which correspond to the tracks of reclustering Golgi elements (c) in this region are represented by solid lines. The tracks of several reclustering Golgi elements (1–7 in c) follow the course of the microtubules numbered 1–7 in d. ○: The location of the Golgi elements during nocodazole washout. ●: The location of Golgi elements 30 min after nocodazole washout. Shaded areas indicate clusters of Golgi elements. Bar, 10 μm (a–c), 2 μm (d). [See also Ho et al., 1989.]

8) [see also Ho et al., 1989]. Tracks of moving Golgi elements are transferred manually to transparencies as described above (Fig. 8c) [see also Matteoni and Kreis, 1987]. Coverslips with attached labeled cells are quickly removed from the chamber at the end of the period of recording, and soluble cytoplasmic protein (including nonpolymerized tubulin) is pre-extracted prior to fixation. This gives much sharper contrast and significantly improves the visualization of individual repolymerized microtubules (Fig. 8b). In fact, although possible in principle, covisualization of microtubules by labeling with microinjected rhodamine-tagged tubulin [Kreis, 1987] has been hampered by the presence of too large a pool of soluble tubulin during reclustering of Golgi elements to allow resolution of individual microtubules by video-intensified microscopy. The tracks of

moving Golgi elements can now be compared with the microtubules (or any other labeled cytoskeletal structures). A good correlation between tracks and individual microtubules was found, suggesting that Golgi elements move along microtubules toward their minus ends during reclustering after nocodazole washout (Fig. 8d) [see also Ho et al., 1989].

V. IMAGE PROCESSING AND STATISTICAL ANALYSIS OF MOVEMENT OF FLUORESCENT ORGANELLES

The initial image processing during the time-lapse recording procedure may not be sufficient for an optimal quantitative analysis of movement of organelles in vitally stained cells. Further improvements may be achieved by additional computer processing of stored frames. This additional image processing allows statistical analysis and quantitation of the overall movement of organelles within a given period of time (Table I, Figs. 10–12). Experimental modifications of processes of movements within living cells (e.g., by inhibitors, specific drugs, microinjected antibodies, etc.) could then be measured quantitatively. Such quantitation would be done more easily, more rapidly, and would be less biased than by manual methods.

In this last section we discuss briefly an interactive image processing system connected to the fluorescence microscope which is dedicated for quantifying dynamic properties of fluorescently labeled cytoplasmic structures in living cells (see Fig. 1). This image processing system is directed by a personal computer (Compaq 386/20-130) running with MS-DOS as the operating system and all programs are written in C (further details will be published elsewhere; Hung Leo et al., in preparation).

A. Transfer of Images to the Personal Computer

The image processor (e.g., DVS 3000), tape recorder, and electronic shutter system are connected with the PC via a general-purpose interphase bus (GPIB-IEEE; see Fig. 1). This allows computer-directed editing of the recording of time-lapse series and transfer of frames of images from the image processor to the PC for further processing. The PC is furthermore connected via Ethernet to a VAX 8650 for mass storage of data and for access to an image printer and negative writer.

Frames are transferred to the PC either directly from the image processor during the experiment or from a videotape-recorded time series. Transfer of a full frame (i.e., 512 by 688 pixels) takes ~3 sec. This is, therefore, the shortest time interval for direct recording of full frames into the memory of the PC. Windows of defined smaller sizes within the frame may, however, be defined with this program, which will reduce information and thus speed up the transfer of and further calculations with these images to the PC memory (see Fig. 10a,b, for example).

B. Automatic Identification of Organelles

The result of this process is a list of organelles contained in each stored frame. It further reduces the amount of data considerably without losing the essential information. Subsequent calculations require, therefore, less memory space and proceed much faster. Each organelle is characterized by a centroid which contains the essential information about its position, size, and fluorescence intensity. Organelles are distinguished from the noise and background fluorescence by setting of appropriate threshold levels for size and intensity (thresholded organelles are indicated in red in Fig. 9a,b, showing the original image stored in the memory of the PC).

C. Calculation of Tracks of Moving Organelles

Tracks of moving organelles are calculated by comparing the centroids of sequential

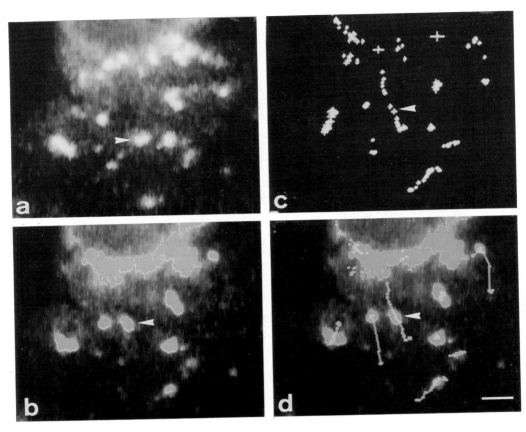

Fig. 9. Image processing steps for visualizing movements of fluorescently labeled organelles in living cells. **a**: The original image of NBD-ceramide-labeled Golgi elements in an area of a Vero fibroblast. **b**: The same area with the thresholded Golgi elements indicated in red. This frame was taken 3 min after the frame shown in a. **c**: Represents the sum of the centroids of the thresholded organelles from 22 frames recorded at 15-sec time intervals. **d**: Displays an overlay of the tracks of moving Golgi elements calculated with the computer over the frame shown in b. The arrowheads indicate one of the Golgi elements. Bar, 10 μm. Figure appears in color in Color Figure Section.

frames (Fig. 9c), correlating the positions of relevant organelles, and calculating the vectors of movement between subsequent time points (frames) for each organelle. It is, however, not a trivial task finding the "clean" tracks of moving organelles immersed in a more or less noisy background, with organelles disappearing and reappearing in focus; therefore, an interactive approach was chosen, allowing, for example, replaying of time-lapse-recorded sequences over the tracks calculated with varying parameters (see, for example, Fig. 9d). The quality of calculation of the tracks of moving organelles depends on the definition of the following parameters: (1) size of the "searching area," (2) distinction of specific features of centroids, and (3) accuracy which defines identity between centroids.

1. Size of searching area. This parameter is used to identify the next position of each centroid in sequential frames among a group of "nearest neighbors" and thus likely candidates. Choosing the best radius

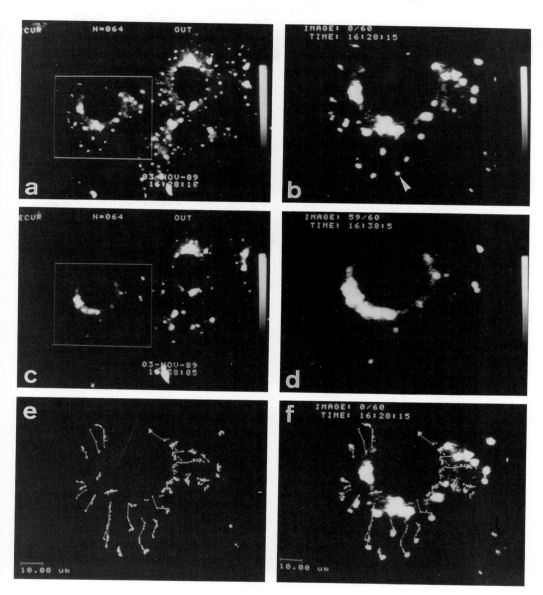

Fig. 10. Computer analysis of movement of NBD-labeled Golgi elements. Golgi elements are labeled with NBD-ceramide in nocodazole-treated cells. Movement of labeled Golgi elements was recorded on videotape and part of the image (window in **a,c**) of a series of frames of this recording (20 min) was transferred to the memory of the PC (**b,d**). Polaroid photographs were taken from the video monitor at the beginning (a,b) and toward the end (c,d) of the reclustering of the Golgi elements. **e:** The tracks of moving Golgi elements calculated by the computer program. **f:** An overlay of the tracks calculated (e) over the initially dispersed Golgi elements shown in b. The arrowhead in b indicates a distinct Golgi element.

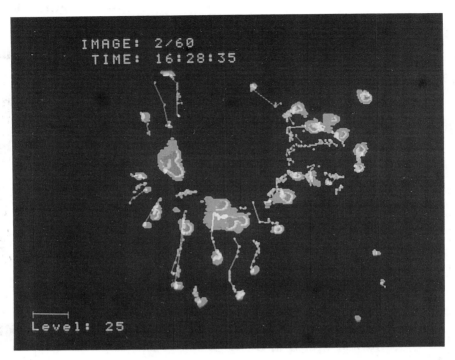

Fig. 11. Overlay of the calculated tracks (green) of moving, fluorescently labeled Golgi elements over the area (shown in Fig. 10a) containing the thresholded NBD-ceramide-labeled organelles at the beginning of the reclustering experiment. Increasing intensity of fluorescence associated with the Golgi elements is indicated by a shift in the color spectrum from red to blue. Bar, 10 μm. Figure appears in color in Color Figure Section.

for the searching area depends on the velocity of movement of organelles and the time period elapsed between intervals.

2. Specific features of organelles. Ambiguity may arise when more than one centroid falls into one searching area. To resolve this ambiguity, a second parameter may be defined, which considers intensity and size of the organelles, or the angle between the previous and new vector of movement calculated.

3. Accuracy. Because organelles move in a three-dimensional environment, their size (and therefore also intensity) may change significantly in a given plane of focus of sequential frames. The limits of "allowed" variation for identifying related centroids are defined by the accuracy criterion.

If the program cannot calculate the full length of tracks, because essential intermediate positions during movement of an organelle have not been resolved (e.g., too fast for the averaging algorithm), consecutive segments belonging to the same track can be concatenated.

D. Statistical Analysis and Quantitation of Movement of Organelles by the Video-Intensified Fluorescence Microscope System

The statistical analysis contains the essential description of the movement of all organelles visible within an area of choice during a defined period of time. The data describing the relevant parameters of movement of each of the organelles identified by the image processing system can be summarized in one table (see Table I; see also Fig. 12).

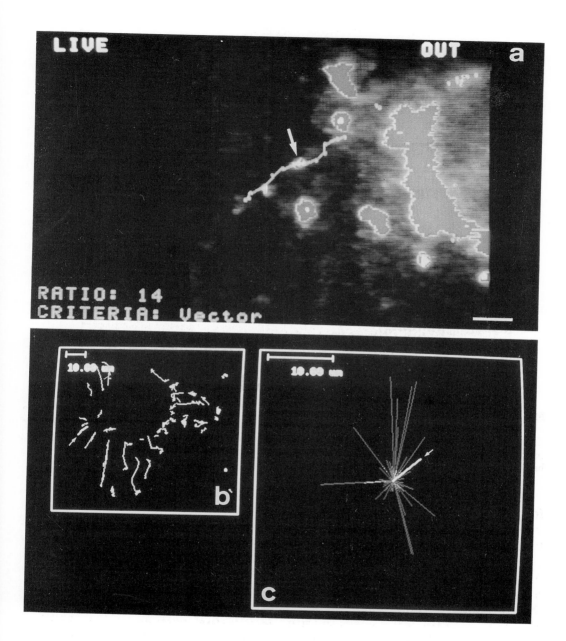

Fig. 12. Statistical analysis of movements of NBD-labeled reclustering Golgi elements. Overlay of the movement of a Golgi element on a track calculated by computer **(a)**, and display of tracks of movement of Golgi elements calculated by computer **(b)** and of the corresponding vectors of movement **(c)**. **a:** The green line corresponds to the track calculated, and red areas are individual Golgi elements or clusters of Golgi elements. The arrow depicts the moving Golgi element. **b,c:** The statistical data of the movement along the track indicated in white and with a small arrow in b and c are given at the bottom; red lines correspond to directed movements of Golgi elements and blue lines to random displacements (c). (a) Bar, 10 μm. Figure appears in color in Color Figure Section.

These parameters allow comparison and quantitation of movement of organelles under variable experimental conditions. They should allow one to distinguish more subtle differences in patterns of movement, differences which could not be determined unambiguously by using the manual approach for characterization of movement of fluorescently labeled organelles in living cells.

ACKNOWLEDGMENTS

We thank Janet Rickard for critically reading the manuscript, and Clemens Storz and Ernst Stelzer for help with the development of the software. K.L. and A.H.L. were recipients of fellowships from the DAAD.

REFERENCES

Arnheiter H, Dubois-Dalcq M, Lazzarini RA (1984): Direct visualization of protein transport and processing in the living cell by microinjection of specific antibodies. Cell 39:99–109.

Gluck S, Kelly S, Al-Awqati Q (1982): The proton translocating ATPase responsible for urinary acidification. J Biol Chem 257:9230–9233.

Hanks JH, Wallace RE (1949): Relation of oxygen and temperature in the preservation of tissues by refrigeration. Proc Soc Exp Biol Med 71:196.

Hart PD, Young MR (1975): Interference with normal phagosome-lysosome fusion in macrophages, using ingested yeast cells and suramin. Nature (Lond) 256:47–49.

Herman B, Albertini DF (1984): A time-lapse video image intensification analysis of cytoplasmic organelle movements during endosome translocation. J Cell Biol 98:565–576.

Ho WC, Allan VJ, van Meer G, Berger EG, Kreis TE (1989): Reclustering of scattered Golgi elements occurs along microtubules. Eur J Cell Biol 48:250–263.

Ho WC, Storrie B, Pepperkok R, Ansorge W, Karecla P, Kreis TE (1990): Movement of interphase Golgi apparatus in fused mammalian cells and its relationship to cytoskeletal elements and rearrangement of nuclei. Eur J Cell Biol 52:315–327.

Johnson LV, Walsh ML, Chen LB (1980): Localization of mitochondria in living cells with rhodamine 123. Proc Natl Acad Sci USA 77:990–994.

Kreis TE (1987): Microtubules containing detyrosinated tubulin are less dynamic. EMBO J 6:2597–2606.

Kreis TE, Matteoni R, Hollinshead M, Tooze J (1989): Secretory granules and endosomes show saltatory movement biased to the anterograde and retrograde directions, respectively, along microtubules in AtT20 cells. Eur J Cell Biol 49:128–139.

Lipsky NG, Pagano RE (1983): Sphingolipid metabolism in cultured fibroblasts: Microscopic and biochemical studies employing a fluorescent ceramide analogue. Proc Natl Acad Sci USA 80:2608–2612.

Lipsky NG, Pagano RE (1985): A vital stain for the Golgi apparatus. Science 228:745–747.

Matteoni R, Kreis TE (1987): Translocation and clustering of endosomes and lysosomes depends on microtubules. J Cell Biol 105:1253–1265.

Maxfield FR, Schlessinger J, Schechter Y, Pastan I, Willingham MC (1978): Collection of insulin, EGF and α_2-macroglobulin in the same patches on the surface of cultured fibroblasts and common internalization. Cell 14:805–810.

Ohkuma S, Poole B (1978): Fluorescent probe measurement of the intralysosomal pH in living cells and the perturbation of pH by various agents. Proc Natl Acad Sci USA 75:3327–3331.

Pagano RE, Sleight RG (1985): Defining lipid transport pathways in animal cells. Science 229:1051–1057.

Rebhun LI (1972): Polarized intracellular particle transport: Saltatory movements and cytoplasmic streaming. Int Rev Cytol 32:93–137.

Robbins E, Marcus PI (1963): Dynamics of acridine orange-cell interaction. I. Interrelationships of acridine orange particles and cytoplasmic reddening. J Cell Biol 18:237–250.

Robbins E, Marcus P, Gonatas NK (1964): Dynamics of acridine orange-cell interaction. II. Dye-induced changes in multivesicular bodies (acridine orange particles). J Cell Biol 21:49–62.

Roth J, Berger EG (1982): Immunocytochemical localization of galactosyltransferase in HeLa cells: Codistribution with thiamine pyrophosphate in trans Golgi cisternae. J Cell Biol 93:223–229.

Schlessinger J, Shechter Y, Willingham MC, Pastan I (1978): Direct visualization of binding, aggregation, and internalization of insulin and epidermal growth factor on living fibroblastic cells. Proc Natl Acad Sci USA 75:2659–2663.

Swanson JA, Yrinec BD, Silverstein SC (1985): Phorbolesters and horseradish peroxidase stimulate pinocytosis and redirect the flow of pinocytosed fluid in macrophages. J Cell Biol 100:851–859.

Taylor DL, Wang YL (1989): "Fluorescence Microscopy of Living Cells in Culture." Meth Cell Biol 30, Academic Press.

Terasaki M, Song JD, Wong JR, Weiss MJ, Chen LB (1984): Localization of endoplasmic reticulum in living and glutaraldehyde-fixed cells with fluorescent dyes. Cell 38:101–108.

Thyberg J, Moskalewski S (1985): Microtubules and the organization of the Golgi complex. Exp Cell Res 159:1–16.

van Deurs B, Ropke C, Thorball N (1984): Kinetics of pinocytosis studied by flow cytometry. Eur J Cell Biol 34:96–102.

van Meer G, Stelzer EHK, Wijnaendts-van-Resandt RW, Simons K (1987): Sorting of sphingolipids in epithelial (Madin-Darby canine kidney) cells. J Cell Biol 105:1623–1635.

van Renswoude JK, Bridges KR, Harford JB, Klausner RD (1982): Receptor-mediated endocytosis of transferrin and the uptake of Fe in K562 cells: Identification of a nonlysosomal acidic compartment. Proc Natl Acad Sci USA 79:6186–6190.

Wang YL, Taylor DL (1989): "Fluorescence Microscopy of Living Cells in Culture." Meth Cell Biol 29, Academic Press.

Willingham MC, Pastan I (1978): The visualization of fluorescent proteins in living cells by video intensification microscopy (VIM). Cell 13:501–507.

Willingham MC, Maxfield FR, Pastan I (1979): α_2 Macroglobulin binding to the plasma membrane of cultured fibroblasts. J Cell Biol 82:614–625.

Color Figure Section

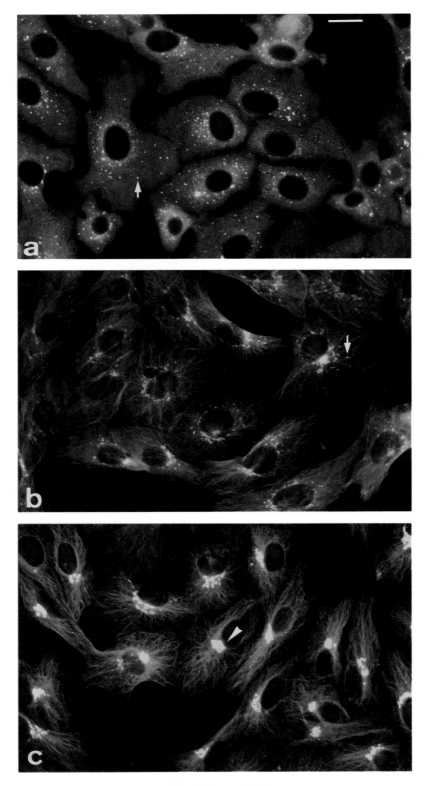

Fig. 7 (page 171)

Vital Staining and Time-Lapse Video-Intensified Fluorescence Microscopy of Organelles in Living Cells

A. Hung Leo, K. Lahteenmaki, and T.E. Kreis

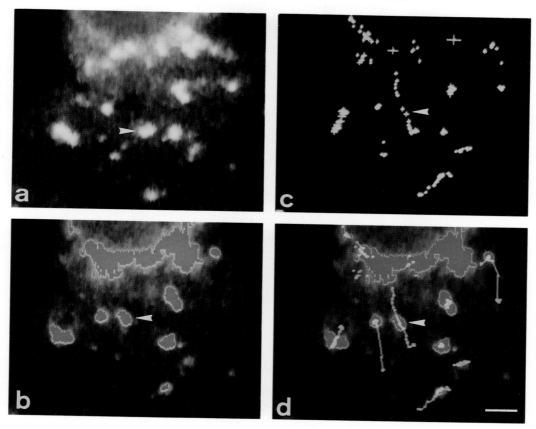

Fig. 9 (page 174). Image processing steps for visualizing movements of fluorescently labeled organelles in living cells. **a:** The original image of NBD-ceramide-labeled Golgi elements in an area of a Vero fibroblast. **b:** The same area with the thresholded Golgi elements indicated in red. This frame was taken 3 min after the frame shown in a. **c:** Represents the sum of the centroids of the thresholded organelles from 22 frames recorded at 15-sec time intervals. **d:** Displays an overlay of the tracks of moving Golgi elements calculated with the computer over the frame shown in b. The arrowheads indicate one of the Golgi elements. Bar, 10mm.

Fig. 7 (page 171). Reclustering of Golgi elements in the region of the microtubule organizing center. The experimental protocol described in the legend to Figure 6 was applied; however, Golgi elements and microtubules were labeled in fixed and permeabilized Vero cells with specific antibodies at 0 min, when the microtubules are completely depolymerized **(a)**, and 15 min **(b)** after removal of nocodazole. **c:** The normal distribution of microtubules (green label, murine monoclonal antibodies against tubulin visualized with fluorescein-labeled secondary antibodies against mouse IgG) [Kreis, 1987] and the Golgi apparatus (orange label, antibodies against galactosyltransferase visualized with rhodamine-labeled secondary antibodies against rabbit IgG) [Roth and Berger, 1982] in untreated cells. An intact Golgi apparatus is indicated with an arrowhead in c, and scattered or reclustering Golgi elements are indicated with arrows in a and b, respectively. Bar, 25mm.

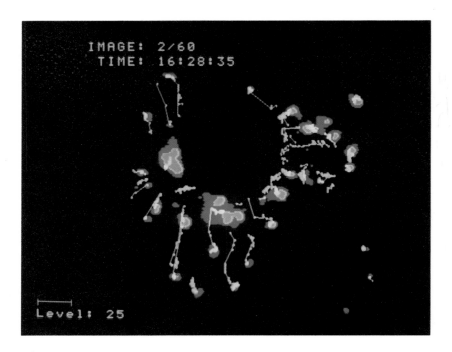

Fig. 11 (page 176). Overlay of the calculated tracks (green) of moving, fluorescently labeled Golgi elements over the area (shown in Fig. 10a) containing the thresholded NBD-ceramide-labeled organelles at the beginning of the reclustering experiment. Increasing intensity of fluorescence associated with the Golgi elements is indicated by a shift in the color spectrum from red to blue. Bar, 10 mm.

Fig. 12 (page 177). Statistical analysis of movements of NBD-labeled reclustering Golgi elements. Overlay of the movement of a Golgi element on a track calculated by computer (a), and display of tracks of movement of Golgi elements calculated by a computer (b) and of the corresponding vectors of movement (c). a: The green line corresponds to the track calculated, and red areas are individual Golgi elements or clusters of Golgi elements. The arrow depicts the moving Golgi element. b,c: The statistical data of the movement along the track indicated in white and with a small arrow in b and c are given at the bottom; red lines correspond to directed movements of Golgi elements and blue lines to random displacements (c). (a) Bar, 10 mm.

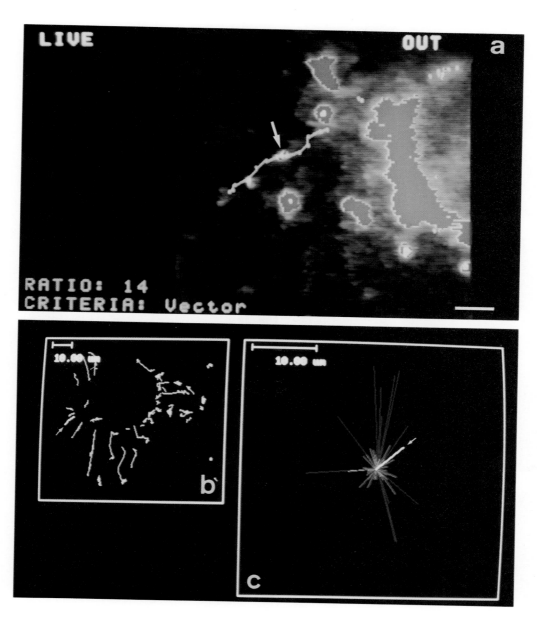

Fig. 12 (page 177)

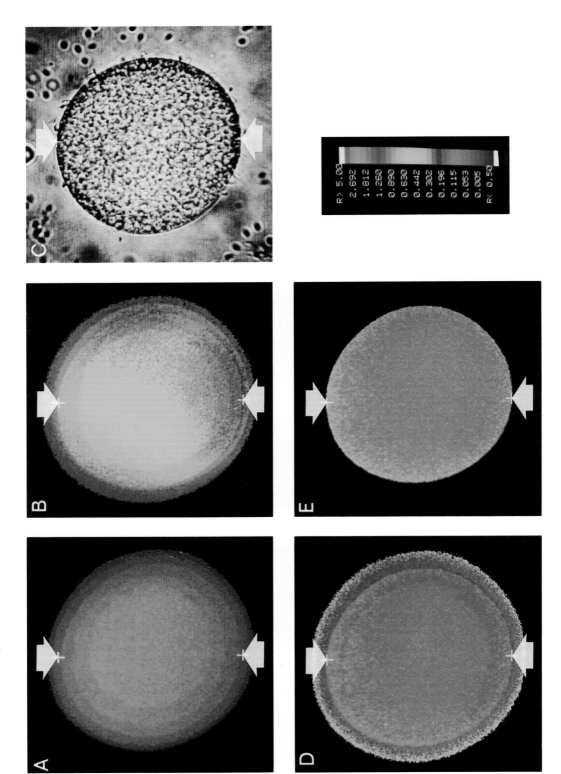

Plate 1 (page 185)

Ratio Imaging Measurement of Intracellular Calcium and pH
Stephen R. Bolsover, R. Angus Silver, and Michael Whitaker

Plate 1 (page 185). Ratio images of $[Ca^{++}]_i$ in a sea urchin (*Lytechinus pictus*) egg. The images illustrate certain problems of ratio imaging, as discussed in the text. The white arrows (tips 117 mm apart) are placed in the same position in each image. The egg was injected with Fura-2 to a final concentration of 50 mM, and with guanosine 5'-thiophosphate (GDPS) to a final concentration of 3 mM. **A:** Before fertilization. The ratio image extends beyond the margins of the cell visible in the transmitted light image **(C)**; this is the *halo*. **B:** After fertilization, during the calcium wave. **C:** Transmitted light image of the cell at the end of the experiment. Although GDPS did not prevent the calcium wave, it blocked exocytosis: The cell has not raised a fertilization membrane. **D:** Peak of postfertilization $[Ca^{++}]_i$. The image shows a bulls-eye pattern of $[Ca^{++}]_i$, with the accurate indicated ion concentration at the center of the cell grading to an entirely erroneous ion concentration in the halo. **E:** As D, but threshold increased from 1/8 to 1/4 camera saturation to remove the halo. $[Ca^{++}]_i$ scale is in mM and was calculated using a viscosity correction factor of 0.5. For methods, see Swann and Whitaker [1986].

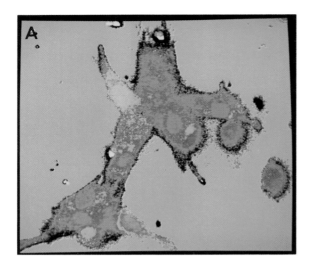

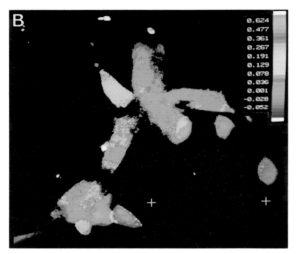

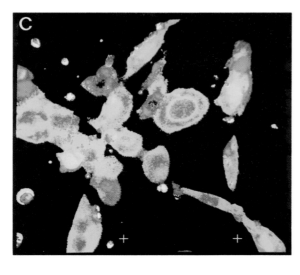

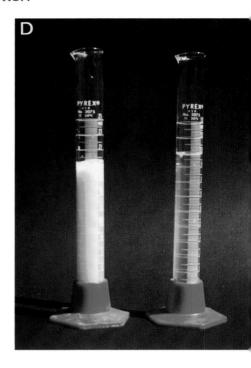

Plate 2 (page 187)

Plate 2 (page 187). A,B: Ratio image of Swiss 3T3 fibroblasts loaded with Fura-2 by incubation in growth medium (DMEM plus 6% fetal calf serum) plus mM Fura-2 and 200 mg/l Pluronic F-127 at 20°C. **A:** Image not thresholded. Ratio values from cell edges are noisy and give incorrect values. Regions of the field outside the cells appear yellow, since this is the hue corresponding to a ratio "value" of $0 \div 0$. **B:** Thresholded at 1/8 camera saturation. The individual cells can now be clearly distinguished and related to the transmitted light image (Fig. 1A). $[Ca^{++}]_i$ is spatially uniform in individual cells, indicating few problems with noncytosolic dye. **C:** Ratio image of Swiss 3T3 fibroblasts loaded with Fura-2 by incubation in growth medium plus 10 mM Fura-2 at 36°C without Pluronic F-127. Measured $[Ca^{++}]_i$ is spatially nonuniform in individual cells, which show the characteristic perinuclear high-$[Ca^{++}]_i$ ring due to Fura-2 loading into endosomes (see Fig. 1B-D). Cursor crosses in B and C are 100 mM apart. $[Ca^{++}]_i$ scale in B (mM) applies to A, B, and C. **D,E:** An egg has approximately equal volumes of yolk and white, but when fried the yolk comprises only 15% of the total area. Estimating mean egg calcium concentration by giving equal weight to equal areas of fried egg will result in a weighting error.

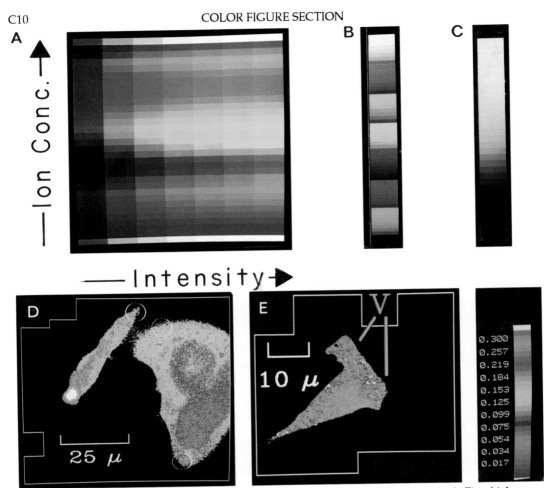

Plate 3 (page 204). A–C: Alternative pseudocolor scales for display of ratio images. **A:** Five-bit hue, 3-bit luminance scale used in our system. Higher concentrations of measured ion are indicated by warmer hues, while the intensity of the fluorescence signal is indicated by the luminance (brightness) of the display. In this plate, hue and luminance interfere, so that, for instance, an ion concentration that appears as green at low intensity appears as yellow at high intensity. This is an artifact of color saturation of the photographic film and does not occur on the monitor display. For this and other reasons, we usually turn the intensity coding off when taking pictures. **B:** A type of scale in which the single parameter—ion concentration—is coded for by both hue and luminance. This type of scale is used in the Spex system but seems inappropriate for a smoothly varying parameter such as ion concentration. **C:** Five-bit (32-shade) gray scale, illustrating the difficulty of recording subtle gray-level changes in monochrome photographic images. **D,E:** N1E-115 mouse neuroblastoma cells. $[Ca^{++}]_i$ scale at right (mM) applies to both images. **D:** loaded by incubation in growth medium (DMEM plus 6% fetal calf serum) plus 20 mM Fura-2 at 36°C without Pluronic F-127. The cell body on the right-hand side of the image shows a clear high-$[Ca^{++}]_i$ ring around the nucleus, indicating that Fura-2 has loaded into endosomes. The neurite, which enters the field at the top center and runs diagonally left to terminate in growth cone, shows a pronounced spatial gradient of $[Ca^{++}]_i$. Figure 4A shows a transmitted light image of this field. Figure 4B is the same ratio image as Plate 3D but using a gray scale to indicate $[Ca^{++}]_i$. The circles in Figure 4B and Plate 3D indicate three regions; $[Ca^{++}]_i$ is approximately the same in the top two but is lower in the bottom circle. However, the relative $[Ca^{++}]_i$ in the three regions is very difficult to assess from the gray-scale image. **E:** Growth cone of N1E-115 cell injected with Fura-2. There are no spatial gradients of $[Ca^{++}]_i$ and mean $[Ca^{++}]_i$ is much lower since only cytoplasmic $[Ca^{++}]_i$ is being indicated. Figure 4C shows a transmitted light image of this growth cone.

Acquisition, Processing, and Visualization of Three-Dimensional Confocal Images

H.T.M. van der Voort and A.W.M Smeulders

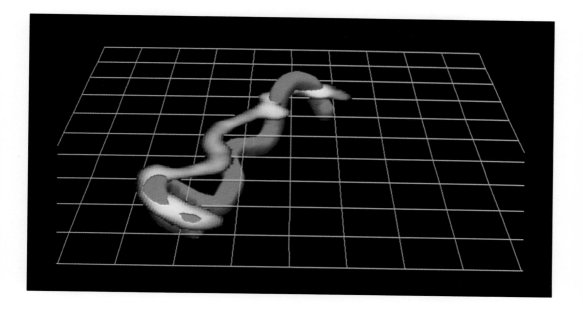

Fig. 12 (page 269). A single chromosome of *Crepis capillaris* as visualized after shape analysis (same data as in Fig. 10). In the analysis, the "spine" of the chromosome was determined by chaining a series of points, each corresponding to a local center of mass (see text). Using building elements resembling the cylindrical and elbow pieces, a stovepipe-like structure was constructed, centered around the spine of the chromosome **(red structure)**. To convey the perspective of the scene, a rectangular grid was added. To show the gain of the combination confocal imaging and 3D analysis, the 2D image of the chromosome as it would have been seen in a conventional microscope was projected, but without out-of-focus blur **(blue-gray structure)**.

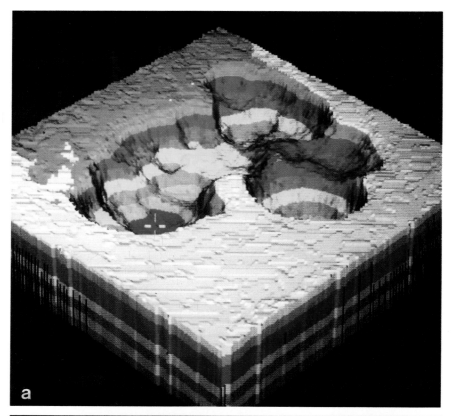

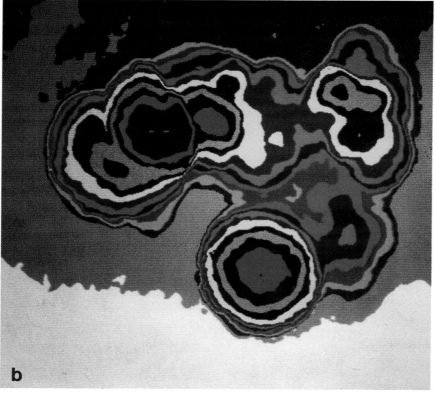

Plate 1 (page 309)

Real-Time Direct-View Confocal Light Microscopy

A. Boyde

Plate 1 (page 309). Complex resorption pit excavated by an osteoclast which differentiated from chick bone marrow in 19-day culture on a slice of sperm whale dentine: Sample prepared as for SEM. A map of the surface was generated by through-focusing over a range of 24 mm—slightly more than the depth of the pit—to find the focal plane at which the brightest signal was recorded for each pixel in a 512 × 512 field. This map was then reprojected to produce the solid model display: Eight contour levels are shown in monochrome in Figure 13a (p. 308) and in pseudocolor in Plate 1a. Plate 1b shows an orthogonal pseudocolor height-coded image. A "wire frame" at 8-pixel interval intersections is shown in Figure 13b (p. 308). Photograph of TV monitor on Tracor 8502 image-analyzing computer. Raw data collected by confocal reflection imaging using a Nikon ×60/1.4NA oil-immersion objective.

CHAPTER 8

Ratio Imaging Measurement of Intracellular Calcium and pH

Stephen R. Bolsover, R. Angus Silver, and Michael Whitaker

Electronic Light Microscopy, pages 181–210 © 1993 Wiley-Liss, Inc.

I. INTRODUCTION

The field of intracellular ion concentration measurement expanded greatly in the 1980s due primarily to the development by Roger Tsien of ratiometric dyes. Since the initial synthesis of the ratiometric calcium indicator Fura-2 [Grynkiewicz et al., 1985], Tsien has developed ratiometric dyes selective for such ions as H^+, K^+, and Mg^{2+}. Tsien has recently written an article in which he describes the hardware and software he has used to obtain ratio images of ion concentrations [Tsien and Harootunian, 1990]. Tsien and Harootunian's review is aimed at people who may want to construct (or improve) their own imaging system, though it is also a very useful general account of how ratio dye imaging systems are put together. Here, we are aiming in a slightly different direction, at people who want to use ratio dye imaging, but do not have at their disposal the large amounts of time, resources, and expertise that are necessary to implement a custom-built ratio dye imaging system. We hope to supply enough information to permit a reasonably informed choice between the various imaging systems that are now commercially available. In order to prevent users from stumbling into some of the more common pitfalls that lie in the path of the unwary, we also discuss a number of problems and artifacts that we have encountered [Silver, Whitaker, and Bolsover, 1992] or that we have noticed in other work.

II. RATIO IMAGING PRINCIPLES

A. Ratiometric Fluorescent Dyes

The property that distinguishes a fluorescence ratio dye is a shift in the maximum of the excitation or emission spectra on ion binding. The usefulness of this sort of a spectral shift is that it can provide a signal related to ion binding and independent of dye amount. A spectral shift is essential for imaging spatial differences in ion concentration within cells: Without the shift, there are no means of distinguishing between spatial differences due to ion concentration and differences due to variations in cell thickness and dye distribution.

To exploit the spectral shift of a ratio dye requires measurement at two excitation wavelengths or two emission wavelengths, depending on the dye in question. Emission ratio dyes have the advantage that both emission wavelengths can be measured simultaneously, while excitation ratio dyes require alternating presentation of the two excitation wavelengths. This is cumbersome. It might appear, therefore, that emission ratio dyes would be preferred for imaging applications. In practice, real or perceived problems in the alignment of two video cameras (one looking at each emission wavelength) have led most workers, and all commercial suppliers, to use excitation ratioing systems. In this review, we concentrate on excitation ratioing systems, although most of what we say is also applicable to emission ratio imaging.

The most widely used fluorescent ratio dyes for imaging are the excitation ratio dyes Fura-2 (for calcium) and BCECF and SNAFL (for pH). Other appropriate dyes are Mag-Fura-2, for magnesium, and the emission ratio dyes Indo-1 (for calcium), SNARF (for pH), and Mag-Indo-1 (for magnesium). These dyes are usually either introduced directly into intact cells through a glass micropipette, or applied as the membrane-permeant acetoxymethyl ester [Tsien, 1981]. Alternatively, cells can be rendered temporarily leaky by techniques such as electroporation, ATP^{4-}, or scraping, so that dye in the extracellular fluid can diffuse in [Moore et al., 1990].

B. Applications of Ratio Dye Imaging

Ratio dye imaging can be used for two distinct purposes. The first and most obvious use of imaging is to detect spatial differences in ion concentration within a single cell. This is an exacting application that is prone to several artifacts and to limits in spatial resolution that we discuss below. However, a second and equally common use of ratio imaging is to measure mean ion concentration in each of a number of cells in the field of view. The advantages of measuring ion concentration in single cells rather than measuring a population mean ion concentration is that if each cell responds to a particular stimulus with a different latency, or if the intracellular ion concentration changes are oscillatory, the population mean is a poor indicator of the actual ion concentration being experienced by each cell. Why use an imaging system rather than single-cell photometry? Some reasons are: (1) Simultaneous measurement of responses in 6 or 10 cells ensures that each of the cells has the same history, as well as allowing data to be gathered several times faster. (2) Many a single-cell photometry experiment has to be discarded when one looks down the microscope at the end of the experiment and finds that the cell has moved out of the measurement aperture. Such problems are far less acute in imaging systems. (3) Investigations of intercellular signaling may require simultaneous observation of ion concentration in more than one cell. An example is the elegant experiments of Cheek [1990], where an increase in the intracellular free calcium ion concentration, $[Ca^{++}]_i$, in a chromaffin cell was seen to trigger secretion which in turn caused $[Ca^{++}]_i$ changes in neighboring fibroblasts. Measurement of mean ion concentration in each of a number of cells in the field of view is a less exacting application than detection of spatial differences in ion concentration within a single cell. Less is required of an imaging system making these sorts of measurements.

C. Calculating Ratio Images from Fluorescence Images

Raw data from a ratio imaging system consist of a pixel-by-pixel record of the emission signals at the two excitation wavelengths. The ion concentration at each pixel can be calculated from the ratio equation, but before this can be done, the raw data have to be corrected for other sources of light and for any differences in the spatial distribution of the two excitation beams. The mathematical manipulations required are shown in Box 1.

1. Background subtraction. The background is all light measured by the camera that is not emitted from the ratio dye. The most likely sources are camera dark noise, incompletely blocked excitation light, fluorescence of optical components, slide or coverslip, or extracellular medium, and cell autofluorescence. The optimal approach to background subtraction is to move the cells of interest into the field of view and acquire background images at each wavelength before introducing dye. This optimal approach is only possible if the cells do not change shape during dye introduction; we have successfully used this approach in neurons at room temperature loaded from a whole-cell patch pipette. However, it is usually not possible to correct for autofluorescence in this way, and one must therefore ensure that one uses a quantity of dye that gives a signal sufficient to swamp any autofluorescence. In this latter case, background images are obtained from regions of the slide or coverslip that do not contain cells; such background images will correct for camera dark noise and incompletely blocked excitation light plus fluorescence of optical components, slide or coverslip, and

BOX 1   PIXEL-BY-PIXEL CALCULATION OF RATIO IMAGE

Acquisition of a ratio image begins with the acquisition of images at each of two excitation wavelengths, λ_1 and λ_2. Let the fluorescence intensities at a particular pixel be $I_{\lambda 1}$ and $I_{\lambda 2}$. The first step is to subtract backgrounds:

$$I'_{\lambda 1} = I_{\lambda 1} - B_{\lambda 1}$$

$$I'_{\lambda 2} = I_{\lambda 2} - B_{\lambda 2}$$

where the backgrounds $B_{\lambda 1}$, $B_{\lambda 2}$ are from images of light emitted from the preparation in the absence of ratio dye.

The ratio image is then calculated as

$$R = \frac{I'_{\lambda 1} S_{\lambda 2}}{I'_{\lambda 2} S_{\lambda 1}}$$

where the shading values $S_{\lambda 1}$, $S_{\lambda 2}$ for each individual pixel are obtained from background-corrected images of a uniform field of dye (usually, at zero $[Ca^{++}]$) held between coverslips. One can either store two images, consisting of values $S_{\lambda 1}$ and $S_{\lambda 2}$, respectively, or one can store a single "shading ratio image," consisting of values $S_{\lambda 1}/S_{\lambda 2}$.

---

extracellular medium. Backgrounds must be measured at intervals throughout the experiment, because many of the components of the background, such as camera dark noise, will change. Inadequate or excessive background subtraction will have the greatest effect at the edge of the cell, where the signal from the dye is weakest. Images subject to this error therefore show a "bull's-eye" pattern, with one (relatively accurate) ion concentration indicated at the center of the cell that grades to an entirely erroneous ion concentration at the periphery (Plate 1D). Such images are regrettably common in published work.

An increase of $[Ca^{++}]_i$ can reveal an error produced by inadequate or excessive background subtraction that was not apparent at lower $[Ca^{++}]_i$. Plate 1 illustrates this point. At low $[Ca^{++}]_i$ (image A), both 350 and 380 nm signals are of large amplitude, and a slight error in background has little effect. The edges of the cell therefore show

roughly the same $[Ca^{++}]_i$ as the rest of the cell. At high $[Ca^{++}]_i$ (image D), the 380 nm signal is small, and the effect of a slight error in background is correspondingly larger. This image shows the characteristic bull's-eye pattern, with an incorrect $[Ca^{++}]_i$ at the edge of the cell.

2. Shading corrections. These compensate for any differences in the spatial distribution of the two excitation beams. Such differences are more likely to occur in systems using two light sources to provide the two excitation beams. To make a shading measurement, one uses a uniform field of dye held between coverslips and acquires images at each of the two excitation wavelengths. The way in which these "fluorescence shading images" are used is different in different systems, although the final result is the same. Some systems store both fluorescence shading images, and use them to correct the individual fluorescence images of cells acquired later. Corrected fluorescence

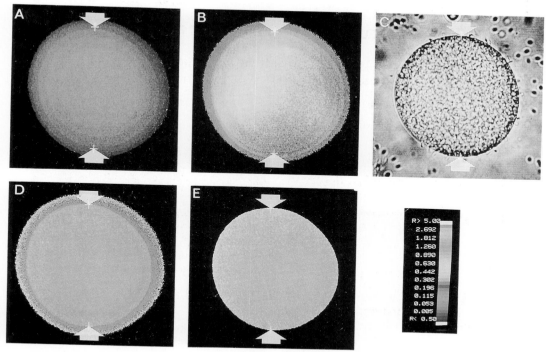

```
R> 5.00
   2.692
   1.812
   1.260
   0.890
   0.630
   0.442
   0.302
   0.196
   0.115
   0.053
   0.005
R< 0.50
```

Plate 1. Ratio images of $[Ca^{++}]_i$ in a sea urchin (*Lytechinus pictus*) egg. The images illustrate certain problems of ratio imaging, as discussed in the text. The white arrows (tips 117 μm apart) are placed in the same position in each image. The egg was injected with Fura-2 to a final concentration of 50 μM, and with guanosine 5'-thiophosphate (GDPβS) to a final concentration of 3 mM. **A:** Before fertilization. The ratio image extends beyond the margins of the cell visible in the transmitted light image (**C**); this is the *halo*. **B:** After fertilization, during the calcium wave. **C:** Transmitted light image of the cell at the end of the experiment. Although GDPβS did not prevent the calcium wave, it blocked exocytosis: The cell has not raised a fertilization membrane. **D:** Peak of postfertilization $[Ca^{++}]_i$. The image shows a bulls-eye pattern of $[Ca^{++}]_i$, with the indicated ion concentration at the center of the cell grading to an entirely erroneous ion concentration in the halo. **E:** As D, but threshold increased from ⅛ to ¼ camera saturation to remove the halo. $[Ca^{++}]_i$ scale is in μM and was calculated using a viscosity correction factor of 0.5. For methods, see Swann and Whitaker [1986]. Figure appears in color in Color Figure Section.

images are then divided to give a ratio image. Our system uses an alternative and equally valid approach: The two fluorescence shading images are divided to yield a "shading ratio image." Subsequent ratio images, from cells, are divided by this shading ratio image. This approach means that only one shading image needs to be stored in memory, and reduces the number of calculations that need to be performed for each measurement.

3. Thresholding. The accuracy of a ratio measurement falls off toward the edges of a cell as the amplitude of the fluorescence signal declines. Ratio measurements on regions outside cells are meaningless. The results shown in Figure 1A and Plate 2A,B illustrate this. Figure 1A shows a transmitted light picture of a field of Fura-2-loaded fibroblasts, while Plate 2A shows the ratio values from the entire field. Ratio values from cell edges are noisy and give incorrect values since they are produced by the division of small integers ($2 \div 1$, $1 \div 1$, and so on). Regions of the field outside the cells appear yellow, since this is the hue corresponding to a ratio "value" of $0 \div 0$. In order that the observer not be confused or actually misled by these ratio values, the ratio

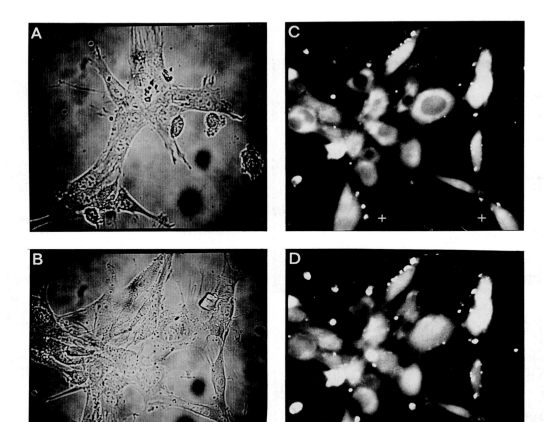

Fig. 1. Swiss 3T3 fibroblasts. **A:** Transmitted light companion image to Plate 2A,B. **B:** transmitted light image of cells loaded with Fura-2 by incubation in growth medium plus 10 μM Fura-2 at 36°C without Pluronic F-127. **C:** Fluorescence image of cells shown in B acquired upon 350 nm excitation, showing prominent bright rings about the nuclei. **D:** Fluorescence image of same cells acquired upon 380 nm excitation, showing relatively uniform intracellular fluorescence. During the nonoptimal loading pro-tocol, a large fraction of the dye has entered endo-somes, which are clustered in the endosomal region around the nucleus. $[Ca^{++}]$ is relatively high in the endosomes, so this dye fluoresces at 350 nm, but not at 380 nm. The presence of noncytoplasmic dye causes apparent spatial gradients of $[Ca^{++}]_i$, with high-$[Ca^{++}]_i$ rings about the nucleus (Plate 2C). The small spots fluorescing brightly at 380 nm (D) but dimly at 350 nm (C) are droplets of Fura-2 AM. Cursor crosses in C are 100 μm apart.

Plate 2. **A,B:** Ratio image of Swiss 3T3 fibroblasts loaded with Fura-2 by incubation in growth medium (DMEM plus 6% fetal calf serum) plus 10 μM Fura-2 and 200 mg/l Pluronic F-127 at 20°C. **A:** Image not thresholded. Ratio values from cell edges are noisy and give incorrect values. Regions of the field outside the cells appear yellow, since this is the hue corresponding to a ratio "value" of 0 ÷ 0. **B:** Thresholded at ⅛ camera saturation. The individual cells can now be clearly distinguished and related to the transmitted light image (Fig. 1A). $[Ca^{++}]_i$ is spatially uniform in individual cells, indicating few problems with noncytosolic dye. **C:** Ratio image of Swiss 3T3 fibroblasts loaded with Fura-2 by incu-bation in growth medium plus 10 μM Fura-2 at 36°C without Pluronic F-127. Measured $[Ca^{++}]_i$ is spa-tially nonuniform in individual cells, which show the characteristic perinuclear high-$[Ca^{++}]_i$ ring due to Fura-2 loading into endosomes (see Fig. 1B–D). Cursor crosses in B and C are 100 μm apart. $[Ca^{++}]_i$ scale in B (μM) applies to A, B, and C. **D, E:** An egg has approximately equal volumes of yolk and white, but when fried the yolk comprises only 15% of the total area. Estimating mean egg calcium concentra-tion by giving equal weight to equal areas of fried egg will result in a weighting error. Figure on p. 187 appears in color in Color Figure Section.

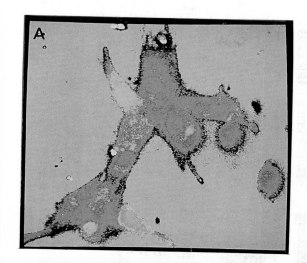

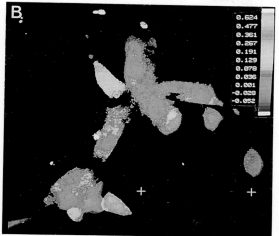

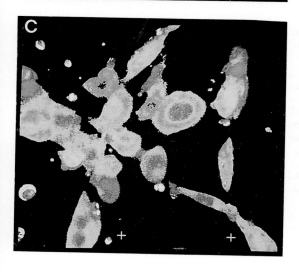

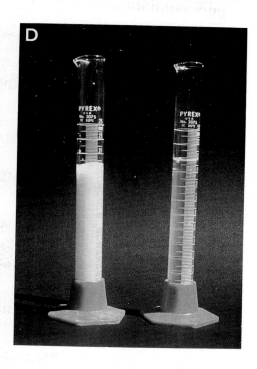

image should be thresholded: Pixels at which the background-corrected fluorescence is lower than a user-defined threshold should be displayed as black or some neutral color. Plate 2B has been thresholded: Pixels where the average background-corrected fluorescence ($\{I_{\lambda 1} + I_{\lambda 2}\}/2$) is less than ⅛ camera saturation are shown as black. The image is both easier to understand and quantitatively more meaningful. Ratio values from pixels at which the background-corrected fluorescence is lower than a user-defined threshold should not be used in any subsequent calculations.

Thresholding can either be done during data acquisition or during postexperiment analysis. The advantage of the latter option is that the appropriate level of thresholding can be decided at leisure. For instance, Plate 1D shows an example where a threshold of ⅛ camera saturation did not entirely remove a "halo" around the cell; only in image E, in which a threshold of ¼ camera saturation was applied, is the diameter of the ratio image equal to the diameter of the cell in the transmitted light image C. The origin of the halo is discussed below (Section IIIA 2). Postexperiment thresholding requires that the stored data contain intensity information, a point that is discussed further in Section VE.

D. Calculating Ion Concentration from Ratio

Figure 2 shows the relationship between the ratio signal (R) and [Ca^{++}] for Fura-2 in our system. The theoretical basis for the smooth line which relates R and [Ca^{++}] is shown in Box 2. While the dissociation constant K_d is an intrinsic property of the dye at any particular pH, ionic strength, and temperature, the parameters $S_{f,\lambda 1}$, $S_{f,\lambda 2}$, $S_{b,\lambda 1}$, $S_{b,\lambda 2}$, R_{min}, R_{max}, and $K_{1/2}$ depend both on the dye spectra and on the intensities of light at the two excitation wavelengths. The excitation light intensities are determined by the light source and the transmission characteristic of the system's opti-

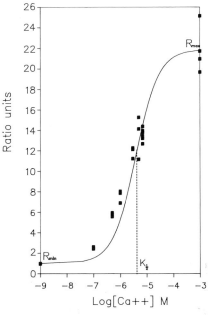

Fig. 2. Relationship between the ratio $R = I_{\lambda 350}/I_{\lambda 380}$ and free calcium concentration at 36° for our system. Squares: Ratio values measured from Ca:EGTA buffers containing 100 μM Fura-2, 19.1 mM EGTA, 20 mM HEPES, and 70 mM KCl, pH 7.0, plus CaCl$_2$ to give free [Ca^{++}] as indicated [calculated as Bartfai, 1979]. Our method of shading correction (see Box 1) causes R_{min} to be normalized to 1. Continuous curve: calibration curve

$$[Ca^{++}] = K_{1/2} \frac{R \times R_{min}}{R_{max} \times R}$$

where $K_{1/2} = 3.9$ μM, $R_{min} = 1$, and $R_{max} = 21.9$.

cal components and will therefore vary from one system to another.

All indicator dyes, including ratio dyes, are intrinsically most sensitive when the concentration of the measured ion is equal to the K_d of the dye, since at that ion concentration a fractional increase of ion concentration will produce the largest increase in the concentration of bound dye and the largest fall in the concentration of free dye. To take full advantage of this sensitivity, the two excitation intensities should be adjusted so that the fluorescence intensity at each excitation wavelength is the same when the ion concentration is equal to the dye K_d.

BOX 2 DERIVATION OF THE CALIBRATION EQUATION FOR A FLUORESCENT RATIO DYE [Grynkiewicz, et al., 1985].

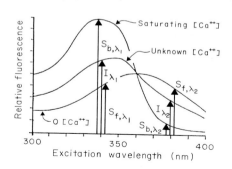

$I_{\lambda 1}$ = total fluorescence intensity upon excitation at λ_1
= $S_{f,\lambda 1} C_f + S_{b,\lambda 1} C_b$

>   Where $S_{f,\lambda 1}$ = intensity of fluorescence of a standard amount of free dye when excited at λ_1
> $S_{b,\lambda 1}$ = intensity of fluorescence of a standard amount of bound dye at λ_1
> C_f = concentration of free dye
> C_b = concentration of bound dye

$I_{\lambda 2}$ = total fluorescence intensity upon excitation at λ_2
= $S_{f,\lambda 2} C_f + S_{b,\lambda 2} C_b$

>   Where $S_{f,\lambda 2}$ = intensity of fluorescence of a standard amount or free dye when excited at λ_2
> $S_{b,\lambda 2}$ = intensity of fluorescence of a standard amount of bound dye at λ_2

but, from law of mass action:

$$C_b = C_f [Ca^{++}] / K_d$$

so $\quad I_{\lambda 1} = S_{f,\lambda 1} C_f + S_{b,\lambda 1} C_f [Ca^{++}] / K_d$

$\quad I_{\lambda 2} = S_{f,\lambda 2} C_f + S_{b,\lambda 2} C_f [Ca^{++}] / K_d$

thus $\quad R = I_{\lambda 1}/I_{\lambda 2} = \dfrac{S_{f,\lambda 1} + S_{b,\lambda 1} [Ca^{++}] / K_d}{S_{f,\lambda 2} + S_{b,\lambda 2} [Ca^{++}] / K_d}$

because the C_f terms cancel.

so $\quad R S_{f,\lambda 2} + R S_{b,\lambda 2} [Ca^{++}] / K_d = S_{f,\lambda 1} + S_{b,\lambda 1} [Ca^{++}] / K_d$

$\quad R S_{f,\lambda 2} - S_{f,\lambda 1} = [Ca^{++}] \{ S_{b,\lambda 1} - R S_{b,\lambda 2} \} / K_d$

$$[Ca^{++}] = K_d \dfrac{R S_{f,\lambda 2} - S_{f,\lambda 1}}{S_{b,\lambda 1} - R S_{b,\lambda 2}}$$

(continues)

BOX 2 *(continued)*

$$= K_d \frac{R - S_{f,\lambda1}/S_{f,\lambda2}}{S_{b,\lambda1}/S_{b,\lambda2} - R} \frac{S_{f,\lambda2}}{S_{b,\lambda2}}$$

$S_{f,\lambda1}/S_{f,\lambda2}$ is the limiting value of R at zero $[Ca^{++}]$, and can be written as R_{min}.

$S_{b,\lambda1}/S_{b,\lambda2}$ is the limiting value of R at saturating $[Ca^{++}]$, and can be written as R_{max}.

Substituting:

$$[Ca^{++}] = K_d \frac{S_{f,\lambda2}}{S_{b,\lambda2}} \frac{R - R_{min}}{R_{max} - R}$$

or:
$$[Ca^{++}] = K_{1/2} \frac{R - R_{min}}{R_{max} - R}$$

where $K_{1/2} = [Ca^{++}]$ for which $R = (R_{min} + R_{max})/2$.

R_{min}, R_{max}, and $K_{1/2}$ are parameters unique to each imaging setup, and depend on the light source and optical system used.

E. Confocal Ratio Imaging

The aim of ratio imaging techniques is to measure the spatial distribution of ion concentrations. Unfortunately, the conventional fluorescence microscope limits the spatial resolution of a ratio imaging system to between 0.7 and 1.4 times cell thickness (see Section IIIA). Similar problems of spatial resolution have been overcome in other fluorescence applications by using confocal fluorescence microscopes, described in chapters later in this volume. The principles and resolution of confocal fluorescence microscopes are discussed by Wilson in Chapter 10, this volume. Confocal ratio imaging offers such an enormous improvement in spatial resolution over conventional fluorescence ratio imaging, at similar cost, that it is likely to supersede conventional imaging for measuring the spatial distribution of ion concentrations across single cells.

Only laser-based confocal microscopes can deliver a sufficient intensity of exciting light to be useful to ratio imaging. Commercially available laser confocal microscopes equipped with visible wavelength argon ion lasers and the facility to measure emission at two wavelengths can in principle be used to measure pH$_i$ using the emission ratio dye SNARF. However, the only visible wavelength calcium-sensitive dyes presently available are not ratiometric. Use of the emission ratio dye Indo-1 (or, indeed, use of Fura-2) requires UV excitation, introducing two problems. First, continuous UV argon ion lasers are cumbersome and expensive. Second, microscope optics cannot be corrected for chromatic aberration in the UV, resulting in differing planes of focus for exciting and emitted light. At the moment, confocal ratio imaging of calcium is possible [and, indeed, is being attempted: see Niggli and Lederer, 1990; Tsien, 1990], but difficult. A visible-excitation ratio emission calcium dye would solve these problems and may soon become available.

III. KNOWN ARTIFACTS AND THEIR CORRECTION

A. Optical Artifacts

1. General properties of fluorescence microscopes. Ratio imaging systems produce two-dimensional images from a three-dimensional object, the dye-filled cell. The image produced by a fluorescence microscope contains light emitted from the whole thickness of the cell. Figure 3A illustrates this point: Pixel P' in the image plane receives light not only from point P in the plane of the cell that is in focus but also from a solid cone of cell of half-angle A. If the cone is regarded as a series of horizontal slices of equal thickness, then each slice makes the same total contribution to the intensity at pixel P', because although the intensity from each volume element decreases as $1/r^2$, the area of each slice increases as r^2 (where r is the axial distance from point P to the slice). This basic property of the fluorescence microscope is responsible for three artifacts or problems: the halo, unexpectedly low horizontal resolution, and the absence of vertical resolution.

2. The halo. The diameter of the ratio image of thick cells often appears larger than the true diameter as seen in transmitted light images. Plate 1 shows a particularly clear example. The white arrows are placed at the edge of the transmitted light image in C; the arrows appear at the same position on the image in A, B, and D, but the ratio image extends considerably further out. Figure 3B demonstrates the origin of the halo. Pixel P', which corresponds to the point P outside the cell, nevertheless receives light emitted from dye within the cell. Since the halo is of relatively low intensity, one can often remove it from the final image by adjusting the threshold, as described above. Plate 1E has been produced by raising the threshold to $\frac{1}{4}$ camera saturation; the halo has disappeared. The indicated calcium concentration in the halo of Plate 1D is different from the indicated calcium concentration of the cell itself. Although complicated optical effects might cause the halo to show ion concentrations that differ from the true intracellular value, a more likely origin of the error seen in Plate 1D is inappropriate background subtraction.

3. Horizontal resolution. Figure 3C demonstrates the problem of horizontal resolution. The microscope is assumed to be focused on the equatorial plane E-E' of a flat cell of thickness x; light from points above and below the equatorial plane is not brought to a focus on the image plane but forms blur circles. Figure 3C is drawn for the worst case, light emitted from the very top portion of the cell. A change of ion concentration at a single point P will cause a change in indicated ion concentration at pixels a, b, and all pixels between. The effective horizontal resolution is therefore d, given by

$$d = x \tan A.$$

(The corresponding formula for a spherical cell is $d = \chi \sin A$, where χ is the cell diameter.) In order to measure the effective collection angle A, we suspended a 0.3 μm-diameter fluorescent bead in standard extracellular fluid plus 20 g/l agar. When we focused up and down with a $100 \times$ Nikon Neofluor, NA 1.3, we observed a diffraction pattern that was contained within a cone of half-angle 32°. The acceptance angle is therefore 32°, a value much less than the value of 77° expected from the numerical aperture. Fay, Fogarty, and Coggins [1986] report a similar narrow acceptance cone for a lens of NA 1.4. The reason for the discrepancy between measured and theoretical acceptance angle is not known. For the situation shown in Figure 3C, an acceptance angle of 32° gives a horizontal resolution equal to 0.53 times the cell thickness. When working on flattened cells or cell processes, it is probably difficult to do better than focus somewhere between the upper and lower surfaces of the cell, so that in the worst case (microscope focused on one surface of cell, ion concentration change at other surface of cell) resolution becomes half as good as in the

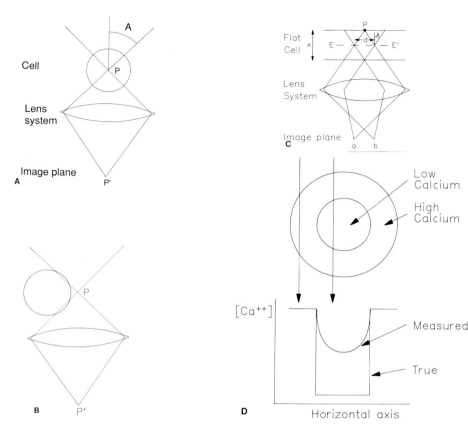

Fig. 3. Illustration of artifacts or problems created by the optical properties of the fluorescence microscope. **A:** Light collection path of a fluorescence microscope. Although the microscope is focused on the equatorial plane of the cell, pixel P' receives light from the entire double cone, half-angle A, with apex at P. **B:** Origin of the halo. Although pixel P' corresponds to point P, outside the cell, it receives light emitted from intracellular dye. **C:** Loss of horizontal resolution. The microscope is focused on the equatorial plane of the cell E——E'. A change of ion concentration at point P causes a change of intensity at pixels a, b, and all pixels in between. **D:** Lack of vertical resolution. The concentric circles represent a cross section through a stimulated skeletal muscle cell in which t-tubule conduction has failed. $[Ca^{++}]_i$ is high close to the cell membrane but low at the cell center. Below is a graph of $[Ca^{++}]_i$ across the cell. Although true $[Ca^{++}]_i$ falls to a low level at the center of the cell, $[Ca^{++}]_i$ measured from the center of the cell does not fall as low, because pixels in this region receive light from out-of-focus regions at high $[Ca^{++}]_i$ above and below the plane of focus. For simplicity, the effect of the horizontal resolution problem (C) has not been considered.

situation of Figure 3C, that is, 1.06 times the cell thickness for the Nikon neofluor. An example of this problem is seen in neuroblastoma growth cones. Calcium ions enter the growth cones at discrete sites where channel density is high, giving rise to "hot spots" at which $[Ca^{++}]_i$ concentration is high [Silver et al., 1990]. In our images, these hot spots have a diameter of 7 μm, close to the predicted horizontal resolution of 6 μm. It is therefore likely that the true size of the hot spots is very much smaller.

4. Vertical resolution. The fluorescence intensity at each pixel of the image contains contributions from the entire thickness of the cell. If there is a vertical gradient of ion concentration, the indicated concentration at any point will be the average value along the vertical axis through that point, rather than the concentration at a unique point. (For the sake of simplicity, this argument ignores the problem of horizontal blurring discussed above.) We encountered this problem when studying $[Ca^{++}]_i$ in skeletal muscle fibers during electrical stimulation [Westerblad et al., 1990]. Figure 3D represents a cross section through a fiber during electrical stimulation in low-sodium solution. Low-sodium solution prevents the action potential propagating down the transverse tubules to excite calcium release in the center of the fiber, so that while $[Ca^{++}]_i$ rises to a high level in the outer regions of the cell, it remains low at the center. The lower graph shows how true $[Ca^{++}]_i$ on the equatorial plane falls to a low level at the fiber center, but measured $[Ca^{++}]_i$ falls relatively little because pixels in the image that correspond to points in the center of the cell receive light emitted from the outer regions of the cell above and below the plane of focus.

Confocal microscopes, which eliminate the majority of out-of-focus fluorescence, are subject to none of the above problems. Ratio imaging on confocal microscopes, when possible, will yield images of ion concentration that correspond to a true optical section through the cell.

5. Movement artifact. Ratio images are generated by illuminating a cell with excitation light of first one, then another, wavelength. Gross movements of the cell during the sequential acquisition of the two raw images used to calculate a ratio image therefore produce a characteristic artifact. At the leading edge of the cell, more dye is present during illumination with the second excitation wavelength than is present during illumination with the first wavelength. Similarly, at the trailing edge of the cell, more dye is present during illumination with the first excitation wavelength than is present during illumination with the second wavelength. The result is that the ratio is distorted downward at one edge and upward at the other. Gross movement artifacts are instantly recognizable and are an annoyance rather

than a danger. More subtle movements, such as the slow movement of cytoplasm into a growth cone, might produce apparent ion concentration gradients that are not recognized as artifact. In such cases, wisdom dictates reversing the order of presentation of the excitation wavelengths. If ion gradients persist with the same orientation, they may be real; if the gradients reverse, they are a movement artifact.

B. Hardware/Software Artifacts

1. Camera nonlinearity. The ratio technique for indicating ion concentration requires that the photometric system be linear. A nonlinear input–output relation will distort the ratio measurement: Thicker regions of the cell will give a ratio value different from that recorded from thin regions. Linearity of our COHU 5000 series silicon intensifier target (SIT) camera and Photonic Science Extended ISIS-M intensified charge-coupled device (CCD) camera are excellent. Tsien and Harootunian [1990] describe how to ensure linearity of the DAGE-MTI model 66 SIT camera. Cooled CCD cameras, which are photon-counting devices, should be perfectly linear.

2. Weighting error. This is a trivial artifact, but one that appears to have been overlooked in the design of most imaging systems. For didactic purposes, we illustrate it with the egg in Plate 2D,E, the shape of which approximates well to the shape of a flattened cultured fibroblast. An egg comprises approximately equal volumes of white, of free $[Ca^{++}]$ (say) 2 mM, and yolk, of free $[Ca^{++}]$ (say) 0 mM. The mean $[Ca^{++}]$ in an egg is therefore 1 mM. However, the area of yolk in the image of the fried egg comprises only 15% of the total area. Any system that gave equal weight to equal area in the image would report a mean free $[Ca^{++}]$ of 1.7 mM. The correct value, 1 mM, is obtained by weighting the $[Ca^{++}]$ value from each pixel of the image by the thickness of the fried egg at that pixel. In measurements of actual dye-filled cells, an easy estimate of the thickness of cytoplasm at each pixel is given by the intensity of the background-corrected fluorescence at that pixel. This approach assumes that the dye is present at equal concentration throughout the cytosol, an assumption that appears to be true for Fluo-3 in neurons [Hernandez-Cruz et al., 1990]. During calculation of mean ion concentration for a defined area, our system therefore weights the contribution of each pixel by the mean fluorescence ($\{I_{\lambda 1} + I_{\lambda 2}\}/2$) at that pixel. A further advantage of this weighting is that thin regions of the cell, where the ratio value is prone to large errors from small errors in background subtraction, contribute relatively little to any calculation of mean cell ion concentration. Our system is able to calculate weighted averages because the stored ratio image contains information about the fluorescence intensity at each pixel (see Section VE).

C. Dye Artifact

Dye artifact can introduce two types of error into the final ratio image. *Spatially homogeneous* errors distort the ratio by an equal amount over the whole image. Spatially homogeneous errors are not usually dangerous, since they will not lead to any very important biological misinterpretation. Furthermore, spatially homogeneous errors can be eliminated by relatively simple procedures, as described in Section IIIC2 below. *Spatially inhomogeneous* errors can create artifactual intracellular ion gradients, or apparent differences in mean intracellular ion concentration in two types of cell. Spatially inhomogeneous errors are therefore potentially much more serious.

1. Protein error. Proteins can bind to many indicators, changing some or all of their properties [Clark, 1928; Lakowicz, 1983]. Konishi et al. [1988] found that the spectrum of Fura-2 was altered by certain soluble cytoplasmic proteins at concentrations that are likely to be found in cells. As long as the proteins are truly cytoplasmic, and therefore at equal concentration in all

regions of the cell, errors produced by these proteins will be spatially homogeneous, except, perhaps, in the region of the nucleus. Binding of dye to a cytoskeletal or membrane protein that is restricted to one region of the cell, thus creating an apparent ion concentration gradient, would be a potentially more serious problem. Such an error has been reported for the calcium indicator dye Antipyrylazo III, which changes its optical properties upon binding to myofibrils [Baylor et al., 1986] but not, as yet, for any of the dyes used in ratio imaging.

2. Viscosity effect. Poenie [1990] has reported that the excitation spectrum of Fura-2 is affected by the viscosity of the medium. When compared to the spectrum measured *in vitro* in solutions of low viscosity, spectra measured in solutions of high viscosity show enhanced fluorescence when excited at long wavelengths. If a calibration based on low-viscosity *in vitro* measurements is applied to intracellular measurements, then the measured $[Ca^{++}]_i$ will be lower than the true $[Ca^{++}]_i$. The fluorescence of the Ca:free and Ca:bound forms of the dye is affected equally, while the K_d of calcium binding is unaffected. These results imply that the effect of viscosity is on the fluorophore alone, not on the part of the Fura-2 molecule that binds calcium, and make possible a relatively easy mathematical correction that can be applied to the *in vivo* data to produce valid $[Ca^{++}]_i$ values: The *in vivo* ratio values are simply divided by a viscosity correction factor V (typically 0.85 for 340/380 nm ratioing) (see Box 3, part A).

Although the theoretical basis of this correction is the viscosity effect, it can in fact be used to correct for a variety of spatially homogeneous errors, such as a protein error or absorption of one or other excitation wavelength by intracellular or extracellular components. In the face of these problems, three options are available. First, do nothing: Simply apply the *in vitro* calibration, since any errors that can be corrected for by this procedure are spatially homogeneous and therefore not dangerous. Second, calibrate the dye in the intracellular environment by artificially setting the cytoplasmic ion concentration to a range of levels [Almers and Neher, 1985]. Third, use the *in vitro* calibration but divide the measured ratio values by the viscosity correction factor V. Many papers simply choose a value of V that gives reasonable ion concentrations! This somewhat questionable approach has been used in Plate 1; no other figures in this chapter use viscosity correction. Poenie gives a method for directly measuring V, based on a comparison of the increase of fluorescence at the short excitation wavelength and the fall of fluorescence at the long excitation wavelength during an increase of $[Ca^{++}]_i$ (Box 3, part B).

3. **Noncytoplasmic dye.** Although exceptions may exist, most of us are interested in cytoplasmic (and nucleoplasmic) ion concentrations. Any dye that is present inside organelles will indicate a concentration that is different from the cytoplasmic value; spatial gradients of organelle density within cells will produce spatially inhomogeneous errors. The presence of noncytoplasmic dye can often be detected by careful use of digitonin to permeabilize only the cell's plasma membrane; dye that remains is noncytosolic [Poenie et al., 1987]. Each cell type appears to produce its own brand of noncytoplasmic dye error, so that the following review is unlikely to cover every problem that will be encountered. The papers in *Cell Calcium*, Vol. 11, No. 2/3, 1990, contain much useful discussion of these problems in many cell types.

a. *Loading problems.* Indicators are often "AM-loaded" into cells by bathing the cells in a suspension of the acetoxymethyl ester of the indicator [Tsien, 1981]. The acetoxymethyl ester can pass through lipid membranes, and therefore in principle can enter each and every intracellular compartment before being hydrolyzed to the membrane-impermeant, ion-indicating form. However, depending on the cell and the conditions, dye can accumulate in secretory granules [Almers and Neher, 1985], sarcoplasmic reticulum [Williams et al., 1985], or mito-

chondria [Roe et al., 1990]. One of the commonest manifestations of this artifact is the appearance of a perinuclear high-$[Ca^{++}]_i$ (or low-pH) ring, visible in many of the 3T3 cells of Plate 2C. Malgaroli et al. [1985] have traced this artifact to endocytosis of acetoxymethyl ester micelles. Hydrolysis by lyososmal enzymes produces the membrane-impermeable, ion-indicating form, which then indicates the high-$[Ca^{++}]$, low-pH environment of the endosome. Endosomes localize in a perinuclear region close to the Golgi body [Marsh et al., 1986], creating the characteristic perinuclear ring. To avoid this artifact, the acetoxymethyl ester should be dispersed using a nonionic detergent such as Pluronic F-127 (a detergent specifically designed to be nontoxic to cells; see Molecular Probes data sheets), and loading should be carried out at low temperatures (e.g., room temperature for a mammalian cell). Plate 2B shows a field of 3T3 cells loaded at room temperature in the presence of 200 mg/l Pluronic F-127. No perinuclear high-$[Ca^{++}]_i$ rings are visible, and mean indicated $[Ca^{++}]_i$ is much lower since the signal from Fura-2 in the high-$[Ca^{++}]_i$ endosomes has been eliminated (or at least reduced).

Incomplete hydrolysis of acetoxymethyl esters can produce fluorophores that are trapped within the cell, but which are not ion-sensitive. Most users incubate loaded cells for at least 90 min at 36°C after AM loading in the hope of achieving full hydrolysis. Errors due to incomplete hydrolysis of dye will usually be spatially homogeneous, and therefore nondangerous. Serious problems might however arise if one type of cell in a mixed population hydrolyzed acetoxymethyl esters more rapidly than a second type of cell in the same field of view.

b. *Compartmentalization.* Brownlee and Pulsford [1989] have shown that cytosolic Fura-2 and Indo-1 are taken up avidly by the vacuole of plant cells, and suggested that this uptake occurred through an anion transport system. Di Virgilio et al. [1990] have shown a similar uptake, inhibited by the

BOX 3   VISCOSITY CORRECTION

A. Theoretical Basis for the Correction Factor V

Peonie [1990] has found that increased viscosity changes the fluorescence of Ca:free and Ca:bound Fura-2 to the same extent. K_d is unaffected. Thus, the calculations in Box 2 must be adapted as follows:

$I_{\lambda 1}$ = total intracellular fluorescence intensity upon excitation at λ_1
$= A\, S_{f,\lambda 1}\, C_f + A\, S_{b,\lambda 1}\, C_b$

> Where $S_{f,\lambda 1}$ = fluorescence efficiency of free dye at λ_1
> $S_{b,\lambda 1}$ = fluorescence efficiency of bound dye at λ_1
> C_f = concentration of free dye
> C_b = concentration of bound dye
> A = viscosity effect upon fluorescence at λ_1

$I_{\lambda 2}$ = total intracellular fluorescence intensity upon excitation at λ_2
$= B\, S_{f,\lambda 2}\, C_f + B\, S_{b,\lambda 2}\, C_b$

> Where $S_{f,\lambda 2}$ = fluorescence efficiency of free dye at λ_2
> $S_{b,\lambda 2}$ = fluorescence efficiency of bound dye at λ_2
> B = viscosity effect upon fluorescence at λ_2

but C_b $= C_f [Ca^{++}] / K_b$

so $I_{\lambda 1}$ $= A\, S_{f,\lambda 1}\, C_f + A\, S_{b,\lambda 1}\, C_f [Ca^{++}] / K_d$

$I_{\lambda 2}$ $= B\, S_{f,\lambda 2}\, C_f + B\, S_{b,\lambda 2}\, C_f [Ca^{++}] / K_d$

thus R $= I_{\lambda 1}/I_{\lambda 2}$ $= \dfrac{A\, S_{f,\lambda 1} + A\, S_{b,\lambda 1}\, [Ca^{++}] / K_d}{B\, S_{f,\lambda 2} + B\, S_{b,\lambda 2}\, [Ca^{++}] / K_d}$

because the C_f terms cancel.

so $B R\, S_{f,\lambda 2} + B R\, S_{b,\lambda 2}\, [Ca^{++}] / K_d = A\, S_{f,\lambda 1} + A\, S_{b,\lambda 1}\, [Ca^{++}] / K_d$

$B R\, S_{f,\lambda 2} - A\, S_{f,\lambda 1} = [Ca^{++}] \{ A\, S_{b,\lambda 1} - B R\, S_{b,\lambda 2} \} / K_d$

$[Ca^{++}]$ $= K_d\, \dfrac{B R\, S_{f,\lambda 2} - A\, S_{f,\lambda 1}}{A\, S_{b,\lambda 1} - B R\, S_{b,\lambda 2}}$

$= K_d\, \dfrac{B R - A\, S_{f,\lambda 1}/S_{f,\lambda 2}}{A\, S_{b,\lambda 1}/S_{b,\lambda 2} - B R}\, \dfrac{S_{f,\lambda 2}}{S_{b,\lambda 2}}$

Dividing top and bottom by A:

$[Ca^{++}]_i$ $= K_d\, \dfrac{R/V - S_{f,\lambda 1}/S_{f,\lambda 2}}{S_{b,\lambda 1}/S_{b,\lambda 2} - R/V}\, \dfrac{S_{f,\lambda 2}}{S_{b,\lambda 2}}$

> Where $V = A/B$ = viscosity correction factor

(continues)

BOX 3 (continued)

Written as:

$$[Ca^{++}] = K_d \frac{S_{f,\lambda 2}}{S_{b,\lambda 2}} \frac{R/V - R_{min}}{R_{max} - R/V}$$

or:

$$[Ca^{++}] = K_{1/2} \frac{R/V - R_{min}}{R_{max} - R/V}$$

Where
R_{min} = R value recorded from a solution at zero $[Ca^{++}]$
R_{max} = R value recorded from a solution saturated with calcium
$K_{1/2}$ = $[Ca^{++}]$ for which $R = (R_{min} + R_{max})/2$.

R_{min}, R_{max}, and $K_{1/2}$ are parameters unique to each imaging setup, and depend on the light source and optical system used.

B. Direct Measurement of V [Poenie, 1990]

The method requires that you be able to induce a $[Ca^{++}]_i$ change in the cytosol under conditions in which total intracellular dye concentration does not change. You do not need to know the value of the $[Ca^{++}]_i$ change. The fluorescence at each excitation wavelength will change, by $\Delta I_{\lambda 1}$ and $\Delta I_{\lambda 2}$, respectively. Call the ratio $\Delta I_{\lambda 1}/\Delta I_{\lambda 2}$ measured in this intracellular environment Z_i. The process is now repeated in vitro, under the conditions in which you measure the in vitro R_{max} and R_{min}. One approach would simply be to add calcium to your Ca^{++}-free standard. Call the ratio $\Delta I_{\lambda 1}/\Delta I_{\lambda 2}$ measured in this extracellular environment Z_e. Now V is simply Z_i/Z_e.

---

anion transport inhibitors probenecid and sulfinpyrazone, into macrophage endosomes, and suggest that routine use of these inhibitors may help remove this particular artifact.

c. Release of dye into the extracellular medium. The cell membrane is not completely impermeable to dyes such as Fura-2 and BCECF [Di Virgilio et al., 1990]. Cytoplasmic dye is therefore slowly lost to the extracellular medium. This can be a serious problem in cuvette measurements, where signal from extracellular dye contaminates the cell signal, but is not a problem in imaging systems. Almers and Neher [1985] have investigated a potentially much more serious artifact in AM-loaded cells. Much

of the Fura-2 in AM-loaded mast cells is located within secretory vesicles, and is released into the extracellular fluid upon stimulation. In an imaging system, such release would give the appearance of a transient high-$[Ca^{++}]$ ring around the periphery of the cell, and might well be misinterpreted as a submembrane rise of $[Ca^{++}]_i$ due to calcium influx through the membrane. In this, as in other situations of apparent $[Ca^{++}]_i$ changes upon stimulation, the critical test is to examine the raw fluorescence data. A true $[Ca^{++}]_i$ increase will cause an increase of whole-cell fluorescence at excitation wavelengths <360nm, while loss of Fura-2 from the cell on stimulation will, once the released dye has diffused away, reduce whole-cell fluorescence at all wavelengths.

4. Buffering of ion concentrations by the indicator. All indicators that bind ions will necessarily affect the amplitude or time course of concentration changes. The distortion will be significant if the additional buffering capacity due to the indicator is comparable to the intrinsic buffering capacity of the cytosol. A simple measure of buffering capacity is the ratio of bound to free ion. For instance, when the intracellular concentration of Fura-2 is 100 μM and free $[Ca^{++}]_i$ is 1 μM, 88 calcium ions are bound to Fura-2 for every one ion free in solution (at 20°C). Unfortunately, estimates of the intrinsic calcium-buffering capacity of cytosol vary widely, from 1:40 [Ahmed and Connor, 1988] to 1:4,000 [Bolsover, 1986]. There is therefore no real alternative to checking that a known $[Ca^{++}]_i$-dependent physiological process is not significantly affected by the presence of the dye. Using this criterion, intracellular Fura-2 concentrations of 25, 30 to 100, and 130 μM do not significantly affect, respectively, neutrophil movement, smooth muscle excitation–contraction coupling, and neurotransmitter exocytosis [Marks and Maxfield, 1990b; Yagi et al., 1988; Brethes et al., 1987]. However, the calcium buffer BAPTA at intracellular concentration of 5 mM *does* block neutrophil movement [Marks and Maxfield, 1990b]. Fast $[Ca^{++}]_i$-dependent processes can be affected by concentrations of Fura-2 as low as 10 μM [Timmerman and Ashley, 1986].

IV. BIOLOGICAL APPLICATIONS OF RATIO IMAGING

This brief review is not intended to be exhaustive, but simply to indicate some of the significant advances and ongoing debate. The first two ratio imaging papers appeared in 1985. Sawyer et al. used Quin-2 to study $[Ca^{++}]_i$ in neutrophils, while Roger Tsien and co-workers used Fura-2 to study $[Ca^{++}]_i$ in smooth muscle cells [Williams et al., 1985]. Each of these papers opened a chapter in

the study of $[Ca^{++}]_i$ and motility in non-muscle and muscle cells, respectively. Relatively few papers on pH_i imaging have appeared. The first was by Roger Tsien's group, who used BCECF to examine the different mechanisms of pH_i regulation in two cell types in intact gastric glands [Paradiso et al., 1987].

Sawyer et al.'s [1985] paper opened a question that is still in debate: Is there a role for $[Ca^{++}]_i$ gradients in nonmuscle motility, and is $[Ca^{++}]_i$ higher or lower than average at the leading edge of motile cells? Sawyer et al. saw no $[Ca^{++}]_i$ gradients in motile cells in the absence of targets, but neutrophils that had oriented to and were approaching a chemotactic target showed raised $[Ca^{++}]_i$ at the leading edge. Poenie et al. [1987], studying another motile leukocyte, the killer T cell, again found no $[Ca^{++}]_i$ gradients in the absence of targets. However, in contrast to Sawyer et al.'s finding, cells that had oriented to a target showed raised $[Ca^{++}]_i$ at the trailing edge. This study also demonstrated the advantages of ratio imaging for whole-cell $[Ca^{++}]_i$ measurement, since $[Ca^{++}]_i$ could be simultaneously but independently measured in cells known to be killer T and cells known to be targets.

The growth cone, the motile growing tip of the nerve cell axon, has been the subject of a series of imaging papers. In 1986, Connor introduced cooled CCD camera-based $[Ca^{++}]_i$ imaging, and reported that $[Ca^{++}]_i$ was high at the growth cone tip, even in the absence of stimuli. In a series of papers, Connor, Kater, and co-workers developed the model that a rise of $[Ca^{++}]_i$ from a basal level first promotes, and then inhibits, axonal growth [Cohan et al., 1987; Kater et al., 1988]. We have recently disputed some aspects of this model; in particular, the high-$[Ca^{++}]_i$ region at the tip of the growth cone seems to be an artifact of AM loading [Silver et al., 1989]. When Fura-2 is directly injected into neuroblastoma cells, $[Ca^{++}]_i$ is remarkably uniform in unstimulated growth cones (Plate 3E). However, images of Fura-2 AM-loaded growth cones often show apparent high-$[Ca^{++}]_i$ regions

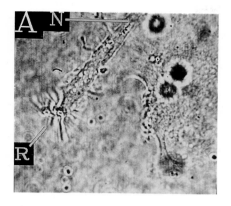

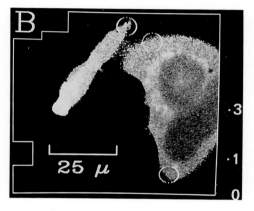

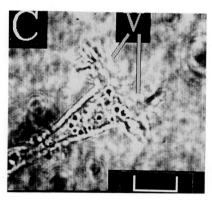

Fig. 4. N1E-115 mouse neuroblastoma growth cones. See legend to Plate 3 for details. Bar in **C** is 10 μm. N: neurite; R: ruffle; V: extending veil at leading edge of growth cone.

at the tip (Plate 3D). This is likely to be due to endocytosis of dye micelles: The tip of the growth cone is a site of active endocytosis [Bunge, 1977]. Growth cones of neuroblastoma cells, therefore, appear to fit the general pattern seen in leukocytes: no $[Ca^{++}]_i$ gradient in the absence of stimuli. Connor has also studied mature nerve cells. In particular, two extremely elegant papers have described measurements of $[Ca^{++}]_i$ in single nerve cells *in situ* within brain slices [Tank et al., 1988; Regehr et al., 1989]. The outgrowth of a *Fucus* rhizoid from the fertilized egg is superficially similar to axon outgrowth. Recently Brownlee and Pulsford [1989] have overcome the considerable problems involved in intracellular measurements on plant cells to show a clear, steady $[Ca^{++}]_i$ gradient in the absence of stimuli, with the highest $[Ca^{++}]_i$ at the advancing tip.

Muscle cells would be expected to show homogeneous $[Ca^{++}]_i$, even during stimuation, in order that the contractile apparatus be uniformly activated. Williams et al.'s [1985] paper confirmed that cytosolic $[Ca^{++}]_i$ was uniform, but raised the unexpected and exciting possibility that intranuclear $[Ca^{++}]_i$ was regulated independently of cytosolic $[Ca^{++}]_i$. Two more recent papers, on calcium-overloaded cardiac cells and fatigued skeletal muscle cells, have shown that cytosolic $[Ca^{++}]_i$ gradients can appear under conditions of physiological stress [Weir et al., 1987; Westerblad et al., 1990].

Many cells respond to steady levels of agonist with a series of discrete $[Ca^{++}]_i$ oscillations, "spikes," or spatial waves. Although such oscillations can be studied using non-imaging methods, ratio imaging has been instrumental in, for instance, demonstrating unequivocally that oscillations in neighboring cells are not in phase [Wilson et al., 1987]. Study of the spatial distribution of $[Ca^{++}]_i$ during oscillations requires high time resolution. Roger Moreton and Michael Berridge have pushed the time resolution of excitation ratio imaging to the limit imposed by video frame rate, taking one ratio image every 2 video frames (i.e., every 80 msec, European standard). They have shown that, in chromaffin cells, activation of cell membrane calcium channels is more effective at raising $[Ca^{++}]_i$ close to the membrane and therefore triggering transmitter release than is calcium release from intracellular stores [Cheek et al., 1989]. Takamatsu and Weir [1990] have achieved a rate of one ratio image every half video frame (17 msec, North American standard). They achieve this by emission ratioing, using two cameras to look at the two emission wavelengths, and by extracting individual half-frames (fields) from the interlaced video signal. A major advantage of Takamatsu and Weir's technique is that it requires no moving parts, making it the system of choice for electrophysiological applications.

High time resolution imaging has revealed waves of high $[Ca^{++}]_i$ propagating at high speed across cells. Such waves have been observed, for instance, in calcium-loaded cardiac cells [Weir et al., 1987] and endothelial cells [Jacob, 1990]. These results should make one cautious about the interpretation of single time-point calcium images acquired from cells that show $[Ca^{++}]_i$ oscillations, particularly if the integration time used to acquire such images is significantly longer than 100 msec.

V. HARDWARE CONSIDERATIONS

A. General Points

Figure 5 summarizes the layout of an imaging system. Images are acquired by the camera and passed to the image processor, where they are digitized. After the manipulations described in Box 1, the ratio image is displayed on a monitor. A host computer tells the image processor what to do. The right-hand column shows various storage options. The image processor itself may have memory available for image storage. Analogue videotape or videodisk can be used to store either the fluorescence images or

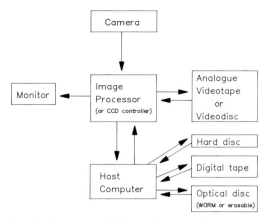

Fig. 5. Image acquisition and processing components of a generalized ratio imaging system.

the computed ratio image. The host computer can store the data on hard disk, tape, or optical disk. Most imaging systems work on images containing 512×512 pixels, each an 8-bit number, so that each image contains 0.25 Mbyte of information. Analyzing, archiving, and retrieving this sort of experimental data once it is acquired can be very time-consuming for the operator. The first question to ask when thinking of using a ratio imaging system is whether a simpler alternative, for example, single-cell photometry, might be a better way of approaching the question.

The trend in imaging systems is toward higher time resolution. However, the problems that arise in trying to achieve high time resolution are not merely financial. In order to calculate accurate ratios from cells in shorter and shorter time periods, the light emitted from the cells must be correspondingly brighter to achieve the same signal/noise performance. This can only be achieved by either injecting more dye, and therefore running the risk of buffering the ion one wishes to measure, or increasing the excitation beam intensity, and running the risk of both bleaching the dye and damaging the cell.

The practical limit to time resolution is often set by the data storage capacity of the system. Only in systems in which analogue video images are stored on videotape or videodisk can all the data from experiments lasting more than a few seconds be collected. The next possibility is storage in the image processor. Thirty-two Mbyte of RAM could store all the ratio images produced in a 10 sec period or could, for instance, store one image every 5 sec during an experiment that lasted 10 min. If the host computer is responsible for even short-term storage, then time resolution is limited by the time taken to send 0.25 Mbyte from the image processor to the host, a time between 1 and 10 sec in most systems. Even when the data has arrived at the host, storage capacity is by no means limitless. If one wishes to store one image every 5 sec then one would fill a 40-Mbyte hard disk in 11 min. It may thus be advantageous to store information only from a small area of interest, rather than the whole field, as discussed in another context by De Brabander et al. (Chapter 6, this volume).

B. Cameras

With cameras, more money buys more sensitivity, but not necessarily more usable signal. Two physical constraints must be added to the financial one. *First*, there is no point in using a camera more sensitive than that needed to visualize the autofluorescence of the cells being studied. For ratio imaging results to be meaningful, the signal from the intracellular dye must be considerably larger than the autofluorescence, so if the camera can detect autofluorescence it will certainly detect the dye signal. If the purpose of the imaging system is to measure ion concentrations in cell bodies, an SIT camera will be adequate [Tsien and Harootunian, 1990]. However, imaging of thinner cell regions may require more sensitivity. *Second*, photon shot noise represents an absolute limit on how dim the fluorescent object can be. For example, the Photonic Science extended ISIS-M camera will at maximum sensitivity saturate when four photons hit

a particular pixel in one video frame time. If we wish to estimate the light intensity at that pixel to within 1%, we must wait until at least 10,000 photons are incident (since error = square root of 10,000 = 100), that is, we must wait for 2,500 video frames or 1 min 40 sec (in Europe). Most of us would, in such circumstances, choose to reduce camera sensitivity and increase the light intensity. Three types of camera have sufficient sensitivity: SIT cameras, cooled CCD cameras, and intensified CCD cameras.

1. SIT cameras. In a silicon intensifier target (SIT) camera, the incoming image falls on a photocathode, liberating electrons that are then accelerated by a high voltage toward a silicon target. Each electron hitting the target generates many electron–hole pairs, which are in turn read out by a scanning electron beam. SIT cameras are relatively cheap but show lag, responding to a change of light intensity with a time constant of 100 to 200 msec. They also suffer from limited dynamic range, geometric distortion, substantial internal electronic noise, and susceptibility to damage from exposure to high light levels. Nevertheless, SIT cameras remain the sensible choice for many applications.

2. Cooled CCD cameras. In a charge-coupled device (CCD) camera, the incoming image falls on an array of silicon photodiodes. The image is not read off by a scanning beam. Rather, an external signal causes the charge built up on the photodiodes to jump serially from diode to diode until it reaches an amplifier that reads out the charge. Quantum efficiency is high: maximally 70%, and typically 45% at 600–700 nm. In a cooled CCD camera, thermal noise is reduced by cooling the array to below $-45°C$, resulting in a true photon-counting device. At the low temperature of operation, charge cannot be made to jump from photodiode to photodiode quickly enough to output the whole image at video frame rate. Cooled CCD cameras are therefore slow-scan devices, interfaced to control systems that permit more rapid readout from particular image regions if required, allowing a trade-off between spatial and temporal resolution. Cooled CCD cameras suffer from none of the disadvantages of SIT cameras listed above. One of the most productive ratio imaging systems, in the laboratory of John Connor (Roche Institute), uses a cooled CCD system.

Cooled CCD cameras thus offer the ability to image whole fields at low time resolution, then collect data from restricted regions of that field at a higher time resolution. A team headed by Jim Gillespie of the University of Newcastle, UK, is using this facility to create, with the intention of marketing, a low-priced (<£20,000) system that will function as an off-line ratio imager (see Section VC) but give a real-time readout of mean ion concentration in each of the many cells in a field of view. Many laboratories, concerned with the response of cells to agonists, drugs, or toxins, would find such a system as useful as more expensive imagers.

3. Intensified CCD cameras. At room temperature a CCD camera can output the entire image at video rate. The simplicity and ruggedness of the room-temperature CCD camera has led to its becoming the standard cheap video camera. At room temperature, however, the CCD camera is not a photon-counting device; since the amplitude of thermal noise is much greater than at $-45°$, the image amplitude must be equivalently brighter.

An intensified camera is simply one in which the incoming light is brightened before being imaged on the camera. Although any type of camera could in principle be used (intensified SIT cameras, for instance, are available) cheap room-temperature CCD cameras are the norm. Most intensified CCD cameras use a single microchannel plate intensifier; the Extended ISIS from Photonic Science uses two intensifiers in series and can be used as a photon-counting device. Unlike SIT cameras, intensified CCDs show no lag and respond instantly to a step change in light intensity. A second advantage is dynamic range: For instance, the sensitivity of the Extended ISIS from Photonic Science can be adjusted over a 1,000-fold range, compared with the 40-fold adjustment possible on the COHU 5000 series SIT camera. The same camera can therefore be used for routine work on thick cell bodies and demanding applications such as measurement of $[Ca^{++}]_i$ in thin cell processes. Criteria for choosing an intensified CCD are that its sensitivity peak is in the visible, not in the infrared, and that its sensitivity can be adjusted to be sufficiently *low* to just detect autofluorescence in the thickest cells studied. Because an intensified CCD camera has a higher time resolution, photon shot noise is more apparent, and one must be careful to avoid signal clipping: That is, the gain of the camera will often need to be adjusted so that the mean signal is very much less than camera saturation, to avoid camera saturation during the peaks of the noisy signal.

C. Image Processing Hardware

Ratio imaging systems can be graded into three performance bands. *Off-line* systems cannot generate a ratio image rapidly, so that one is limited to storing fluorescence images during the experiment and generating the ratio images in a low-priced computer after the experiment is complete. This approach is cheap and can be very effective. It has been used, for instance, to image $[Ca^{++}]_i$ in cardiac myocytes [Weir et al., 1987]. *Blinking* systems use image processors such as the Gould FD5000 that can perform only one arithmetical operation in one video frame time. Generation of a ratio image requires at least three operations, and there is therefore a perceptible delay of ~120 msec, a blink, between acquisition of the fluorescence images and display of the ratio image. This is adequate for all but the most rapidly changing signals. *Real-time* systems use pipelined processors such as the Synoptics Image. The speed of these processors allows one ratio image to be generated from a pair of fluorescence images while the next pair of fluorescence images is being read in from the camera. Ratio images are therefore updated every second video frame (for an excitation ratioing system). Real-time imaging will be required, for instance, to follow spatial gradients of $[Ca^{++}]_i$ during single action potentials. For most purposes, however, such high time resolution is a handicap, since ratio images calculated from two single video frames are very noisy. For instance, O'Sullivan et al. [1989] reduced the time resolution of their real-time imager to 200 msec to average out some of this noise. In order to take full advantage of the capabilities of a fast processor, real-time systems chop between the two excitation wavelengths at video rate. A CCD camera, which responds instantly to a change of intensity, is therefore essential.

D. Data Transfer and Storage

The sheer volume of data created by ratio imaging systems creates problems of data handling. Figure 5 summarizes the possibilities. One solution is to store data in an analogue form, before it reaches the host computer. Only this approach will allow storage of all, rather than selected portions of an experiment. Roger Tsien uses a video-disk to store a pair of averaged fluorescence images. After the experiment, these images are read back into the image processor and divided to produce the ratio image. This approach has the enormous advantage that it is the raw data that is being stored, before any mathematical processing other than averaging. Drs. Berridge and Moreton use videotape to store the ratio image as shades of gray in an analogue picture. This image is read back into the image processor for further analysis. Both systems record a gray-scale image beside each stored image so that any nonlinearities in the analogue storage device can be automatically compensated for.

The alternative is to store data in digital form. An image can either be stored in spare memory in the image processor or downloaded to the host computer. For maximal time resolution, images must be stored in the image processor, since this can occur at video rates. Most processors, such as the Gould FD5000 in our lab, have enough memory for tens of images. Once the image processor memory is full, images must be downloaded to the host computer. This takes 7 sec per image and in the majority of cases is the process that limits time resolution in our system. The Applied Imaging system, in contrast, can be supplied with up to 32 Mbyte of memory, enough for up to 128 images at 512×512 resolution, giving an enormous advantage in terms of time resolution.

Permanent storage of digital images is performed by the host computer. There are two possibilities: storage of the two fluorescence images, or storage of the calculated ratio image. As in all scientific work, there are enormous advantages in storing the raw data. For instance, one can check later that dye is not lost from the cell during stimulation (see Section IIIC3). Having the two fluorescence images makes postexperiment analysis much easier; for instance, one can measure the background in each of a pair of images before calculating the ratio image, and one can adjust the threshold to the most appropriate level. However, storing the two fluorescence images requires twice as much space on the final storage medium.

The final medium can be a hard disk, digital tape, or an optical disk. Internal hard disks are for temporary storage only; removable hard disks, although a possible storage medium, are probably not cost-effective. We at present use VAX TK50 magnetic tape, which is unacceptably slow, and are now upgrading to an optical disk system that will write a single ratio image in ten sec. Retrieval of images from permanent media should be simple and fast; again, the VAX TK50 tapes are unacceptably slow. Many tape drives for MS-DOS machines are much faster. An optical disk is the best solution, giving retrieval of any one image from a database of thousands in 10 sec.

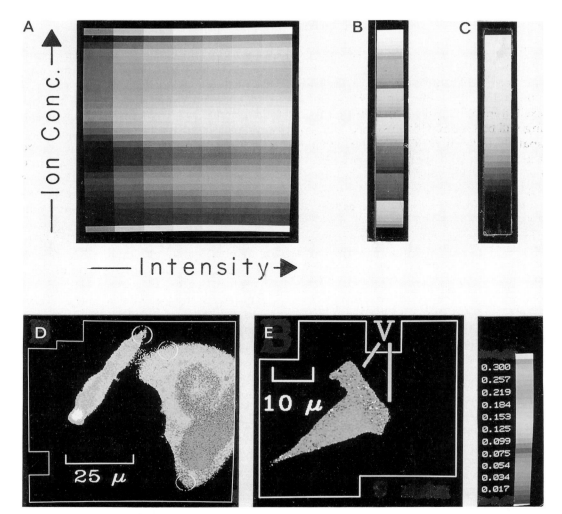

Plate 3. A–C: Alternative pseudocolor scales for display of ratio images. **A:** Five-bit hue, 3-bit luminance scale used in our system. Higher concentrations of measured ion are indicated by warmer hues, while the intensity of the fluorescence signal is indicated by the luminance (brightness) of the display. In this plate, hue and luminance interfere, so that, for instance, an ion concentration that appears as green at low intensity appears as yellow at high intensity. This is an artifact of color saturation of the photographic film and does not occur on the monitor display. For this and other reasons, we usually turn the intensity coding off when taking pictures. **B:** A type of scale in which the single parameter—ion concentration—is coded for by both hue and luminance. This type of scale is used in the Spex system but seems inappropriate for a smoothly varying parameter such as ion concentration. **C:** Five-bit (32-shade) gray scale, illustrating the difficulty of recording subtle gray-level changes in monochrome photographic images. **D,E:** N1E-115 mouse neuroblastoma cells. $[Ca^{++}]_i$ scale at right (μM) applies to both images. **D:** loaded by incubation in growth medium (DMEM plus 6% fetal calf serum) plus 20 μM Fura-2 at 36°C without Pluronic F-127. The cell body on the right-hand side of the image shows a clear high-$[Ca^{++}]_i$ ring around the nucleus, indicating that Fura-2 has loaded into endosomes. The neurite, which enters the field at the top center and runs diagonally left to terminate in a growth cone, shows a pronounced spatial gradient of $[Ca^{++}]_i$. Figure 4A shows a transmitted light image of this field. Figure 4B is the same ratio image as Plate 3D but using a gray scale to indicate $[Ca^{++}]_i$. The circles in Figure 4B and Plate 3D indicate three regions; $[Ca^{++}]_i$ is approximately the same in the top two but is lower in the bottom circle. However, the relative $[Ca^{++}]_i$ in the three regions is very difficult to assess from the gray-scale image. **E:** Growth cone of N1E-115 cell injected with Fura-2. There are no spatial gradients of $[Ca^{++}]_i$ and mean $[Ca^{++}]_i$ is much lower since only cytoplasmic $[Ca^{++}]$ is being indicated. Figure 4C shows a transmitted light image of this growth cone. Figure appears in color in Color Figure Section.

E. Displaying Ratio Images

Roger Tsien makes a good case for displaying ratio images on a pseudocolor scale, on the grounds that this scale best suits the contrast discrimination of the human visual system [Tsien and Harootunian, 1990]. A gray scale is a much inferior alternative, as a comparison of Figure 4B and Plate 3D demonstrates. Pseudocolor displays are readily available from most image processors. Plate 3A shows the scale we use routinely, 32 hues running from blue through red to violet code for increasing Ca^{++} or H^+. A white color at the top and bottom of the scale is used to warn the operator that the scale should be expanded. Marks and Maxfield [1990a] describe how to create such a scale. An alternative color scale is shown in Plate 3B. Here the lowest ion concentrations are indicated by low luminance violet. As one moves up the scale toward higher ion concentrations, the same violet color increases in luminance until a sudden change to low-luminance blue, and so on up the scale. This type of color scale, which is used by Spex, is inappropriate for a smoothly varying function like ion concentration. Some systems do not display ratio images in pseudocolor during an experiment, but only during postexperiment analysis. The extent to which this is a disadvantage will depend on the type of experiment being performed. If choice of experimental protocol depends on knowing the intracellular ion concentrations during an experiment, then a system that displays pseudocolor ratio images in real time is a definite advantage.

Tsien and Harootunian [1990] point out that there is little advantage in using more than 5 bits per pixel to code for ratio, since noise in the ratios usually exceeds a 32nd part of the total indicated range. In systems with 8 available bits per pixel, the 3 additional bits can be used to store information about the fluorescence intensity at each pixel. This allows postexperiment thresholding (see Section IIC3) and prevents weighting errors (see Section IIIB2). The monitor display can use the fluorescence intensity infor-mation to vary the luminance of individual pixels, as was done in Plate 1A, or this facility can be turned off, as was done in Plate 1E. We find that some users are confused by the luminance variation and turn it off, while others find it useful. The properties of photographic film mean that the luminance variation must normally be turned off for publication photography. Plate 3A illustrates why. In the original video display each horizontal band had a constant hue, but increased in luminance toward the right. However, in the photographic reproduction a hue that appears green at low luminance appears yellow at high luminance. (The problem appears mainly during the reversal printing; color slide film is relatively innocent in this respect.) Examples occur in the literature of cells that appear green, while the calcium calibration scale (which is generally displayed at high luminance) does not show green as an option!

In systems that store the ratio image, rather than the two fluorescence images, as the permanent record of the experiment, the ratio image must contain intensity information, since otherwise postexperiment thresholding is impossible, and measurements of mean ion concentration will be subject to the weighting error (Section IIIB2).

F. Advantages and Disadvantages of Specific Systems

We have looked at commercial ratio imaging systems produced by Spex, Applied Imaging, Improvision, and Photon Technology, and have compared them with the systems in our laboratory and the laboratories of Drs. Berridge and Moreton in Cambridge and Ashmore and Kolston in Bristol. We discuss each system briefly below. To aid comparison, we state the approximate price of the image processor plus host. Any ratio imaging system would of course contain other components, such as a camera, excitation light source, and permanent storage medium.

1. UCL system. Our system is a blinking imager using a Gould FD5000 image processor and DEC MicroVax computer. The software is based upon the system written by Roger Tsien, although it has been entirely rewritten by Angela Lamb. The most important feature that we inherited from Dr. Tsien is the coding system used in ratio images and discussed in Section VE, which we strongly recommend. This system includes intensity information in the ratio image, and has two major advantages. First, it allows calculation of true mean ion concentration in a cell or region of a cell, eliminating the weighting error (Section IIIB2). Second, it allows thresholding (Section IIC3) to be carried out on stored data after the experiment, rather than during the experiment. We store all data digitally. During experiments, up to 13 images can be stored in the image processor. Downloading to the MicroVax hard disk takes ~7 sec and limits time resolution in all but the briefest of experiments. The facilities for quantitative analysis of ratio images are more comprehensive than in any other system we have seen. We use TK50 tape cartridges for permanent data storage. As noted above, these are unacceptably slow, and we are upgrading to a WORM (optical disk) system. The approximate hardware cost, including the TK50 tape back-up system, is £95K. We are willing to make our software available.

2. Cambridge system. This is a real-time imager using a Synoptics Imagine image processor and DEC MicroVax computer. Excitation wavelengths of 340 and 380 nm are alternated at video rate. Each new ratio image is calculated and displayed during the time that the next pair of fluorescence images is being read in; that is, the time resolution is 80 msec. Thresholding is performed at this stage; regions of the image of lower than a criterion intensity are blanked to black; other than this thresholding, the ratio image contains no intensity information. During the experiment, ratio images are viewed as gray-scale images only: That is, regions of cells with low ion concentration appear dark, regions of cells with high ion concentration appear bright. Ratio images (as gray-scale

images) are stored as a continuous record on videotape. This inevitably degrades the image somewhat, but means that during the experiment the experimenter does not have to be concerned with choosing specific images for collection. After the experiment, ratio images are read back for analysis, and can be displayed as pseudocolor images if so desired. The approximate hardware cost is £85K. Dr. Moreton is willing to make his software available.

3. Bristol system. This is a real-time imager comprising a Quintek frame grabber and six Inmos T800 transputers in an MS-DOS host. The excitation filter is changed at video frame rate. Although a pair of fluorescence images therefore takes 80 msec to acquire, the ratio image display is updated every 40 msec using the last two fluorescence images acquired. The image processor reduces the data in the raw 340 nm and 380 nm images to an amount that can be stored in real time on the hard disk of the PC. For instance, of the entire field of view of 512×256 pixels, the operator might choose a region 200×100 pixels around the cell. Spatial averaging would then reduce this to 100×50 values, or 5 KBytes, which can be stored on the hard disk in the 40 msec available. A 100 Mbyte disk can therefore store information about every video frame in a 13 min period. The total cost was approximately £15K. We were very impressed with this system, which we recommend to anyone prepared to program it to fit their particular needs.

4. PTI (Photon Technology International). This is a blinking imager with no separate image processor; images are fed straight into an MS-DOS computer and all further manipulation is performed in the computer. For this reason the system is slower than systems with dedicated image processors, taking approximately 10 sec to produce a ratio image of 512×512 pixels. Software is convenient and easy to use. Intensity information can be included in the ratio image. In the base system, permanent data storage is on floppy disk, but a tape streamer or optical disk can easily be fitted. The approximate cost (hardware and software) is £40K.

5. **Spex.** This is a blinking system designed as an add-on to Spex fluorimeters. The user's camera is interfaced to a Spex image processor and MS-DOS computer. The software is easy to use but contains no analysis protocols: That is, the end product is a color ratio image, with no facility for measuring ion concentration. The color scale used is of the unusual type shown in Plate 3B. Thresholding is performed, but the ratio image does not contain intensity information. Permanent data storage was on tape streamer, which appeared fast and convenient. Approximate cost: £55K.

6. **Applied Imaging "Magical."** This system is based on an Applied Imaging image processor and MS-DOS computer. Long-term data storage is by a fast tape streamer. In its base configuration, with 2 Mbyte of image processor memory, the system is a blinking ratio imager. Thresholding is performed, but the ratio image does not contain intensity information. Image size is variable, so that if one chose, for instance, a resolution of 128×128 pixels, over 100 ratio images could be stored in the image processor. Software is extremely easy to use; for instance, areas of interest for averaging are defined by drawing around them on the screen with a light pen. The unique and very attractive feature of the Applied Imaging system is the availability of an enormous image processor memory—up to 32 Mbyte at present. Users of top-of-the-line Applied Imaging systems tend, paradoxically, to use off-line ratioing, since in this way the problems caused by the lack of intensity information in the ratio image can be overcome. During the experiment, the fluorescence images themselves are stored in the image processor memory. If one chose, for instance, a resolution of 128×128 pixels, over 1,000 image pairs could be stored in the image processor. After the experiment, the image pairs are divided to yield the ratio images; at this stage, the experimenter can

vary the threshold applied. Since the raw fluorescence data is available, mean ion concentrations in areas of interest can be calculated, with no weighting error, by measuring total fluorescence intensity in the area of interest at each excitation wavelength and dividing. The disadvantage of this approach is that the experimenter does not have a ratio image to look at during the course of the experiment. Applied Imaging intends to offer a real-time ratioing system soon. In such a system, the lack of intensity information in the ratio image would be a more serious problem. The approximate cost of the base system is £60K, while the 32 Mbyte system is £80K.

7. **Improvision.** This is a blinking imager with no separate image processor; images are fed straight into an Apple Macintosh Quadra 900 computer and all further manipulation is performed in the computer. For this reason the system is slower than systems with dedicated image processors, taking approximately 10 sec to produce a ratio image of 512×512 pixels. Alternatively, the system can be used as an off-line imager, acquiring pairs of fluorescence images at 1 sec intervals for subsequent ratioing. The operator has complete freedom to increase speed by reducing resolution and/or collecting data from a window of any size placed anywhere on the image. For instance, a 128×128 pixel ratio image can be generated in approximately 1 sec. Software is convenient and easy to use. Thresholding is performed; there is no intensity information in the ratio image. The product is very new, and the system we saw had teething problems: For instance, the software required one to take a new measure of background fluorescence before each series of ratio images. These problems should be cured soon. The approximate cost is £48K, while an additional £20K buys a rapid deconvolution package from Vaytek that allows the system to emulate a confocal microscope.

8. **Conclusions.** While we cannot suggest a best buy, we offer the following comments. We discuss only those systems that we have

seen, since only for these can we judge such factors as ease of use.

1. The system at Bristol, based on transputers, is both the fastest and the cheapest of those that we have seen. Anyone willing to do some of their own programming should consider this route.

2. Of commercial systems, the top-of-the-line Applied Imaging system is both the most expensive and the most powerful. The operating system on this machine is very user-friendly and has clearly benefitted from considerable input from working scientists.

3. The cheapest commercial system, from PTI, works well and the user we spoke to was very pleased with the support he has received from PTI.

4. Of commercial systems, only that from PTI offers the option of storing both ion concentration and intensity data in the ratio image. Other systems will be subject to the weighting error (see Section IIIB2) when calculating mean ion concentrations in a cell or region of a cell, although one can always get around this problem by storing and subsequently analyzing the two fluorescence images.

5. Users of nonimaging fluorimeters who wish to upgrade to ratio imaging should not assume that they must buy their imaging system from the manufacturer of their fluorimeter, but should shop around: Most imaging systems will interface with most fluorimeters.

6. The time resolution of ion imaging systems is often determined by the rate at which data can be stored and the total capacity of the storage medium. Furthermore, storage of regions of the image outside the cell of interest is unnecessary and expensive. The Improvision system allows the user to define a window within the main image from which subsequent measurements are taken, and then to choose just the appropriate resolution. The Applied Imaging Magical is not quite as versatile: It only allows collection from the entire field of view and allows choice between a limited number of resolutions such as 512×512 or 64×64, etc. Other systems, such as the PTI one, allow no choice at all: Collection is from the whole field of view at 512×512 resolution.

7. Rapid and quantitative deconvolution packages offer the potential for ion imagers based on conventional fluorescence microscopes to measure intracellular ion concentration with the spatial resolution available at present only from confocal microscopes. Such systems would have two advantages over confocal microscopes: First, they could easily use dyes excited in the UV, and could therefore use ratiometric dyes such as Fura-2. Second, they would be more optically efficient, using all the light collected by the objective lens rather than rejecting the majority of light emitted by the dye. They would therefore require less intense illumination of the specimen. A rapid deconvolution package is supplied by Vaktek for the Improvision system. However, it is not yet clear whether this or other deconvolution packages preserve the low spatial frequency information essential for accurate measurement of ion concentrations.

ACKNOWLEDGMENTS

We are especially grateful to Tim Cheek, Graham Collingridge, George Duncan, Anthony Galione, Morris Hallett, Bill Mason, Roger Moreton, and George Schofield for showing us their systems, and John Connor, Jim Gillespie, George Reicher, and Roger Tsien for help and advice. The experiment illustrated in Plate 1 was performed by Ian Crossley. Our work is supported by The Wellcome Trust, the MRC, the SERC, and the Smith-Kline (1982) Foundation.

REFERENCES

Ahmed Z, Connor JA (1988): Calcium regulation by, and buffer capacity of, molluscan neurons during calcium transients. Cell Calcium 9: 57–59.

Almers W, Neher E (1985): The calcium signal from Fura-2 loaded mast cells depends strongly on the method of dye loading. FEBS Lett 192:13–18.

Bartfai T (1979): Preparation of metal-chelate complexes and the design of steady-state kinetic experiments involving metal nucleotide complexes. Adv Cycl Nucl Res 10:219–242.

Baylor SM, Hollingworth S, Hui CS, Quinta-Ferreira ME (1986): Properties of the metallochromic dyes Arsenazo III, Antipyrylazo III and Azo I in frog skeletal muscle fibres at rest. J Physiol 377:89–141.

Bolsover SR (1986): Two components of voltage-dependent calcium influx in mouse neuroblastoma cells: measurement with Arsenazo III. J Gen Physiol 88:149–165.

Brownlee C, Pulsford AL (1989): Visualization of the cytoplasmic calcium gradient in Fucus serratus rhizoids: Correlation with cell ultrastructure and polarity. J Cell Sci 91:249–256.

Bunge MB (1977): Initial endocytosis of peroxidase or ferritin by growth cones of cultured nerve cells. J Neurocytol 6:407–439.

Cheek TR (1990): Fura-2 imaging provides new insight into the role of Ca^{++} in triggering exocytosis from adrenal chromaffin cells. Proc Physiol Soc, University College London Meeting, March 30–31, 1990, 4 p.

Cheek TR, O'Sullivan AJ, Moreton RB, Berridge MJ, Burgoyne RD (1989): Spatial localization of the stimulus-induced rise in cytosolic Ca^{++} in bovine adrenal chromaffin cells: Distinct nictonic and muscarinic patterns. FEBS Lett 247:429–434.

Clark WM (1928): "The Determination of Hydrogen Ions." Baltimore: Williams and Wilkins.

Cohan CS, Connor JA, Kater SB (1987): Electrically and chemically mediated increases in intracellular calcium in neuronal growth cones. J Neurosci 7:3588–3599.

Connor JA (1986): Digital imaging of free calcium changes and of spatial gradients in growing processes in single, mammalian central nervous system cells. Proc Natl Acad Sci USA 83:6179–6183.

DiVirgilio F, Steinberg TH, Silverstein SC (1990): Inhibition of Fura-2 sequestration and secretion with organic anion transport blockers. Cell Calcium 11:57–62.

Fay FS, Fogarty KE, Coggins JM (1986): Analysis of molecular distribution in single cells using a digital imaging microscope. Soc Gen Physiol Ser 40:51–63.

Grynkiewicz G, Poenie M, Tsien RY (1985): A new generation of Ca^{++} indicators with greatly improved fluorescence properties. J Biol Chem 260:3440–3450.

Hernandez-Cruz A, Sala F, Adams PR (1990): Subcellular calcium transients visualized by confocal microscopy in a voltage-clamped neuron. Science 247:858–862.

Jacob R (1990): Imaging cytoplasmic free calcium in histamine stimulated endothelial cells and in fMet-Leu-Phe stimulated neutrophils. Cell Calcium 11:241–249.

Kater SB, Mattson MP, Cohan C, Connor J (1988): Calcium regulation of the neuronal growth cone. Trends Neurosci 11:315–321.

Konishi M, Olson A, Hollingworth S, Baylor SM (1988): Myoplasmic binding of Fura-2 investigated by steady-state fluorescence and absorbance measurements. Biophys J 54:1089–1104.

Lakowicz JR (1983): "Principles of Fluorescence Spectroscopy." New York: Plenum Press.

Malgaroli A, Milani D, Meldolesi J, Pozzan T (1985): Fura-2 measurements of cytosolic free Ca^{++} in monolayers and suspensions of various types of animal cells. J Cell Biol 105:2145–2155.

Marks PW, Maxfield FR (1990a): Local and global changes in cytosolic free calcium in neutrophils during chemotaxis and phagocytosis. Cell Calcium 11:181–190.

Marks PW, Maxfield FR (1990b): Transient increases in cytosolic free calcium appear to be required for the migration of adherent human neutrophils. J Cell Biol 110:43–52.

Marsh M, Griffiths G, Dean G, Mellman I, Helenius A (1986): Three-dimensional structure of endosomes in BHK-21 cells. Proc Natl Acad Sci USA 83:2899–2903.

Moore EDW, Becker PL, Fogarty KE, Williams DA, Fay FS (1990): Ca^{2+} imaging in single living cells: Theoretical and practical issues. Cell Calcium 11:157–179.

Niggli E, Lederer WJ (1990): Real-time confocal microscopy and calcium measurements in heart muscle cells: Towards the development of a fluorescence microscope with high temporal and spatial resolution. Cell Calcium 11:121–130.

O'Sullivan AJ, Cheek TR, Moreton RB, Berridge MJ, Burgoyne RD (1989): Localization and heterogeneity of agonist-induced changes in cytosolic calcium concentration in single bovine adrenal chromaffin cells from video imaging of Fura-2. EMBO J 8:401–411.

Paradiso AM, Tsien RY, Machen TE (1987): Digital image processing of intracellular pH in gastric oxyntic and chief cells. Nature 325:447–450.

Poenie M (1990): Alteration of intracellular Fura-2 fluorescence by viscosity: A simple correction. Cell Calcium 11:85–91.

Poenie M, Tsien RY, Schmitt-Verhulst A (1987): Sequential activation and lethal hit measured by $[Ca^{++}]_i$ in individual cytolytic T cells and targets. EMBO J 6:2223–2232.

Regehr WG, Connor JA, Tank DW (1989): Optical imaging of calcium accumulation in hippocampal pyramidal cells during synaptic activation. Nature 341:533–536.

Roe MW, Lemasters JJ, Herman B (1990): Assessment of Fura-2 for measurements of cytosolic free calcium. Cell Calcium 11:63–73.

Sawyer DW, Sullivan JA, Mandell GL (1985): Intracellular free calcium localization in neutrophils during phagocytosis. Science 230:663–666.

Silver RA, Lamb AG, Bolsover SR (1989): Elevated cytoplasmic calcium in the growth cone inhibits neurite elongation in neuroblastoma cells: Correlation of behavioral states with cytosolic calcium concentration. J Neurosci 9:4007–4020.

Silver RA, Lamb AG, Bolsover SR (1990): Calcium hotspots caused by L channel clustering promote morphological changes in neuronal growth cones. Nature 343:751–754.

Silver RA, Whitaker M, Bolsover SR (1992): Intracellular ion imaging using fluorescent dyes: Artifacts and limits to resolution. Pflügers Arch 420: 595–602.

Swann K, Whitaker MJ (1986): The part played by inositol trisphosphate and calcium in the propagation of the fertilization wave in sea urchin eggs. J Cell Biol 103:2333–2342.

Takamatsu T, Weir WG (1990): High temporal resolution video imaging of intracellular calcium. Cell Calcium 11:111–120.

Tank DW, Sugimori M, Connor JA, Llinas RR (1988): Spatially resolved calcium dynamics of mammalian purkinje cells in cerebellar slice. Science 242:773–777.

Timmerman MP, Ashley CC (1986): Fura-2 diffusion and its use as an indicator of transient free calcium changes in single striated muscle cells. FEBS Lett 209:1–8.

Tsien RY (1981): A non-disruptive technique for loading calcium buffers and indicators into cells. Nature 290:527–528.

Tsien RY (1990): Laser scanning confocal microscopy at video rates (30 frames/sec) with dual wavelength emission ratioing for quantitative imaging of intracellular messengers. Proc R Microsc Soc 25:S52.

Tsien RY, Harootunian AT (1990): Practical design criteria for a dynamic ratio imaging system. Cell Calcium 11:93–109.

Weir WG, Cannell MB, Berlin JR, Marban E, Lederer WJ (1987): Cellular and subcellular heterogeneity of [Ca^{++}] in single heart cells revealed by Fura-2. Science 235:325–328.

Westerblad H, Lee JA, Lamb AG, Bolsover SR, Allen DG (1990): Spatial gradients of intracellular calcium in skeletal muscle during fatigue. Pflügers Arch 415:734–740.

Williams DA, Fogarty KE, Tsien RY, Fay FS (1985): Calcium gradients in single smooth muscle cells revealed by the digital imaging microscope using Fura-2. Nature 318:558–561.

Wilson HA, Greenblatt D, Poenie M, Finkelman FD, Tsien RY (1987): Crosslinkage of B lymphocyte surface immunoglobulin by anti-Ig or antigen induces prolonged oscillation of intracellular calcium. J Exp Med 166:601–606.

Yagi S, Becker PL, Fay FS (1988): Relationship between force and Ca^{2+} concentration in smooth muscle as revealed by measurements on single cells. Proc Natl Acad Sci USA 85:4109–4113.

CHAPTER 9

Computer Reconstruction in Three-Dimensional Fluorescence Microscopy

Peter J. Shaw

I. INTRODUCTION

The development of the theory of optical imaging by Abbe and his successors has given microscope designers the information they need to build the best possible optical systems for many different purposes. This development continues today, with one of the most interesting developments for biologists being the present generation of confocal scanning microscopes [Wijnaendts van Resandt et al., 1985; Amos et al., 1987; Brakenhoff et al., 1989; Carlsson and Liljeborg, 1989; see Shotton, 1989, for a recent review]. On the other hand, the rapid increase in power and decrease in cost of digital computers has opened another avenue for the development of optical micros-

copy—that of postprocessing the optical images. With as full an understanding as possible of the imaging system, the images produced are manipulated in a well-defined way so as to increase or enhance certain desired features ("features" here being used in a general sense), or to remove or decrease unwanted features. This chapter attempts to give some of the theory underlying these image processing methods with specific reference to methods for three-dimensional reconstruction and describes some of the algorithms that have been used for this purpose.

The interest in developing methods for detailed three-dimensional reconstruction in optical microscopy has arisen from the rapid renaissance in optical microscopy as a tool

Electronic Light Microscopy, pages 211–230 © 1993 *Wiley-Liss, Inc.*

in cell biology after its eclipse by electron microscopy for the past decades. There are several reasons for this resurgence; probably the most important is that there are now many very specific probes for all types of biological components. The probes are almost invariably designed to be imaged by fluorescence emission and this makes their detection extremely sensitive. Discussion of the vast range of probes for cellular constituents which is now available is beyond the scope of this chapter, but it includes specific antibodies, modified polynucleotides, and reporter dyes for determination of pH and the concentration of specific cations. (See Tsien [1989] for a recent review. The catalogue produced by Molecular Probes Inc. [P.O. Box 22010, 4849 Pitchford Avenue, Eugene, OR 97402, USA] is also a mine of useful information about this subject.) A second reason is that optical microscopy is inherently relatively nondestructive to biological tissue, at least in comparison with such techniques as electron microscopy. Although high light levels can damage cells, especially when light-absorbing molecules are present, it is often possible to carry out optical imaging experiments on living cells with no detectable harm. The rapid improvement of microinjection and micromanipulation methods has now opened up the possibility of carrying out biochemical and physiological experiments within individual cells.

The major limitation of optical microscopy is of course its resolving power. For the best objective lenses, this is given by classical theory as between 0.2 and 0.3 μm, depending on the wavelength of the light, the definitions adopted, etc. This, however, is the resolution in the image plane. When the microscope is regarded as a three-dimensional imaging device, the resolution in the third dimension, the direction of the optical axis, must be considered. This introduces some problems. The depth of field for a high numerical aperture objective lens (i.e., that portion of the specimen which contributes to the image at a particular focus level) is large compared to the in-plane resolution

limit, but small compared to the thickness of most specimens. This means that the microscope cannot be used to produce true projections—and thus to enable reconstructions to be made from multiple tilt methods such as those used in X-ray tomography or electron microscopy. On the other hand, the image at a given focal plane is a poor estimate of a section through the specimen because of contamination of the image by out-of-focus contribution from adjacent parts of the specimen. The effective depth of field also depends on the spatial frequency, large-scale structure having a larger depth of field than fine image detail. This complicates still further the interpretation of through-focal series of images. These problems are particularly severe for epifluorescence imaging, currently the most useful for biological studies. It is essentially a "dark-field" imaging mode, and with high levels of fluorescence labeling, the out-of-focus low-frequency components from a fairly large distance either side of the plane of focus build up to give a large degree of flare. The in-focus fine image detail is often seen as a relatively small modulation against the large out-of-focus flare. Its contrast is thus reduced. This effect is inherently less objectionable in bright-field techniques; the out-of-focus information adding to the already bright field has a smaller effect on the image contrast. Moreover, there exist differential imaging modes, such as differential interference contrast, which can increase the depth discrimination markedly [in fact, the narrowness of the depth of field which can be produced is still not fully understood; see Inoué, 1989]. The techniques described in this chapter have thus been specifically designed for fluorescence imaging. The extent of their applicability to other imaging modes has yet to be explored.

II. THEORETICAL BACKGROUND

A. Linear, Shift-Invariant Imaging, and the OTF

Before soundly based image processing and reconstruction schemes can be carried out on optical microscope images, we must provide a mathematical formulation of the imaging process. We shall do this in a very condensed manner, following treatments given by Agard et al. [1989] and Young [1989]. For more thorough discussions, see these authors and also Castleman [1979], Born and Wolf [1975], and Oppenheim et al. [1983]. The two key features of the underlying physical phenomenon are that the imaging operation should be linear and shift-invariant.

We will denote the object as a function of position $f(r)$ and the resulting image as $g(r)$. Then representing the imaging operation by an arrow:

$$f(r) \Rightarrow g(r). \tag{1}$$

Linearity means that a linear combination of objects produces a linear combination of images or:

$$k_1 f_1(r) + k_2 f_2(r) + \ldots \Rightarrow$$
$$k_1 g_1(r) + k_2 g_2(r) + \ldots \tag{2}$$

(where k_1, k_2, etc. are constants).

Shift invariance means that the imaging operation has the same effect no matter what part of the field it is applied to. The importance of these conditions is that if they hold, any object may be described as the sum of its parts, and the resulting image is the sum of the images of the parts. In particular, although the object is in general a continuous, indivisible body, at least in the realms of light microscope resolution, we can approximate it as closely as we like by a series of points of suitable intensity. We denote the point function by $\delta(r)$. Then the object $f(r)$ can be represented by the sum:

$$f(r) = \sum_{i=-\infty}^{+\infty} f(s_i)\delta(r - s_i)\Delta s_i \tag{3}$$

or in the limit:

$$f(r) = \int_{-\infty}^{+\infty} f(s)\delta(r-s)ds. \tag{4}$$

Now if we know the way in which a single point $\delta(r)$ is affected by the imaging process, we can describe the image of any object: Let the image of the point be $o(r)$, i.e.,

$$\delta(r) \Rightarrow o(r) \tag{5}$$

then: $\delta(r-s) \Rightarrow o(r-s) \tag{6}$

and

$$f(r) = \int_{-\infty}^{+\infty} f(s)\delta(r-s)ds$$

$$\Rightarrow \int_{+\infty}^{+\infty} f(s)o(r-s)ds = g(r). \tag{7}$$

Thus the image of an object, $g(r)$, is given by Equation 7. This is a familiar mathematical form: a convolution. The image is therefore the convolution of the object with the image of a point. The point image is generally termed the impulse response function in one-dimensional situations or the point spread function (PSF) by microscopists. Once this function is known, then providing the linearity and shift-invariance conditions are met, the image obtained from any object can be predicted by the convolution:

$$g = f \otimes o, \tag{8}$$

using $\otimes$ to denote the convolution operation more concisely. This equation is often more easily handled in terms of Fourier transforms, where the convolution becomes a product of transforms.

$$G = FO, \tag{9}$$

where G and F are the Fourier transforms of the image and object, and O is the transform of the PSF. The function O is commonly called the optical transfer function (OTF) or the modulation transfer function (MTF). It is also somewhat loosely called the contrast transfer function. It is not strictly correct to call it this, using the conventional definition of contrast, although the true contrast transfer function is closely related to the OTF (see Young [1989] for a fuller discussion).

The determination of the original object, f, given the measured image g and the response function o, is the fundamental problem in three-dimensional optical reconstruction. It is a familiar mathematical problem which occurs in several different areas besides optical microscopy. Many different techniques of differing degrees of sophistication have been used to solve the problem, and the extension and improvement of these techniques is a very active area of computational research. We describe below some methods which have so far proved useful in optical microscopy.

B. Three-Dimensional Imaging

All the mathematical formulations so far have been expressed in vector notation for simplicity and generality, without specifying the dimensionality. In fact, optical microscopy is traditionally expressed in two-dimensional terms: All the available image detectors, including the eye, are two-dimensional, and the optical microscope is inherently approximately two-dimensional in that a fairly thin region of the specimen around the focal plane is brought to the final image plane of the microscope. The extension to three dimensions is quite straightforward, however. The third dimension—conventionally denoted the z-axis—is parallel to the optical axis. A three-dimensional image is produced as a set of two dimensional images at successive focal planes. The 3D point spread function is then similarly defined as the image at successive focal planes of a single three-dimensional point object. This then gives a consistent and physically reasonable interpretation of the three-dimensional object as the 3D distribution of light emission.

III. APPLICABILITY OF THE THEORY

The application of Equations 8 and 9 requires consideration of two factors: First, how far is the imaging system linear and shift-invariant; second, what is the form of the PSF. As before, we shall restrict the discussion to fluorescence imaging.

Fluorescence imaging probably approximates linearity well. Each part of the specimen can be thought of as emitting fluorescent light to an extent which depends only on the incident intensity and the local concentration of the fluorochrome. Similarly, if the optical system is adequately corrected for image plane aberrations, such as flatness of field, etc., the imaging should be shift-invariant to a good approximation. However, it should be realized that circumstances can occur in which these conditions do not hold. For example, in a particular fluorescent specimen, nonfluorochrome-specific light scattering may be considerable; thus both incident and emitted light might be affected in some complicated way by the specimen structure, and the imaging would no longer be linear. Similarly, for very strongly fluorescent specimens, secondary absorption and scattering of the emitted light may take place.

Another possible limit to linearity is saturation of the fluorochrome, a situation which is easily attained with laser scanning microscopy. Changes in refractive index within the specimen can also give rise to nonlinear imaging, effectively the focal plane is no longer flat, but may contain specimen-dependent irregularities. With these reservations in mind, the assumption of linear, shift-invariant imaging is still usually justified for epifluorescence.

A. The PSF for Fluorescent Imaging

If the foregoing assumptions are true, the imaging can be characterized by a PSF (or OTF). Several authors have sought to calculate its form [Stokseth, 1969; Castleman, 1979; Agard, 1984; Erhardt et al., 1985]. The question immediately arises of how far the theoretical calculations agree with real measurements of the PSF.

For epifluorescence imaging, the PSF is probably best measured using submicroscopic fluorescent beads as described by Hiraoka et al. [1988, 1990] [see also Agard et al., 1989]. Sisken [1989] gives details of suitable beads and a list of suppliers. A suspension of beads is fixed to the surface of a slide or coverglass, and then a focal series of images of a field of beads is recorded in exactly the same way as for a standard specimen. Provided the bead diameter is well below the resolution limit of the objective, the images of the beads will be good approximations to the PSF. The principal difficulty is that the amount of light recorded from such small beads is often not enough to give an accurately defined image, even when using a highly sensitive, accurate imaging device such as a charge-coupled device (CCD) camera. It is thus necessary to average the data from many individual beads in order to obtain a sufficiently noise-free estimate. Finally, cylindrical averaging about the z-axis can be used to further decrease the noise. (It should be noted, however, that Hiraoka et al. found a degree of rotational asymmetry for certain objective lenses, and so it may not always be desirable to carry out cylindrical averaging.)

A PSF series generated in this way is shown in Figure 1a as x–y sections and in Figure 2a as x–z sections. The resultant OTF, derived by cylindrical averaging of the 3D Fourier transform of the PSF, is shown in Figure 3a. The central x–y section of the PSF is the familiar Airy pattern, a central disk surrounded by rings of subsidiary maxima. However, in three dimensions the form is considerably more complicated, with maxima extending in an approximately conical form either side of the central plane. The effect of the function is more easily seen by considering the OTF. The most striking feature is a "missing cone"—the OTF is in fact a torus. Information is completely lost in the missing region at low spatial frequencies. This is another way of viewing the lack of

z-resolution at low frequencies, or the out-of-focus flare. Also, although it is not so readily apparent from an intensity plot, there is considerable attenuation of high spatial frequencies, even in the x–y plane. In other words, the fine detail is not present with as great a contribution in the image as it was in the original object.

These results indicate that the agreement with theory was fairly good. The overall shape was as predicted. However, two factors had a large effect on the experimentally determined PSF. First, the PSF was often highly asymmetrical with respect to the focal plane, i.e., it fell off much more rapidly one side of focus than the other. This is due to effective spherical aberration; high-resolution objective lenses are designed to minimize aberrations for a particular optical path length. Oil-immersion lenses are intended in general to image objects immediately beneath a coverglass of the specified thickness using an immersion oil of the specified refractive index. However, biologists often need to image deep within a relatively thick specimen. Thus the optical path is increased by an additional thickness of whatever mounting medium is being used. Hiraoka et al. [1990] have shown that it is possible to correct this by changing the refractive index of the immersion oil to compensate. The best resolution was obtained when the PSF was as nearly symmetrical about the focal plane as possible. A surprising finding was that the field of illumination had an effect on the depth of field; when an effective aperture of a few micrometers at the specimen was used, the PSF showed better z-resolution. The authors explained this as being partial confocal behavior. These effects, and the effect of inhomogeneity in the illumination source, need further investigation. It is clearly necessary to pursue theoretical calculations so as to be sure we have an adequate understanding of the imaging process. However, for processing of real experimental data, it seems important to use the experimentally determined PSF for the imaging system. If it is not possible to deter-

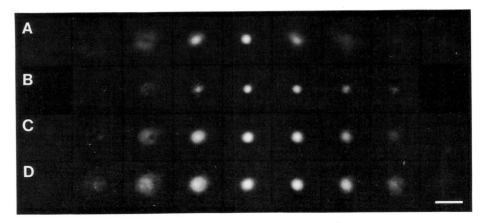

Fig. 1. Comparison of point spread functions for conventional (**A**), confocal with 1 mm pinhole (**B**), confocal with 3.5 mm pinhole (**C**), confocal with 6 mm pinhole (**D**). Data collected with Leitz 63× oil-immersion objective lens (NA 1.4); cooled CCD camera (a); or Biorad MRC500 (b–d). Focal step between sections = 0.4 μm. Bar = 1 μm.

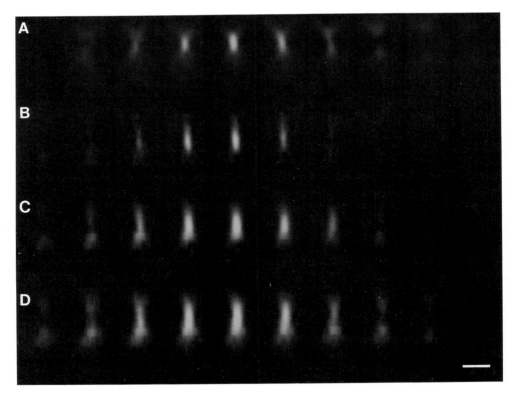

Fig. 2. As Figure 1, but rotated to show x–z sections. (The z-axis is vertical, x is horizontal.) Step in y (between successive sections) = 0.4 μm. Bar = 1 μm.

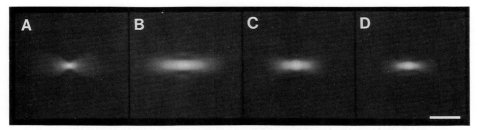

Fig. 3. OTFs for PSFs shown in Figures 1 and 2. The functions have been cylindrically averaged. The z^*-axis is vertical, the radial axis is horizontal. Bar $= 1/0.24\,\mu m^{-1}$.

mine the PSF for the apparatus in use, it would seem to be preferable to use an experimentally determined PSF for a different objective, wavelength, etc., with appropriate corrections, rather than a theoretical PSF.

In the case of a confocal scanning microscope, the situation is a little better. The OTF approximates an axially symmetric ellipsoid rather than a torus, and thus the missing cone problem is eliminated. Figures 1B and 2B show a PSF measured with a Biorad MRC500 confocal microscope with minimum detector aperture. In Figures 1C,D and 2C,D, the detector aperture has been opened to 3.5 mm and 6 mm, respectively. Opening of the detector aperture is often necessary to detect enough light in weakly fluorescent specimens. However, it is clear from these PSFs that doing so considerably degrades the resolution. In Figure 3B–D, the resulting OTFs are shown for the confocal data. The objectionable out-of-focus flare is largely absent, and the different spatial frequencies fall off more evenly with z. However, the z-resolution is still considerably less than the in-plane resolution, and the high spatial frequencies are still markedly attenuated. It seems likely that the image processing schemes described in this chapter will also find a useful application in confocal microscopy, especially when low light levels make the use of a relatively large detector aperture necessary. An illustration of this is shown later in this chapter.

The PSF is a real function, in the mathematical sense, and in general therefore the OTF is Hermitian symmetric. However, in the ideal case the PSF is both cylindrically symmetric and has mirror symmetry about the x–y plane. Thus the function is also centrosymmetric. This means that the ideal OTF is also a real function. Any departure from symmetry about the x–y plane causes it to become complex.

IV. HARDWARE CONSIDERATIONS

A. Image Recording

The methods described in this chapter are to some extent applicable to all 3D fluorescent microscope image data, no matter how the data sets were acquired. So, for example, a good deal of useful work has been done using images relayed to a video camera and digitized by a video-linked computer frame store. In general, at least a silicon intensifier target (SIT) video camera is required for the low light levels encountered in fluorescence imaging, and for some applications another level of intensification is needed (SIT). More recently, video-rate CCD cameras have become available, together with various types of image intensifier—for example, microchannel plates. Generally, video frame averaging is required to reduce the image noise level. In our experience the chief problems with video cameras are image defects and blemishes, either on the target itself or in the image intensifiers. These are generally not stable enough to be corrected completely, and they are accentuated by the image enhancement procedures. Another problem is the limited dynamic range of a video signal. This is usually not much better than 6 bits. In fluorescence imaging, a very great range of intensities is produced, and three-dimensional processing often depends on accurate measurements of differences between closely similar images. For these reasons video cameras will never be ideal for this type of imaging, and it is probably not worthwhile to use any but the simplest restoration schemes on this type of data.

There is little doubt that the highest quality image data currently obtainable from an optical microscope is provided by the new generation of high-sensitivity, cooled CCD cameras (see, e.g., Aikens et al. [1989]). They are nearly ideal imaging devices: high detection efficiency (up to 80% in the best cases, routinely 30–40%); very low noise; excellent geometrical accuracy and linearity of response; and a very large dynamic range. The fact that these devices can be read after any required illumination period means that the inevitable electronic noise can be reduced far below that obtained in any video-rate camera—the image integration is performed on the CCD chip itself and thus electronic noise is introduced only once during readout, rather than every video frame. Video frame averaging cannot restore the accuracy lost in this way. Earlier CCD cameras of this type required liquid nitrogen cooling to achieve very low dark currents. Some current cameras use a technique in the driving circuitry known as MPP (multi-pinned phase) or inverted clock mode, which reduces the dark current to such an extent that electrical cooling to $-40°C$ is sufficient for virtually any microscopy application. Many earlier CCD chips contained a fairly large variation in sensitivity from well to well on the chip; this variability is much less in current devices. In any case, the behavior of each well is highly reproducible, and so it is possible to correct for any discrepancies to a high degree. In this way extremely accurate data of a high dynamic range can be obtained.

B. Computing

The scale of computer facilities required for this type of work would have been judged very substantial a few years ago. Today little more than a high-end graphics workstation is required. The time taken for 2D and 3D Fourier transforms is a good benchmark for all of these algorithms. When this type of work was initiated in our laboratory, the only computer available was a multiuser VAX 11/750, which made all but the simplest image processing schemes too time-consuming. The nearest neighbor approach (see below) requires two 2D Fourier transforms per image section. For a 512 by 512 pixel image, a Fourier transform took about 3 min, and thus a typical 3D image of 30 focal sections took about 3 h central processing unit time. This work has now been transferred to an Ardent Titan graphics super-workstation. On this machine, a 512 by 512 pixel 2D transform takes about 2 sec and thus the neighbor approach takes only a few minutes. A 3D Fourier transform of this size data set takes about 3 min on the latter machine, and the 3D iterative constrained method requires two 3D transforms per cycle, giving reasonable refinement in a couple of hours. Thus approaches which were formerly out of the reach of all but the best equipped laboratories should now be easily accessible.

Other important considerations are fast input of images, from whatever source, and fast and high-quality image display, preferably with color capabilities. In our experience, considerable interaction of the user with the display is very important in visualizing and understanding the vast amount of data in a 3D image, and a very tight and efficient coupling of display hardware to the computing power and the main computer memory is essential. This organization is provided by many of the current generation of graphics workstations. Finally, it is virtually axiomatic that in this type of application there is never enough disk storage or main memory. We currently have 500

Mbyte of disk storage and 32 Mbyte of memory. Even so, this can easily become a serious limitation.

V. METHODS FOR DEBLURRING

We shall present results of three methods we have used for removing the out-of-focus contamination introduced by the OTF. Technically, to the extent that these methods reverse the convolution of the object with the PSF, they could all be termed deconvolutions. More precisely, however, we should restrict that term to algorithms that specifically use a convolution or its inverse. We shall use a single data set to illustrate the methods. This is an image of a metaphase chromosome set measured from an intact larva of *Drosophila melanogaster*, stained with the DNA dye DAPI [Mitchison and Sedat, 1983]. The data were collected on a Zeiss Axiomat microscope using a water-immersion, coverglass-free objective lens

(Zeiss, $63 \times / 1.2\,NA$). The image was relayed to a liquid nitrogen-cooled CCD camera (Photometrics Inc.); (see Shaw et al. [1989] for more details). Some sections from the unprocessed data (corrected for inhomogeneities in the camera, however) are shown in Figure 4A. The x–y pixel spacing was 0.06 μm and the z focus movement between successive sections in the original data set was 0.2 μm; every fourth section is shown in Figure 4. These mitotic figures are very small (about 5 μm in diameter), and so extracting the maximum amount of information from the images poses a considerable challenge for optical microscopy.

A. Nearest Neighbor Subtraction

Methods involving transforms of the entire 3D data sets require a large amount of computation and can be very time-consuming. The nearest neighbor approach attempts to reduce the 3D problem to a series of 2D com-

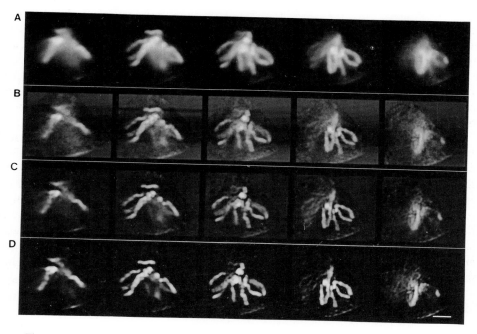

Fig. 4. Comparison of sections from reconstructions calculated by the various methods. **A:** Original data; **B:** nearest neighbor method; **C:** Wiener filtering; **D:** iterative constrained deconvolution. Bar = 2 μm. Focal step between sections = 0.2 μm, every fourth section shown.

putations by correcting each plane in the stack individually. It was the first method to be used in 3D optical microscopy [Castleman, 1979; Agard, 1984]. The basic assumption is that the effect of the 3D OTF can be considered as having two components: First, a roll-off of high-frequency information in the plane of each section—convolution with the Airy pattern; second, addition of extraneous out-of-focus information from nearby focal planes:

$$g_j = f_j \otimes o_0 + f_{j-1} \otimes o_{-1} + f_{j+1} \otimes o_{+1}$$
$$+ f_{j-2} \otimes o_{-2} + f_{j+2} \otimes o_{+2} + \ldots \quad (10)$$

In the simplest method we assume that the out-of-focus addition is strongest from the immediately adjacent planes f_{j-1} and f_{j+1}, and ignore contributions from any planes further away. Furthermore, we assume the measured data planes either side of the plane of interest are a sufficiently good approximation to the real object, and subtract them after blurring by an additional focus step (o_Δ):

$$f_j \otimes o_0 = g_j - f_{j-1} \otimes o_{-1} -$$
$$f_{j+1} \otimes o_{+1} - \ldots \text{(other planes)}$$
$$\approx g_j - g_{j-1} \otimes o_\Delta - g_{j+1} \otimes o_\Delta \quad (11)$$

or

$$f_j \approx [g_i(g_{j-1} + g_{j+1}) \otimes o_\Delta] \otimes p_0 \quad (12)$$

where $p_0 = [o_0]^{-1}$.

However, this immediately shows the limits to the validity of the approximations. The terms subtracted from g_j in Equation 12 are too large, especially at low spatial frequencies, and a great deal of the large-scale structure in the object is removed. Castleman originally suggested subtracting only half of this correction. A more general form is to subtract a fraction c of the correction:

$$f_j = [g_j - c[g_{j-1} + g_{j+1}] \otimes o_\Delta] \otimes p_0. \quad (13)$$

Agard et al. [1989] and Gruenbaum et al. [1984] have suggested $c = 0.45$ as giving a better result, as judged by lack of negative ripples around large peaks in the reconstructed image. More detailed analysis shows that the situation is considerably more complicated [Agard et al., 1989], and suggests that the constant c be replaced by a function of frequency β, which starts at a value of 0.45 and increases to about 0.9 at high spatial frequencies. (In fact, it is at high frequencies that the contributions fall off most rapidly with z, and thus that the nearest neighbor approximation is best.) For the function β, Agard et al. suggest the inverse transform of B, given by:

$$B = c O_0 O_\Delta / (O_0 + O_1 O_\Delta). \quad (14)$$

In spite of these problems, this method, even in its simplest form, is very effective at removing out-of-focus components and has been used in several published studies, both with CCD and video camera data [see, for example, Lloyd et al., 1987; Traas et al., 1987; Hiraoka et al., 1988; Rawlins and Shaw, 1988; Agard et al., 1989; Flanders et al., 1989]. The result of this algorithm is shown in Figure 4B. The main problems are noise amplification and generation of ripples around intense features. It also suffers from problems due to any scaling differences between successive sections.

B. Wiener Filtering

Equation 9 suggests that F could be recovered simply by dividing G by O; inverse Fourier transformation would then give the reconstructed object f. Unfortunately, this type of approach suffers from inevitable problems in regions where O becomes very small. In theory, G should become correspondingly small, but in practice it always contains a noise component distributed over its entire space. Dividing by the extremely small components of O serves merely to amplify the noise, which can then easily dominate the final reconstruction. There is an extensive literature concerning linear filters to overcome this problem. The best that can be done in general is some form of Wiener filter [see, e.g., Castleman, 1979]. We have used:

$$F = GO/(O^2 + K). \qquad (15)$$

In general, K should be a function of spatial frequency, whose value depends on the variance of the data with respect to frequency. However, we have found that simply using a constant value for K works well in practice.

Erhardt et al. [1984] used a similar approach, but calculated a theoretical approximation to the OTF, then rolled it off smoothly to zero in regions where it was very small. The inverse filter was then only applied in nonzero regions of the OTF.

The chief problem with this type of method, as with the previous method, is that negative ripples tend to surround large peaks. In order to reduce these ripples to acceptable levels, especially with noisy data, the constant K often has to be increased to such a degree that little sharpening is achieved. Figure 4C shows the Wiener-filtered result (with $K = 0.001$).

C. Nonlinear Methods

Although it can be shown that a Wiener or closely related filter is the best that can be done in the absence of any other information and in the presence of a particular noise level, it does not necessarily provide the best reconstruction if some degree of prior knowledge can be assumed and incorporated as a constraint in the deconvolution process.

The wavelength of the radiation used, the numerical aperture of the objective lens, and the imaging mode used clearly affect the maximum spatial frequencies that can be transferred by the optical system. But this is not the same as the resolution available in a particular image, or indeed as to the level to which a particular image can be interpreted. The level of detail interpretable from particular images can be either more or less

than the classical resolution limit, depending upon circumstances. A simple example is the presence of noise in the recorded image; since the high-frequency image components are generally much smaller in intensity than the low-frequency components, noise will tend to limit the resolution available. Conversely, to obtain the highest resolution it is important to reduce the noise component to as low a level as possible, and noise level is always a fundamental limitation to the accuracy attainable in a reconstruction. On the other hand, the availability of prior knowledge about an object can increase the useful resolution. This type of knowledge is often used in a complicated way in the final interpretation of images rather than in processing; for example, fine filamentous structures may be interpreted in terms of single microtubules, although in themselves they are far below the resolution limit of the optical microscope.

We almost always have some prior knowledge of the object of a fairly general nature. For example, we may know that it is spatially bounded. Another reasonable assumption is that the object density is nowhere negative. However, it is important to realize that these assumptions should always be critically examined as applied to the light emission of the object in question. (Negative emission—in the form of secondary absorption—may at least theoretically be possible.) Another piece of "knowledge" about the object is the idea of smoothness. This has been formalized in various ways as information content, image entropy, maximum likelihood, etc. A detailed discussion of this is beyond the scope of this chapter. (See Holmes et al. [1991] for an interesting recent application of maximum likelihood restoration to microscopy.) In a general way it is easy to see that information content is similar to lack of smoothness in an image. We might assume that a totally uninformative picture would be uniformly gray—and thus completely smooth. Introducing information into the image must cause departures from this uniformity. When this constraint is used in reconstruction methods, choosing the smoothest reconstruction may be thought of as deriving the minimum information consistent with the available data; i.e., drawing the "safest" or most likely inference. However, in this field, computational, mathematical, and philosophical pitfalls abound.

Several schemes have been devised for incorporating this type of information as a constraint in the deconvolution. Unfortunately they all turn out to give nonlinear solutions and must therefore be calculated by iteration. They are invariably very computationally intensive. We shall examine in more detail the algorithm originally devised by Van Cittert and Jansson [Jansson et al., 1970], which has been used most extensively in optical reconstruction so far.

The algorithm starts with a trial solution (often the observed data), then successively modifies it according to the difference between the blurred trial and the observed data. Since this is done in iterative stages, modifications can be made to the trial solution along the way so as to impose constraints such as positivity. The solution is also smoothed, generally every few cycles, via a Gaussian filter on the Fourier transform. In more detail, the algorithm is:

i. $g^k = f^k \otimes o$
ii. $f^{k+1} = f^k + \gamma(g)(g - g^k)$
iii. where $(f^{k+1} < 0)$ set equal to 0
iv. $k = k + 1$
     go to (i) until f^{k+1} and f^k converge
     where $\gamma = 1 - [g^k - A]^2/A^2$ (16)

(f^k is the current guess, g^k the blurred current guess, g the observed data, and A is a constant set to the maximum value of $g/2$). This method often requires 50–100 cycles for satisfactory convergence in three dimensions. A modification of this scheme due to Gold [1964] uses a different modification step ii):

ii. $f^{k+1} = f^k(g/g^k),$ (17)

which often converges faster. One reason for the slow convergence is that high-frequency components in the correction are damped down by the convolution in Equation 16i. An ideal correction term ϵ would be one that would exactly add the difference between the current blurred trial and the observed data, so that:

$$(f^k + \epsilon) \otimes o = g \qquad (18)$$

$$\text{or } (F^k + E)O = G \qquad (19)$$

$$\text{giving } E = (G - F^k O)/O. \qquad (20)$$

Thus E is the difference between the blurred trial and the observed data filtered with the inverse of the OTF. This, of course, brings us directly back to the problems of inverse filtering and noise amplification. However, we have found that using a Wiener filter for this inverse filtering of the update is quite successful. Using this update for the first two or three cycles greatly increases the rate of convergence. After the first cycles, the conventional Jansson or Gold filter is used. With this acceleration, satisfactory convergence is often obtained in 5–10 cycles. This algorithm is illustrated in Figure 4D. Ten cycles of iteration were used, incorporating the accelerated convergence method for the first three cycles. Figure 5 shows x–z sections of the original data and each reconstruction. See Jordan et al. [1992] for a recent application of this algorithm.

VI. RECONSTRUCTION FROM MULTIPLE TILTED VIEWS

The image processing methods of the previous section are very successful at improving the clarity and ease of interpretation of 3D images. This is accomplished primarily by changing the relative weighting of spatial frequencies according to the optical transfer function so as to reserve their differential attenuation in the image. However, it is not clear that any of the methods is capable of restoring the missing spatial frequency data and thus truly increasing the available resolution. Linear methods certainly cannot do this. There is considerable debate about the extent to which nonlinear, constrained meth-

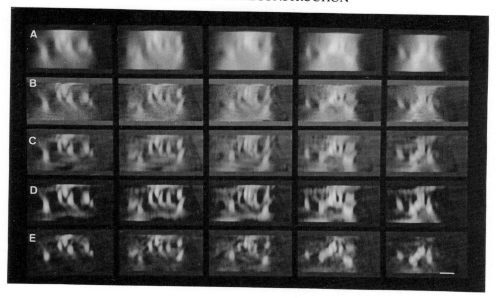

Fig. 5. The same data sets as Figure 4 **(A–D)** and the combined tilt reconstruction **(E)**, rotated to show x–z sections. Bar = 2 μm.

ods can restore missing data. An alternative approach which definitely can fill in the missing regions is to use multiple tilted views. This method is used extensively in X-ray crystallography, electron microscopy, and medical tomographic imaging, but so far has been very little applied to optical microscopy. There are one or two examples of limited tilting being used for the generation of stereo views [Skaer and Whytock, 1975; Inoué, 1986]. Recently we demonstrated the feasibility of large degrees of tilting coupled with through-focal sectioning significantly to increase the z-resolution [Shaw et al., 1989].

The overall image processing scheme is shown diagrammatically in Figure 6. Two key problems need to be solved for this approach. First, a suitable method of mounting the specimen so that it can be rotated through large angles must be devised. Second, the data sets obtained must be placed accurately in a common coordinate system so that they can be combined.

The first problem was solved by constructing a special rotating specimen holder which carried the specimen adhered to the outside of a fine glass capillary tube. The rotation was adjusted under computer control. Translational adjustment was provided to allow the area of interest to be brought to the center of the field of view. Other adjustments allowed the rotational axis to be set to intersect the optical axis, so that rotation would be eucentric. This holder allowed rotations of up to 360° of suitable specimens. In practice, the maximum tilt angles that could be used were determined by the nature of the specimen; at large angles the region of interest was obscured by other parts of the specimen. The specimen was imaged by a coverglass-free water-immersion objective lens (Zeiss, 63 × /1.2 NA). In one case five separate data sets were collected. More usually, two or three data sets at angles of 45° to 90° were collected.

To bring the data sets to a common origin, a cross-correlation procedure was used. The standard cross-correlation is given by:

$$x(s) = FT[X(S)]$$

$$\text{where } X(S) = F_1(S)F_2(S)^*. \qquad (21)$$

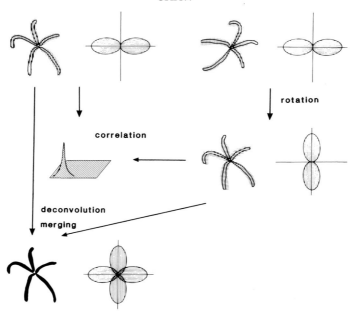

Fig. 6. Image processing scheme used for the tilted view reconstruction.

$F_1(S)$ is the Fourier transform of one data set $f_1(r)$, and $F_2(S)^*$ the complex conjugate of the transform of a second data set $f_2(r)$, after rotation to the same orientation as $f_1(r)$. This correlation function has a very broad peak, since it is dominated by a few very low frequency components. The sharpness of the peak was considerably improved by using instead:

$$X'(S) = \frac{F_1(S)F_2(S)^*}{(|F_1| + K_1)(|F_2| + K_2)}. \quad (22)$$

This function effectively replaces each Fourier term in the transform of the cross-correlation by its phase; the terms K_1 and K_2 are chosen to limit the noise contribution in the same way as for a Wiener filter.

Figure 7 shows two data sets—one at zero tilt (Fig. 7a) (the same as illustrated in the previous sections), and another at a tilt angle of 45° (Fig. 7b). In Figure 7c, the 45° tilt data set has been rotated back to zero tilt for comparison with the true zero tilt. The data sets are clearly very similar, but are affected dif-

ferently by the OTF. It became apparent during this work that the distance between the focal sections was less than the physical focus movement (0.16 μm as opposed to 0.2 μm). This discrepancy can be explained as being due to a difference in refractive index at the specimen (chromatin) and the objective lens (water). The correct value was determined by refinement using the modified cross-correlation method and searching for the section spacing which maximized the height of the cross-correlation peak. The relative orientation angles were also determined in this way [for details, see Shaw et al., 1989].

Stereo projections of the original data sets, processed by the constrained, iterative procedure, and of the combined reconstruction are shown in Figure 8, viewed from each of the original tilting angles. The combined reconstruction effectively gives the best resolution available in each of the original data sets. The increase in resolution in z is seen more clearly in the x–z sections shown for comparison with the previous methods in Figure 5E.

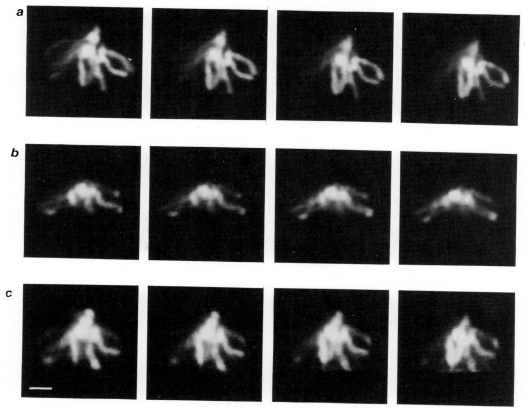

Fig. 7. Comparison of tilted data sets. **a:** The same data set as the previous figures (0°); **b:** the object tilted through 45°; **c:** the 45° data set rotated back to 0° for comparison with 0° data in a. Bar = 2 μm.

VII. FUTURE PROSPECTS

This chapter has concentrated mainly on images obtained by "conventional" epifluorescence microscopy. It might be assumed that confocal microscopy would solve all the problems arising in conventional imaging from the highly anisotropic OTF. However, the confocal OTFs (see Fig. 3) show that even in the best case (minimum pinhole) there is still considerable anisotropy, with the z-resolution 3–4 times worse than that in x and y, and high-frequency attenuation as there is for conventional, wide-field microscopy. When the detector pinhole is enlarged, as it often must be to image weakly fluorescent specimens, the OTF becomes markedly worse. In these cases the main

advantage of confocal imaging (still, however, a considerable one) is the presence of measured data in the "missing cone" low-frequency region rather than any overall increase in resolution. The OTF is thus a less objectionable function and the low-frequency out-of-focus contamination is greatly reduced.

We have investigated the use of the deconvolution methods described above to confocal data [Shaw and Rawlins, 1991, Shaw et al., 1992]. For this study, we used data from thick (about 50 μm) sections of roots of *Pisum sativum* cut using a vibratome and labeled in situ with biotinylated ribosomal DNA probe. The labeled DNA was then detected using a fluorescently labeled secondary antibody [see Rawlins and Shaw, 1990a,b,

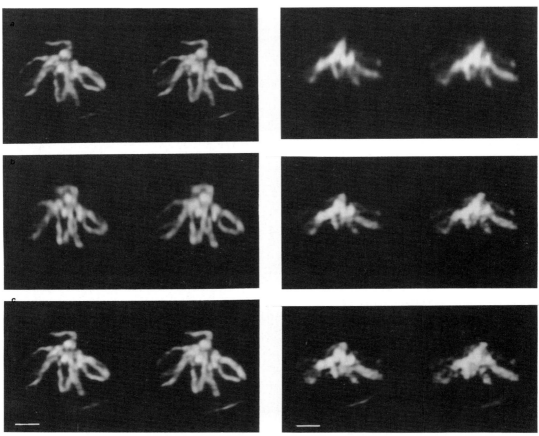

Fig. 8. Stereo projections of the processed 0° data set **(a)**, the processed 45° data set **(b)**, and the combined reconstruction **(c)**. In each case the reconstruction is shown projected at 0° and 45°. The final reconstruction combines the best resolution of each of the tilted data sets. Bar = 2 μm.

and Rawlins et al., 1991, for further details]. This labeling reveals the tandemly repeated rDNA sequences in and around the nucleolus of each cell. Three focal section stacks were measured with different detector pinhole sizes (1 mm, 3.5 mm, and 6 mm) successively from the same field of cells. Each data set was then deconvoluted with the appropriate OTF using the constrained, iterative method described above.

The results of this deconvolution for the smallest detector pinhole are shown in Figure 9, both as x–y and as x–z sections. The deconvoluted image is much crisper and intense features are more clearly defined with less residual blurring in z. This is espe-

cially apparent in the x–z sections. It is evident that even with the smallest detector pinhole the z-resolution in the original image is fairly poor and there is substantial out-of-focus "flare." This is considerably improved in the deconvoluted image.

Finally, an equivalent x–y section from each of the three data sets with the different detector pinhole sizes is shown in Figure 10. The decrease in x–y resolution caused by widening the detector aperture is obvious. However, the deconvolution procedure effectively restores each image to an equivalent final result. In other words, provided the image data has been measured with sufficient accuracy, the deconvolution can

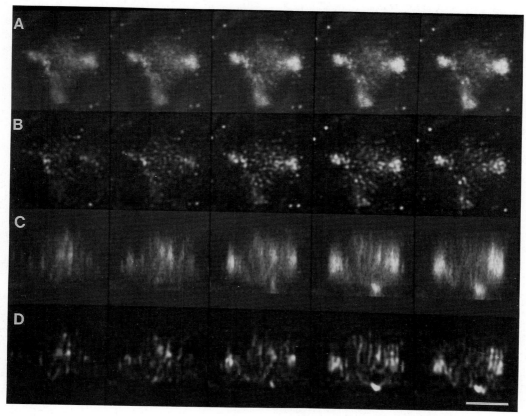

Fig. 9. Sections from a confocal data set of *Pisum sativum* in situ labeled with cDNA probe to ribosomal genes **(a)**, collected with 1 mm pinhole. In **B** the data set has been filtered using the iterative, constrained method and the OTF shown in Figure 3b. In addition to sharpening the image, the processing has removed residual out-of-focus data, and considerably improved the apparent z-resolution. **C,D**: x–z sections of the same data. Bar = 5 μm. (Reproduced from Shaw and Rawlins [1991], with permission from the Royal Microscopical Society.)

restore the different degrees of attenuation of spatial frequencies implied by the different OTFs. Ultimately, the success of this type of restoration depends on the signal-to-noise ratio of the different spatial frequencies in the image, which in turn depends on both the accuracy of image measurement and the form of the imaging OTF.

We regard confocal and conventional microscopy very much as complementary imaging techniques—each has its advantages and disadvantages. Some of the two techniques' relative merits are presented in Table I. The main advantage of confocal micros-

copy is its better OTF, its principal disadvantages the limited laser wavelengths available for excitation and limited sensitivity. Both the latter factors may be considerably improved in the near future. Conversely, the advantages of conventional microscopy coupled with image restoration are very high detection efficiency and the availability of all imaging modes at any excitation frequency obtainable from a conventional light source. The disadvantages are the need for image processing and the requirement for a fairly powerful computer. However, workstations are rapidly decreas-

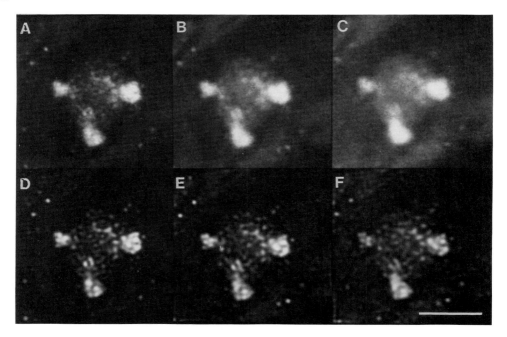

Fig. 10. A single section from each of three data sets of the same nucleus as in Figure 9 imaged with 1 mm, 3.5 mm, and 6 mm confocal detector pinhole **(A–C)**. In **D–F** the equivalent sections from the data sets restored by the iterative, constrained method and the relevant OTF are shown. It is clear that the resolution lost by increasing the detector pinhole size can be restored by the deconvolution procedure, giving very similar results in each case. Bar = 5 μm. (Reproduced from Shaw and Rawlins [1991], with permission from the Royal Microscopical Society.)

TABLE I. Comparison of Confocal Laser Microscopy and Conventional/Image Processing for 3D Imaging

Advantages	Disadvantages
Confocal microscopy	
Improved focal sectioning	Only fluorescence and reflection imaging
Better in-plane resolution (?)	Limited sensitivity
Images available "immediately"	Only certain laser wavelengths available
Simple to use	
Conventional/image processing	
Good focal sectioning possible	Requires considerable computer processing
High sensitivity	Deblurred images not available immediately
Many imaging modes available	Not always simple to use
All conventional wavelengths available	
Bioluminescence imaging possible	

ing in price and increasing in power. Any type of three-dimensional microscopy requires analysis, display, and interpretation of a vast amount of data, and it is becoming necessary to use a powerful computer for these purposes alone; the addition of software for 3D deconvolution and filtering should rapidly become standard.

ACKNOWLEDGMENTS

This work was carried out in part in the laboratories of David Agard and John Sedat at the University of California, San Francisco. I would like to thank them and Yasushi Hiraoka for extensive collaboration and advice. The work was also supported by The Agricultural and Food Research Council of the UK by way of a grant-in-aid to The John Innes Institute.

REFERENCES

Agard DA (1984): Optical sectioning microscopy: Cellular architecture in three dimensions. Ann Rev Biophys Bioeng 13:191–219.

Agard DA, Hiraoka Y, Shaw PJ, Sedat JW (1989): Fluorescence microscopy in three dimensions. Methods Cell Biol 30:353–378.

Aikens RS, Agard DA, Sedat JW (1989): Solid state imagers for microscopy. Methods Cell Biol 29:291–313.

Amos WB, White JG, Fordham M (1987): Use of confocal imaging in the study of biological structures. Appl Opt 26:3239–3243.

Born M, Wolf E (1975): "Principles of Optics." London: Pergamon Press.

Brakenhoff GJ, van der Voort HTM, van Spronsen EA, Nanninga N (1989): Three-dimensional imaging in fluorescence by confocal scanning microscopy. J Microsc 153:151–159.

Carlsson K, Liljeborg A (1989): A confocal laser microscope scanner for digital recording of optical serial sections. J Microsc 153:171–180.

Castleman KR (1979): "Digital Image Processing." Englewood Cliffs, NJ: Prentice-Hall.

Erhardt A, Zinser G, Komitowski D, Bille J (1985): Reconstructing 3D light microscopic images by digital image processing. Appl Opt 24:194.

Flanders DJ, Rawlins DJ, Shaw PJ, Lloyd CW (1989): Computer-aided 3D reconstruction of interphase microtubules in epidermal cells of Datura stramonium reveals principles of array assembly. Development 106:531–541.

Gold R (1964): AEC Research and development report. ANL 6984, Argonne National Laboratory, Argonne, Illinois.

Gruenbaum Y, Hochstrasser M, Mathog D, Saumweber H, Agard DA, Sedat JW (1984): Spatial organization of the Drosophila nucleus: A three-dimensional cytogenetic study. J Cell Sci (Suppl) 1:223–234.

Hiraoka Y, Sedat JW, Agard DA (1988): The use of a charge-coupled device for quantitative optical microscopy of biological structures. Science 238:36–41.

Hiraoka Y, Sedat JW, Agard DA (1990): Determination of three-dimensional imaging properties of a light microscope system: Partial confocal behaviour in epifluorescence microscopy. Biophys J 57:325–333.

Holmes TJ, Liu Y-H, Khosla D, Agard DA (1991): Increased depth of field and stereo pairs of fluorescence micrographs via inverse filtering and maximum-likelihood estimation. J Microsc 164:217–237.

Inoué S (1986): "Video Microscopy." New York: Plenum.

Inoué S (1989): Imaging of unresolved objects, super-resolution, and precision of distance measurement, with video microscopy. Methods Cell Biol 30. New York: Academic Press.

Jansson PA, Hunt RM, Plyler EK (1970): J Opt Soc Am 60:596.

Jordan EG, Zatsepina OV, Shaw PJ (1992): Widely dispersed DNA within plant and animal nucleoli visualized by 3-D fluorescence microscopy. Chromosoma 101:478–482.

Lloyd CW, Pearce KJ, Rawlins DJ, Ridge RW, Shaw PJ (1987): Endoplasmic microtubules connect the advancing nucleus to the tip of legume root hairs but F-actin is involved in basipetal migration. Cell Motil Cytoskel 8:27–36.

Mitchison TJ, Sedat JW (1983): Localization of antigenic determinants in whole Drosophila embryos. Dev Biol 99:261–264.

Oppenheim AV, Willsky AS, Young IT (1983): "Systems and Signals." New York: Prentice-Hall.

Rawlins DJ, Highett MI, Shaw PJ (1991): Localization of telomeres in interphase nuclei by in situ hybridization and 3D confocal microscopy. Chromosoma 100:424–431.

Rawlins DJ, Shaw PJ (1988): Three-dimensional organization of chromosomes of Crepis capillaris by optical tomography. J Cell Sci 91:401–414.

Rawlins DJ, Shaw PJ (1990a): Localization of ribosomal and telomeric DNA sequences in intact plant nuclei by in-situ hybridization and three-dimensional optical microscopy. J Microsc 157:83–89.

Rawlins DJ, Shaw PJ (1990b): Three-dimensional organization of ribosomal DNA in interphase nuclei of Pisum sativum by in situ hybridization and optical tomography. Chromosoma 99:143–151.

Shaw PJ, Agard DA, Hiraoka Y, Sedat JW (1989): Tilted view reconstruction in optical microscopy: Three-dimensional reconstruction of Drosophila melanogaster embryo nuclei. Biophys J 55:101–110.

Shaw PJ, Highett MI, Rawlins DJ (1992): Confocal microscopy and image processing in the study of plant nucleus structure. J Microsc 166:87–97.

Shaw PJ, Rawlins DJ (1991): The point spread function of a confocal microscope: Its measurement and use in deconvolution of 3D data. J Microsc 163:151–165.

Shotton DM (1989): Confocal scanning microscopy and its application for biological specimens. J Cell Sci 94:175–206.

Sisken JE (1989): Fluorescent standards. Methods Cell Biol 30:113–126.

Skaer RJ, Whytock S (1975): Interpretation of three-dimensional structure of living nuclei by specimen tilt. J Cell Sci 19:1–10.

Stokseth PA (1969): Properties of a defocused optical system. J Opt Soc Am 59:1314–1321.

Traas JA, Doonan JH, Rawlins D, Shaw PJ, Watts J, Lloyd CW (1987): An actin network is present in the cytoplasm throughout the division cycle of carrot cells: Actin co-distributes with the four microtubule arrays. J Cell Biol 105:387–395.

Tsien RY (1989): Fluorescent indicators of ion concentration. Methods Cell Biol 30:127–156.

Wijnaendts van Resandt RW, Marsman HJB, Kaplan R, Davoust J, Stelzer EHK, Stricker R (1985): Optical fluorescence microscopy in three dimensions: Microtomoscopy. J Microsc 138:29–34.

Young IT (1989): Image fidelity: Characterizing the imaging transfer function. Methods Cell Biology 30:2–47.

CHAPTER 10

Image Formation in Confocal Microscopy

T. Wilson

I. INTRODUCTION

The conventional optical microscope is a remarkable instrument in the sense that it permits high-resolution images of the entire object field to be obtained instantly. However, if we can relax the requirement of imaging the entire field simultaneously, it turns out that we have more freedom in the design of the optical system. If we take this principle to the limit, all we actually require is that the optical system should be able to produce a good image of *one* point in the object.

A typical arrangement of a reflection scanning optical microscope is shown in Figure 1. The essential components are some form of mechanism to scan the light beam (usually from a laser) relative to the specimen and appropriate detectors to collect the reflected, transmitted, or fluorescent light [Wilson and Sheppard, 1984]. Most of the early systems were analogue in nature [see, for example, Montgomery, 1962], but it is now more usual, thanks to the serial nature of the image formation, to use a computer both to drive the microscope and to col-

lect, process, and display the digital image [Wilson, 1990].

Figure 2a shows the essence of the optical system of a conventional scanning microscope. It consists of an objective lens, a collector lens, and a large-area photodetector. The imaging in this case is identical to that in a conventional, nonscanning instrument. The first lens essentially determines the resolution, while the second lens, the collector lens, plays a secondary role similar to that of the condenser lens in a conventional microscope.

In order to make both lenses contribute to the resolution, we need to turn the collector lens into an *imaging* lens. We can do this simply by using a point detector rather than a large-area one (Fig. 2b). This symmetrical optical arrangement is the form of the *confocal* scanning microscope. Here light from the point source probes a very small region of the object, and the point detector ensures that light from that same small area alone is detected. We might expect that as two lenses are employed simultaneously to image the object, the resolution will be

Electronic Light Microscopy, pages 231–246 © 1993 Wiley-Liss, Inc.

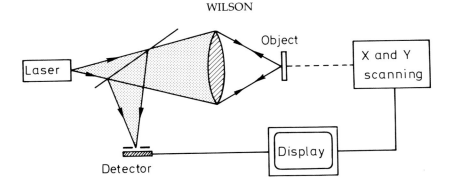

Fig. 1. Schematic diagram of a reflection scanning optical microscope.

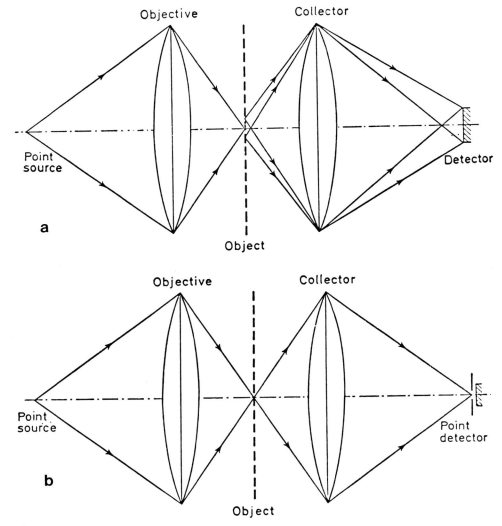

Fig. 2. a: Schematic diagram of a conventional scanning optical microscope. **b:** Schematic diagram of the confocal scanning optical microscope. Note that the only difference in this optical system compared to that of a is the use of a point detector.

improved. This prediction is borne out both in theory and in practice.

It is not appropriate to discuss the theory in great detail here [for more details, see Wilson and Sheppard, 1984, and Wilson, 1990], but we can appreciate the enhanced resolution in the confocal case by considering the image of a single-point object. In the case of a conventional microscope, the image is given by

$$I_{CONV}(v) = \left(\frac{2J_1(v)}{v} \right)^2 \qquad (1)$$

where J_1 is a first-order Bessel function and v is an optical coordinate related to real lateral distance, r, by

$$v = \frac{2\pi}{\lambda} r \sin \alpha \qquad (2)$$

where λ is the wavelength and $\sin \alpha$ the numerical aperture NA of a nonimmersion lens. Equation 1 is the usual form of the Airy disk. In the confocal case, however, both lenses now contribute equally, the image is given by

$$I_{CONF}(v) = \left(\frac{2J_1(v)}{v} \right)^4. \qquad (3)$$

These images are plotted in Figure 3, and although the responses have the same zeros, the central peak of the confocal image has been sharpened up by a factor of 1.4 relative to the conventional image (measured at half-peak intensity). The side lobes are also dramatically reduced, which means that we would expect a reduction in the presence of artifacts in confocal images. Brakenhoff et al. [1979] have confirmed these predictions experimentally.

Figure 4 compares conventional and confocal images of the same object. The confocal image, Figure 4b, is seen to show much more detail than the conventional image. The microscope in both cases was focused on the metallization. In the confocal case, however, we can see additional contrast

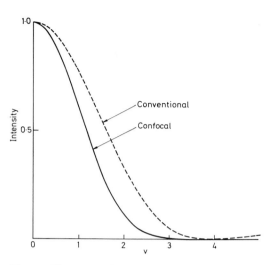

Fig. 3. The image of a point object in a conventional and a confocal microscope.

owing to the variation in surface height of the specimen. The lower regions of silicon which are imaged weakly in the conventional case are not imaged at all in the confocal case. It is the presence of this depth discrimination property which is the key reason for the popularity of confocal microscopes. We discuss this property in detail in the rest of this chapter and point out some of the unique three-dimensional imaging modes which are made possible with confocal microscopes.

II. OPTICAL SECTIONING IN CONFOCAL MICROSCOPES

The origin of the optical sectioning or depth discrimination property may be understood very easily from Figure 5, where we show a reflection-mode confocal microscope and consider the imaging of a specimen with a rough surface. The solid lines show the optical path when an object feature lies in the focal plane of the lens. At a later scan position, the object surface is supposed to be located in the plane of the vertical dashed line. In this case, simple ray tracing shows that the light reflected back

Fig. 4. a: Conventional scanning microscope image of a portion of an integrated circuit. b: Confocal image of the same integrated circuit as in a.

to the detector pinhole arrives as a defocused blur, only the central portion of which is detected and contributes to the image. In this way the system discriminates against features which do not lie within the focal region of the lens. A very simple method of both demonstrating the effect and giving a measure of its strength is to scan a perfect reflector axially through focus and measure the detected signal strength. A simple paraxial theory gives this signal as:

$$I(u) = \left[\frac{\sin (u/2)}{u/2} \right]^2 \qquad (4)$$

where u is a normalized axial coordinate which is related to real axial distance, z, via:

$$u = \frac{8\pi}{\lambda} z \sin^2(\alpha/2). \qquad (5)$$

As a measure of the strength of the sectioning, we can choose the full width at half-intensity of the $I(u)$ curves. Figure 6 shows this value as a function of numerical aperture for the specific case of imaging with red light of a helium neon laser. These curves were obtained using a high-aperture theory [Sheppard and Wilson, 1981], which is more

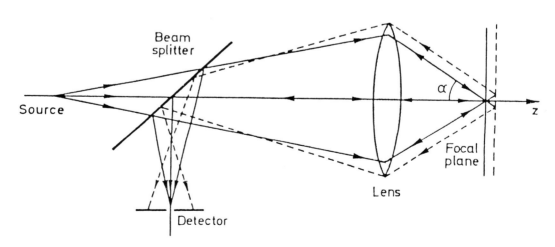

Fig. 5. The origin of the depth discrimination or optical sectioning property of confocal optical systems.

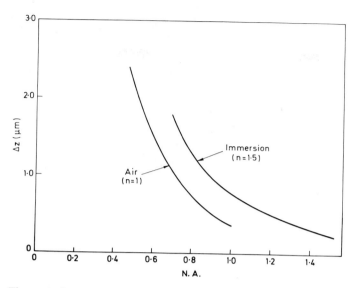

Fig. 6. The optical sectioning as a function of numerical aperture. The curves are for red light (632.8 nm wavelength). Δz is the full width at half-intensity points of the curves $I(u)$ against u.

reliable than Equation 3 at the highest values of numerical aperture.

Although Figure 6 gives a good indication of the degree of sectioning that we might expect, we should remember that these values refer to the imaging of planes. If we consider other object features, such as points and lines, we obtain different degrees of sectioning. Figure 7 shows a comparison of $I(u)$ curves in these cases. In particular, the response for a point object is given by:

$$I(u) = \left[\frac{\sin (u/4)}{u/4} \right]^4, \qquad (6)$$

which shows that the sectioning is weaker than the case for the plane.

It is also worth remembering that these curves refer to bright-field reflection imaging. The image formation in the practically important case of fluorescence imaging is completely different from the bright-field case, and although our remarks apply broadly to both cases, the actual numerical values are different. In the fluorescence case,

for example, the sectioning strength is generally weaker.

Although the degree of optical sectioning is object-dependent, the planar reflector proves to be a very useful object in order to test the alignment and performance of a microscope. It is also useful in determining the quality of objective lenses [Wilson, 1990].

III. IMAGE FORMATION IN FLUORESCENCE MICROSCOPY

We turn now to the area of confocal fluorescence microscopy, which has already proved to be a very useful and powerful tool in many branches of biology and the life sciences. As this is discussed in detail in other chapters of this book, all we do here is emphasize the fact that the image formation in fluorescence microscopes is *completely different* from that in bright-field microscopes. In general, we find that a scanning approach is always to be preferred.

We begin by referring back to the conventional scanning microscope of Figure 2a and imagining that suitable filters are present,

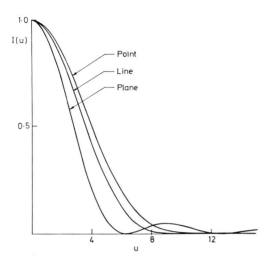

Fig. 7. The variation in detected signal as a point, line, and plane are scanned through focus in a confocal microscope. In the case of the point and line, the microscope is focused on their centers.

such that only the fluorescence radiation is detected. We can immediately see that the *resolution* results essentially from the primary, incident radiation and *not* from the *longer wavelength* fluorescence: The collector lens essentially collects fluorescence radiation onto a detector. In a conventional, nonscanning microscope, this is *not* the case. Here, the primary radiation excites the fluorescence, which is then imaged by the objective lens. Thus, in this case, the resolution results essentially from the *longer wavelength* fluorescence radiation. This leads us to make the very important statement that we expect *scanning* microscopes to be able to produce superior fluorescence images. If we combine these advantages with those concerning dosage and bleaching, together with the optical sectioning of confocal microscopy, we can understand why the technique is so useful [Slomba et al., 1972].

A simple model [Cox et al., 1982] allows us to write the image intensity as:

$$I = |h_{eff}|^2 \otimes f \qquad (7)$$

where the symbol $\otimes$ denotes the convolution operation and f the spatial distribution

of the fluorescence generation. The effective point spread function h_{eff} is given by the product of the point spread functions of the two imaging lenses, $h_1 h_2$, each evaluated at the appropriate primary (λ_1) or fluorescence (λ_2) wavelength. The image formation in the bright-field confocal case is completely different. If t represents transmissivity or reflectivity, then the nonfluorescent image intensity is given by

$$I = |h_1 h_2 \otimes t|^2. \qquad (8)$$

We see, therefore, that the image formation in fluorescence microscopy is always incoherent and linear in f, whereas in the bright-field case it is purely coherent and nonlinear. The linearity of fluorescence imaging makes it an ideal candidate for image enhancement and restoration, as discussed by Shaw in Chapter 9 of this volume. Wiener filters [Gonzales and Wintz, 1987] and Tichonov regularization [Poggio et al., 1985; Bertero et al., 1990] are available. Alternatively, singular value decomposition techniques have been suggested [Bertero et al., 1987; Young et al., 1989].

We finally confirm that a good degree of optical sectioning exists in confocal fluorescent microscopes by considering the signal which we would detect by scanning a uniform fluorescent planar object through focus [Kimura and Munakata, 1990]. We do this by analogy with our plane mirror "brightfield" object. Figure 8 shows that there are no zeros in the response. It also shows that the sectioning becomes coarse as the fluorescent wavelength, λ_2, increases ($\beta = \lambda_2/\lambda_1$). Indeed, in the limit as $\beta \rightarrow \infty$, the sectioning disappears altogether. We can also see that if we detect fluorescent radiation over a range of wavelengths, the final image will contain information from a variety of depths. However, it is clear that in order to obtain optimal sectioning we should try to operate as close to $\beta = 1$ as possible.

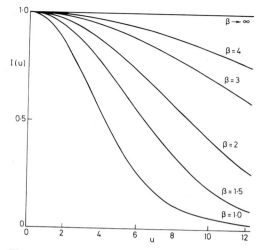

Fig. 8. The detected signal as a perfect planar fluorescent object is scanned axially through focus for a variety of fluorescent wavelengths λ_2. The axial distance, u, is normalized to the primary excitation wavelength, λ_1. We note that if we measure the sectioning by the half-width of these curves, the strength of the sectioning is essentially proportion to β, where $\beta = \lambda_2/\lambda_1$.

IV. APPLICATIONS OF DEPTH DISCRIMINATION

As this property is almost certainly the major reason for the popularity of confocal microscopes, it is worth discussing some of the novel imaging techniques which these microscopes allow. The key point is that the optical sectioning is generally sufficiently strong that any detail which is imaged is imaged *in focus*, that is, the out-of-focus blurring of conventional microscopy is rejected.

This suggests that if we try to image a thick translucent specimen, we can arrange, by the choice of our focal position, to image detail from one specific volume region. In essence, we can section the specimen optically without having to resort to mechanical means. Figure 9 shows an idealized schematic of the process. The portion of the beehive-shaped object that we see is determined by the focus position. In this way it is possible to take a *through-focus* series and obtain data about the three-dimensional structure of the specimen. If we represent

the volume image by $I(x,y,z)$, then by focusing at a position $z = z_1$, we obtain, ideally, the image $I(x,y,z_1)$. This, of course, is not true in practice because the optical section is not infinitely thin.

It is clear that the confocal microscope allows us to form high-resolution images with a depth of field sufficiently small that all the detail which is imaged appears in focus. This suggests immediately that we can extend the depth of field of the microscope by adding together (integrating) the images taken at different focal settings *without* sacrificing the lateral resolutions [Wilson and Hamilton, 1982]. Mathematically, this *extended-focus* image is given by:

$$I_{EF} = \int I(x,y,z)dz. \qquad (9)$$

An example of this technique at high resolution is shown in Figure 10, in which the hairs on an ant's leg, with two hairs projecting to the left, have been imaged. The axial distance between the tips of these hairs is 30 μm, and Figure 10a is an extended-focus image which shows both excellently resolved along their full length, as well as much detail on the leg itself. Figure 10b, in which the microscope has been focused on the tip of the projecting hair, shows an attempt to increase the depth of field in conventional microscopy by using a very low numerical aperture lens, but it is clear that the resolution has suffered dramatically as a result. Even so, nothing approaching the depth of field of the extended-focus image has been achieved.

As an alternative to the extended-focus method, we can form an autofocus or "maximum" projection image by scanning the object axially, and, instead of integrating, selecting the focus at each picture point by recording the maximum in the detected signal. Mathematically, this might be written:

$$I_{AF} = I(x,y,z_{max}) \qquad (10)$$

where z_{max} corresponds to the focus setting giving the maximum signal. The images ob-

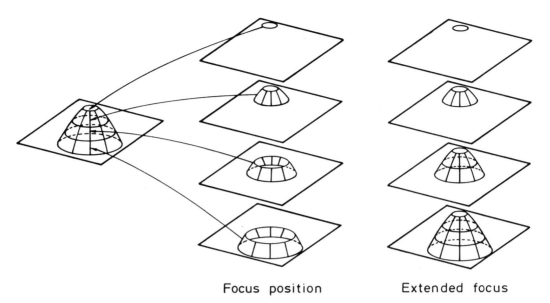

Focus position   Extended focus

Fig. 9. An idealization of the optical sectioning property, showing the ability to obtain a through-focus series of images, which may then be used to reconstruct the original volume object at high resolution.

tained are somewhat similar to the extended focus, and, again, substantial increases in depth of focus may be obtained. Figure 11 shows a digitally scanned autofocus image which has been further improved by edge enhancement, using standard matrix techniques. We can go one step further with the autofocus method and turn the microscope into a noncontacting surface profilometer. Here we simply display z_{max}. This is illustrated in Figure 11c, where we have mapped surface height to gray level. These various techniques may be combined into an isometric image (Fig. 12), where height and surface reflectivity data have been combined.

It is clear by now that the confocal method

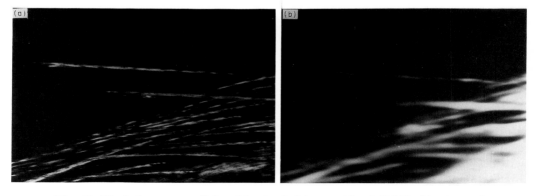

Fig. 10. The hairs on an ant's leg: **(a)** extended focus with a 0.85 NA objective and 632.8 nm He-Ne laser light, and **(b)** conventional scanning image with reduced 1/32 numerical apertures.

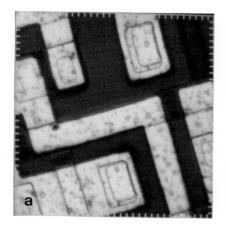

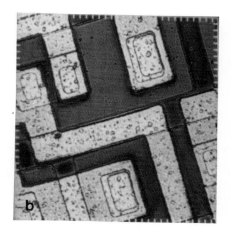

gives us a convenient tool for studying three-dimensional structures in general. We essentially record the image as a series of slices and play it back in any desired fashion. Naturally, in practice it is not as simple as this, but we can, for example, resample and display the 3D data as an x,z vertical section image rather than an x,y image. This is somewhat similar to viewing the specimen from the side. As another example [van der Voort et al., 1989], we might choose to recombine the data as stereo pairs by introducing a slight lateral offset to each image slice as we add them up. If we do this twice, with an offset to the left in one case and the right in another, we obtain, very simply, stereo pairs. Mathematically, we form images of the form:

$$\int I(x \pm \gamma z, y, z)dz \qquad (11)$$

where γ is a constant. In practice it may not be necessary to introduce offsets in both directions in order to obtain an adequate stereo view.

All that we have said so far on these techniques has been by way of simplified introduction. In particular, we have not presented any fluorescence images, because these are dealt with adequately elsewhere. The key point is that, in both bright-field and fluorescence modes, the confocal principle permits the imaging of specimens in three dimensions. The situation is, of course, more involved than we have implied. A thorough knowledge of the image formation process, together with the effects of lens aberrations and absorption, is necessary before accurate data manipulation can take place.

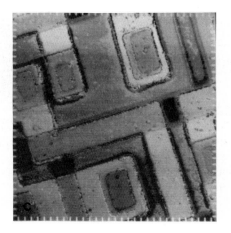

Fig. 11. Digitally scanned and stored images of an integrated circuit: (a) conventional, (b) autofocus with edge enhancements, and (c) a height image where brightness corresponds to surface heights.

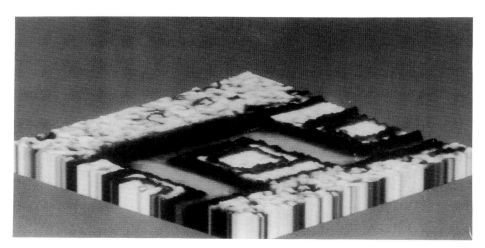

Fig. 12. The **top left-hand** image shows an extended-focus image of a portion of a particularly deep transistor. The **top right-hand** image is the corresponding height image where object height has been coded as image brightness. The **bottom** image is a computer-generated isometric view of the same microcircuit.

V. IMAGING WITH FINITE-SIZED DETECTORS

The only difference between a confocal and a conventional scanning microscope is that the confocal arrangement uses a point detector rather than a large-area one. All the considerable advantages of confocal microscopy derive from the small size of this detector. The success or failure of a particular microscope implementation in achieving true confocal operation depends on the correct choice of pinhole size and shape to approximate as closely as possible the ideal point detector.

In many applications, however, it is not possible to use as small a detector as we would like, because of signal-to-noise problems. This is particularly true in fluorescent imaging. In the following, therefore, we discuss the origins of the optical sectioning property and show how it deteriorates as

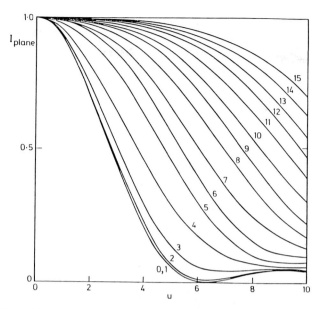

Fig. 13. The variation of $I(u)$ [labeled $I_{plane}(u)$] versus u for a variety of detector pinhole sizes, v_p. $I(u)$, for an ideally small pinhole, is given by Equation 4.

detector geometries of finite size are employed. We begin by considering bright-field microscopes and conclude with a discussion of fluorescent systems. We also mention that use of any finite-sized detector plays an important role in preventing flare and scattered light from being detected.

We will again use the axial scanning of a plane mirror as our measure of optical sectioning. It is clear from Figure 5 that the strength of the sectioning will deteriorate as the detector size increases. Figure 13 illustrates the effect. Here the pinhole size is denoted by a normalized radius, v_p, which is related to real radius, r_p, via:

$$v_p = \frac{2\pi}{\lambda} \frac{r_p}{M} \sin \alpha \qquad (12)$$

where M is the magnification of the lens system between the object and the detector.

We see that the curves corresponding to $v_p = 0$ and $v_p = 1$ are indistinguishable on the scale of Figure 13, which implies that

the strength of the sectioning is insensitive to pinhole size up to some limit [Wilson and Carlini, 1987, 1989]. We can make an estimate of this limit by considering the half-width of the $I(u)$ curves of Figure 13 as a function of detector size, v_p. We plot this metric of resolution, $u_{1/2}$, in Figure 14, and see that the sectioning remains essentially constant until $v_p \sim 2.5$ optical units. If we now claim that we need $v_p \leqslant 2.5$ in order to obtain optimal sectioning, this implies that we need:

$$\frac{M}{NA} \geqslant \left(\frac{\pi}{2.5}\right)\frac{d}{\lambda} \qquad (13)$$

where NA denotes the numerical aperture of the lens and d is the actual diameter of the detector pinhole. This is a very important result, of which we need to be aware if we use a beam-scanning confocal microscope. In these instruments it is necessary to use a variety of objectives in order to cover the entire range of magnification. It is clear

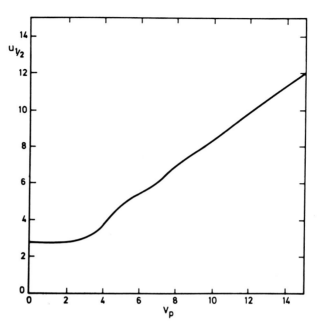

Fig. 14. Half-width of the $I(u)$ versus u curves of Figure 13, $u_{1/2}$ plotted as a function of the normalized detector pinhole radius, v_p.

from Equation 13 that by changing the objective we also change the degree of "confocality." This does not occur in object- or objective-scanning systems, where the same objective may be used to cover the entire range of magnification.

We have concentrated on the axial response because this is the key advantage of confocal microscopy. However, the lateral imaging also deteriorates if finite-sized detectors are used. In particular, the imaging immediately becomes partially coherent, which makes image interpretation and calculation more difficult. As a specific example, we will take a single-point object and ask what happens to the half-width of the image as the pinhole is made larger. We already know that on this basis the ideal confocal system is approximately 1.4 times superior. Figure 15 shows the behavior for intermediate sizes of pinhole. The most important implication, however, is that to obtain the true confocal resolution improvement, a value of v_p less than about 0.5 optical units should be cho-

sen. If we compare this value to that predicted by Figures 13 and 14, we see that the lateral resolution is much more sensitive to pinhole size than the axial resolution. To put these optical units in context, we note that the first dark ring of the Airy disk occurs at a radius of 3.7 optical units.

In the fluorescence case, precisely analogous curves are available [Wilson, 1989, 1990; van der Voort et al., 1990]. Figure 16 shows the half-widths as a function of detector radius. Again we see that optimal sectioning results when the fluorescent wavelength is as close as possible to the primary excitation wavelength. An expression analogous to Equation 13 is also available to determine the degree of confocal behavior. The only difference is in the numerical value of the constant.

We have considered only circularly shaped detectors in our discussion. Similar results are available for other detector shapes, such as slits and squares [Wilson, 1990], which may have advantages in certain applications.

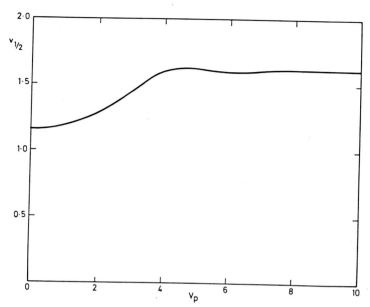

Fig. 15. Half-width of the images of a single-point object, $v_{1/2}$, as a function of the normalized detector pinhole radius, v_p.

Figure 17 compares the degree of sectioning available from confocal microscopes employing ideal point and slit (infinitely narrow) detectors.

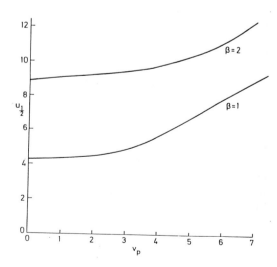

Fig. 16. The half-widths of the fluorescent sectioning curves as a function of normalized detector radius, v_p. The optical units are normalized with respect to the incident wavelength, λ_1, and the parameter $\beta = \lambda_2/\lambda_1$.

VI. DISCUSSION AND CONCLUSIONS

We have not said anything so far about the practical implementation of these microscope systems. This is partly because implementation does not usually change the physics of the image formation, and will usually be application-dependent. The most popular methods of scanning are to scan either the light beam, the object, or the objective. All have their specific advantages and disadvantages. The object-scanning approach may be the most optically desirable, because the beam path is stationary and so can be highly corrected. On the other hand, it may not be convenient, or even possible, to scan the specimen. In the case of the beam-scanning approach, care must be taken in comparing images taken with different numerical aperture and magnification objectives (see Eq. 13).

It might be claimed that a serious drawback to these scanning approaches is that the image is not usually obtained in real time and that an elaborate electrical system is

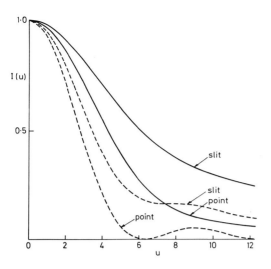

Fig. 17. The axial response of a confocal scanning microscope to appropriate "perfect planar reflector" fluorescent (solid lines) and nonfluorescent (broken lines) specimens using both slit and central point detectors. The fluorescent curves correspond to the limiting best case of $\beta = 1$.

are usually holes in a disk. This disk typically consists of many thousands of holes arranged on interlacing spirals. Finally, the disk is rotated at high speed in order to produce a flicker-free image of the whole field of view. Great care must be taken in the correct choice of size and spacing of these apertures.

The image-formation properties of these instruments are substantially similar to those of single-aperture laser-based confocal microscopes, but have slight differences [Wilson and Sheppard, 1984; Wilson, 1990]. They do have the advantage of not needing a laser source and giving real-time images at frame rates substantially faster than television rate. They do, of course, have drawbacks, the main one being light level. The aperture disk dramatically reduces the amount of light available for imaging, typically by 95% or more. Nevertheless, they can produce good images showing good optical sectioning, and can also operate, with the help of image intensifiers, in the fluorescence mode [Boyde et al., 1990].

We have discussed the origin of the optical sectioning or depth discrimination property of confocal microscopy. This arises from the use of a point detector which essentially serves to eliminate all non-focal-plane-specific detail. If the detector is made sufficiently small, then the optical sectioning effect is sufficiently strong that any detail imaged is imaged in focus. This permits us to build up a through-focus series of images from a thick specimen which may then be processed in many novel ways. In the fluorescence mode, the imaging is linear, and is thus very amenable to further processing and quantification.

required to view the image. These objections are overcome in the tandem scanning microscope [Egger and Petráñ, 1967]. Figure 18 illustrates the principle schematically. A transmission system is shown for ease of explanation, although, in practice, the reflection mode is usually employed. The top diagram shows a microscope with one illumination aperture and one detection aperture. An observer viewing the detection aperture would see a confocal image of the illuminated point. If the apertures were now moved to the position shown in the middle diagram, a confocal image would again be seen of the point that is now illuminated. It is clear now that if we want to form an image of the whole object, an array of a vast number of such apertures would be required, as indicated on the lower figure— each imaging its own illuminated spot on the object. In practice, these systems tend to operate in reflection, and the apertures

In general, then, we can say that there is a great deal to be gained from adopting a scanning approach in microscopy. It opens up the possibility of obtaining a great deal of new information in a relatively straightforward, nondestructive way.

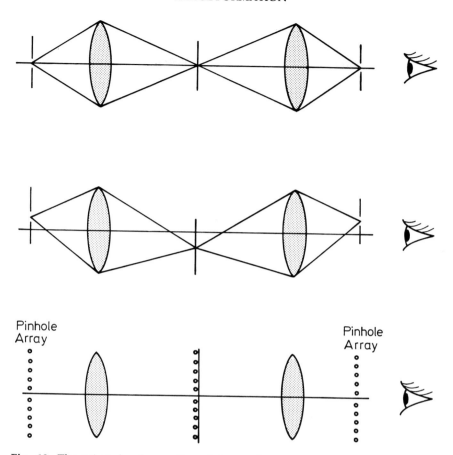

Fig. 18. The principle of operation of the tandem scanning direct view confocal microscope.

REFERENCES

Bertero M, Brianzi P, Pike ER (1987): Super-resolution in confocal scanning microscopy. Inverse Problems 3:195–212.

Bertero M, Boccacci P, Brakenhoff GJ, Malfanti F, van der Voort HTM (1990): Three-dimensional image restoration and super-resolution in fluorescence confocal microscopy. J Microsc 157:3–20.

Boyde A, Jones SJ, Taylor ML, Wolfe LA, Watson TF (1990): Fluorescence in the tandem scanning microscope. J Microsc 157:39–49.

Brakenhoff GJ, Blom P, Barends P (1979): Confocal scanning light microscopy with high aperture immersion lenses. J Microsc 117:219–232.

Cox IJ, Sheppard CJR, Wilson T (1982): Super-resolution by confocal fluorescence microscopy. Optik 60:391–396.

Egger MD, Petràň M (1967): New reflected light microscope for viewing unstained brain and ganglion cells. Science 157:305–307.

Gonzales RC, Wintz P (1987): "Digital Image Processing." New York: Addison Wesley.

Kimura S, Munakata C (1990): Depth resolution of the fluorescent confocal scanning microscope. Appl Opt 29:489–494.

Montgomery PO'B (ed) (1962): Scanning techniques in biology and medicine. Ann NY Acad Sci 97:329–526.

Poggio T, Torre V, Koch C (1985): Computational vision and regularization theory. Nature 317:314–319.

Sheppard CJR, Wilson T (1981): Effects of high angles of convergence on V(z) in the scanning acoustic microscope. Appl Phys Lett 38:858–860.

Slomba AF, Wasserman DE, Kaufman GI, Nester

JF (1972): A laser flying spot scanner for use in automated fluorescence antibody instrumentation. J Assoc Adv Med Instru 6:230–234.

Wilson T (1989): Optical sectioning in confocal fluorescent microscopes. J Microsc 154:143–156.

Wilson T (ed) (1990): "Confocal Microscopy." London: Academic Press.

Wilson T, Carlini AR (1987): Size of the detector in confocal imaging systems. Opt Lett 12:227–229.

Wilson T, Carlini AR (1989): Three-dimensional imaging in confocal imaging systems with finite sized detectors. J Microsc 154:243–256.

Wilson T, Hamilton DK (1982): Dynamic focussing in the confocal scanning microscope. J Microsc 128:139–143.

Wilson T, Sheppard CJR (1984): "Theory and Practice of Scanning Optical Microscopy." London: Academic Press.

van der Voort HTM, Brakenhoff GJ, Boarslag MW (1989): Three-dimensional visualization methods for confocal microscopy. J Microsc 153:123–132.

van der Voort HTM, Brakenhoff GJ (1990): Three-dimensional image formation in high-aperture fluorescence confocal microscopy: A numerical analysis. J Microsc 158:43–54.

Young MR, Davies RE, Pike ER, Walker JG, Bertero M (1989): Super-resolution in confocal microscopy: Experimental confirmation in the 1D coherent and incoherent cases. Europhys Lett 9:773–775.

CHAPTER 11

Acquisition, Processing, and Visualization of Three-Dimensional Confocal Images

H.T.M. van der Voort and A.W.M. Smeudlers

I. INTRODUCTION

The confocal microscope is the first instrument which allows three-dimensional (3D) imaging of intact microscopic objects. As a result, it is of great importance to biology and medicine, where it enables the possibility of studying the 3D structures of tis-

Electronic Light Microscopy, pages 247–271 © 1993 *Wiley-Liss, Inc.*

sues, cells, or cell nuclei [Brakenhoff et al., 1985; Shotton, 1989]. This possibility arises from the unique capability of the confocal arrangement to measure the density of fluorochromes in a small diffraction-limited sampling *volume* around the confocal point [Wijnaends van Resandt et al., 1985], rejecting contributions from any other area regardless of their size and orientation in 3D space. As confocal fluorescent images consist of arrays of these sampling volume elements (voxels), a confocal data set always represents a 3D space in the specimen.

It is essential for 3D microscopy that the data set produced by the confocal scanner is visualized in such a way that the spatial structures in the object can be understood by the operator. In this way, the visualization system, generally a computer, assumes the role of the eyepiece in conventional microscopy, and, as a result, the visualization system must be considered as an *integral part* of the microscope.

In this chapter, several important techniques for the handling and visualization of confocal data are discussed, ranging from data acquisition to graphic techniques for visualization. Because of its key role, special attention is given to the requirements of the computer and the hardware/software organization thereof. In Section II, aspects of the instrumentation are discussed from the point of view of the integration of the digital and opto/mechanical parts. In addition, a description of a confocal microscope based on a standard Unix workstation is given. In Section III, problems associated with data acquisition are discussed. Also in this section, a technique for improvement of the raw data sets is presented. Section IV contains an introduction to the basic principles of 3D image processing. This includes the topics of image enhancement, mathematical morphology, segmentation, and measurement. Finally, in Section V, techniques for visualization of confocal images are discussed.

II. INSTRUMENTATIONAL ASPECTS

A key factor in the efficiency of a confocal microscope is the time required to locate a suitable object in the specimen. Searching for an object in conventional microscopy is greatly facilitated by the instantaneous character of the imaging, together with the fact that out-of-focus objects can easily be found, as they continue to contribute to the image. As mentioned in the Introduction, the image in the confocal microscope always represents a volume in the specimen. Moreover, in contrast to conventional microscopy, areas outside this volume (especially above and below) do not contribute to the image. This property is the basis for the 3D image-forming capabilities of confocal systems, but is a complicating factor when searching for an object.

As a result of its 3D imaging properties, the conventional actions of objective selection, focussing, and x–y movement of a mechanical stage are replaced in the confocal microscope by a manipulation of the size, shape, and location of a "viewing volume." As no clues about the location of objects outside the viewing volume are to be found in the confocal image, the ideal confocal system should be able to generate a representation of a new 3D image quickly (within half a second or less) after an adjustment of the viewing volume. To do so, the computer system must be able to collect data, control the mechanical and electronic parts of the microscope, and visualize the scanned data at a rate sufficient to permit the viewing volume to be manipulated interactively. The following discussion considers the computer system from the point of view of a functional description regarding data acquisition, control and monitoring, and visualization.

A. Functional Demands of the Computer System

1. The data acquisition function. For interactive work, the requirement to generate a new 3D image in less than half a second puts a lower limit on the average data rate of the data acquisition system. The most important factors determining this rate are the smallest possible, but still acceptable, size of the data set, the time available for collection of this data, and the line-scan frequency. The smallest acceptable size depends on factors such as the voxel volume, the type and dimensions of the object, and the type of visualization method used. For our present purpose it is sufficient to concentrate on the information processing problem, leaving the matter of optimal sampling density to Section III.

When simple objects are to be visualized, acceptable results can be generated from a limited data set $D(k,l,m)$, $(k,l = \{0, 1, \ldots , N - 1\}$, $m = \{0, 1, \ldots , M - 1\}$ where k,l represent lateral coordinates perpendicular to the optical axis and m represents coordinates in the direction of the optical axis), adding to a total of $N_v = N^2M \approx 2^{17}$ voxels, with $N = 128$ and $M = 8$. This low value of M is explained in Section IIIA. This minimally-sized data set consists of $N.M$ scan lines, resulting in a lower bound on the line-scan frequency of $f_s \approx 2$ kHz when the entire image is to be scanned within 0.5 sec. The corresponding voxel data rate is $f_d = N.f_s.d^{-1}$, where d is the fraction of a scan period which can be effectively used for scanning a line. With $d = 0.3$ we have $f_d \approx 850$ kHz.

Using beam-scanning devices based on resonant mirrors, rotating polygon mirrors, or acoustic deflectors, scan frequencies as high as $f_s = 15$ kHz (video line frequency) are feasible. As a result, the scanning mechanism in itself need not be the limiting factor in a system with interactive capabilities. However, at this point it must be noted that an increase of the scan frequency must be accompanied by a proportional increase of the laser power, so as to collect an image with the same signal-to-noise ratio, the total energy delivered to the specimen must remain constant. The resulting increase of the dose-rate may cause severe bleaching effects, which may already be a major problem at scan rates as low as 150 Hz. There-

fore, very high speed *single*-beam scanning systems may not be a practical solution for a confocal fluorescence system with interactive capabilities. A possible solution to this problem is the use of multiple beams, thus parallelizing the imaging process. In that way the dose-rate of each individual beam can be reduced while the total laser power delivered is kept constant (see Chapters 13 and 14 on Nipkow disk confocal microscopes).

The number of bits needed in digitizing the signal from the detector depends on the imaging mode and the object type. In transmission or reflection mode, confocal systems can in principle produce dynamic ranges up to 10^4 [van der Voort et al., 1985]. Lossless digitization of such signals requires a 14-bit analogue-to-digital converter (ADC). In the fluorescence imaging mode, the digitization accuracy demands are much less, due to the lower signal-to-noise (S/N) ratio in most fluorescence images. Although there are exceptions, in a typical fluorescence image we find a S/N ratio of 30 dB, corresponding to only 6 bits per voxel. As a result, standard 8-bit commercial hardware can be used satisfactorily in digitizing the signal.

2. The control and monitoring function. The control and monitoring function in a confocal microscope is dominated by two important issues: performance requirements and quality of the user interface. Performance requirements are directly related to the line-scan frequency: The end of each line marks an event which must be reacted upon within a specified amount of time. In a system where flexibility is an important design criterion, this event must be reacted upon by software. The reaction time to the end-of-line event is determined by the line-scan frequency, and can be estimated as roughly $\frac{1}{10}$ of a scan period, i.e., 50 μsec in the case of $f_s = 2$ kHz. As a consequence, a system based on handling the end-of-line events in software must have real-time capabilities. This demand severely limits the choice as to the optimal computer system, in particular the operating system.

Workstations based on the Unix operating system have many properties and facilities which make them particularly attractive for use in instrumentation.[1] Unfortunately, Unix does not comply with real-time performance demands as specified above[2] and as a consequence, the control and monitoring function cannot be fully implemented on a Unix workstation. To solve this problem, the time-critical parts of the control and monitoring process must be delegated to a special front end. Only on the basis of a detailed study of the problems resulting from this added complexity can it be decided whether or not it is feasible to use a standard Unix workstation.

To allow for selection of optimal imaging conditions by nonexpert users, a high-quality user interface is necessary. Such an interface must present the status of the microscope (including the viewing volume being scanned) and allow for modification of the control functions in ways which are intuitively obvious. This is especially relevant for those operations necessary for manipulation of the viewing volume.

3. The visualization function. In the case of two-dimensional (2D) images, visualization by means of a computer is trivial: The 2D data sets can be directly mapped to a monitor using a suitable look-up table to convert measured intensities to gray-values color. To create representations of 3D data sets suggesting depth on 2D devices like CRTs or photographic film, it is necessary to generate 2D images from 3D data sets using special visualization techniques. Detailed descriptions of algorithms for this purpose are given in Section V; here we discuss the consequences for the instrumentation.

In general, there is a direct trade-off between the effectiveness of visualization techniques and computational speed. A useful number to characterize the efficiency of a given visualization algorithm is the average number of standard instructions[3] K_{visu} per voxel. Typical values for K_{visu} are: K_{stereo} = 15 for an simple method to generate a stereoscopic pair (see Section VA2) and K_{SFP} = 180 for a more sophisticated algorithm. The total computing time can be expressed as:

$$T_{visu} = \frac{K_{visu}N^2M}{S}, \tag{1}$$

with N,M the size of the voxel array $D(k,l,m)$ as defined before, and S the average number of standard instruction per second.

For interactive usage, we require $T_{visu} \leqslant$ 0.5 sec, so the workstation must be capable of a performance of

$$S \geqslant 2K_{visu} N^2M. \tag{2}$$

A simple small object can be represented by a data set with N = 128 and M = 8, and effectively visualized by the stereoscopic pair-generating algorithm. In that case we find with K_{stereo} = 15, $S \geqslant$ 0.4 mips, a figure attainable by any current workstation. Interactive visualization of a more complex object represented by a typical data set (N = 256 and M = 16) by the simulated fluorescence process (SFP) method (see Section

---

[1]The excellent facilities for software development, such as packages to generate device drivers, high-level user-interface definition languages, and project management tools, together with the possibility to use software developed for the many other Unix machines, reduce software development time and increase the flexibility of the end result.

[2]Several manufacturers claims to offer "real-time Unix" systems. Whether these claims are justified depends on the definition of real time (especially in connection with the maximum response time), and whether or not nonstandard system calls have to be used to achieve such performance.

[3]By definition, a DEC VAX 11-780 has an average execution speed of 1 million (standard) instructions per second (mips). After measuring the execution speed of other computer architectures in terms of VAX 11–780 equivalents, their performance can be expressed in mips, disregarding the (very architecture-dependent) actual number of instructions needed to achieve this performance. The accuracy of mips figures as given by manufacturers is generally on the order of ±25%.

VA3) requires a workstation with $S \geqslant 360$ mips, a figure out of range of current standard workstations. Interactive 3D visualization on standard workstations is therefore currently only feasible using simple visualization methods.

B. A Workstation-Based Confocal Microscope

The use of a Unix workstation as an integral part of a confocal microscope has several important advantages: (1) The operating system is an industry standard and, as a consequence, a great driving force is present behind the availability of computer programs; (2) an excellent and standardized graphics environment is available to create high-quality user interfaces; and (3) it has very good networking capabilities, resulting in the capability to separate the image acquisition from the image analysis. Off-line analysis can thus be carried out using another workstation without any complications to the user, freeing the microscopic facility for other use. To summarize, use of a Unix workstation would allow for realization of a confocal microscope with many important features with modest development effort and short development time.

The following demands must be met by the workstation:

1. Sufficient performance to visualize 3D images interactively. In the case of a scanning device with a data rate insufficient for this task (for instance, a stage scanning confocal microscope), the system should be able to visualize the data as soon as it is generated.

2. Instrumentation bus. For reasons of physical and electronic robustness, the digital and analogue interfaces between computer and the opto/mechanical system are preferably located in a separate housing, and controlled by a separate instrumentation bus.

3. Real-time aspects. The complexity of the special hardware needed to control the scanner should not be too great, as this would defeat the purpose of a short development time. For the same reason it should be possible to use commercial hardware for the data acquisition unit. Hardware solutions must be found for data acquisition and scanner control, especially for handling of the end-of-line events.

4. To create a high-quality user interface, a single screen should be used. Therefore, it must be possible to use the workstation CRT for the combined display of the visualized images, of the microscope status, and of menu-driven control functions.

To investigate the feasibility of a Unix-based confocal microscope, an experiment was conducted by linking a stage scanning confocal system [Marsman et al., 1983; van der Voort et al., 1985] as the opto/mechanical part with an HP/Apollo DN 4000 workstation.

1. Practical realization. The most critical and difficult item in the list above is number 3, which can only be solved by the addition of special hardware to avoid or handle the end-of-line events and to buffer the incoming data stream. The solution to the first problem was to use a semiautonomous x-y-z scan generator, with the possibility of overriding the automatic mechanism in order to scan a specific line or plane. The second problem was solved by way of a buffered data acquisition unit.

We opted for the VMEbus for use as the instrumentation bus. All interfaces were mounted on this bus with the exception of the data acquisition unit, which was mounted on the peripheral control bus of the workstation (see Fig. 1). Solving the real-time problem by the use of such specially designed electronics requires the development of software (so-called device drivers) to control this hardware. In general, writing Unix device drivers is not a trivial task and requires in-depth knowledge of the Unix operating system. Fortunately, some workstation manufacturers offer toolboxes which greatly reduce this problem. For the HP/ Apollo workstations, such tools were available, making them suitable for use in instrumentation.

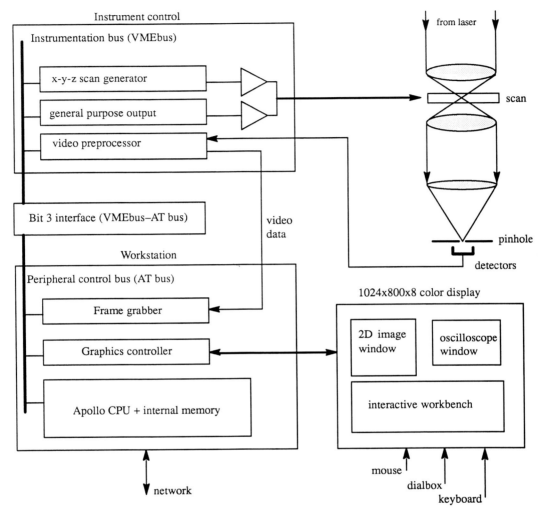

Fig. 1. Workstation-based confocal scanning laser microscope. **Top right:** Confocal arrangement with scanning object stage. The instrumentation bus (VMEbus) is controlled by way of a PC-AT bus to VMEbus interface (Bit 3 Computer Corporation, Minneapolis, MN). The data acquisition is carried out by a "slow-scan" frame grabber (see text), directly mounted on the PC-AT peripheral control bus of the workstation. In this way throughput of image data is optimized. The semiautonomous x-y-z (in order of decreasing repetition rate) scan generator is realized by synchronizing an oscillator driving the x-actuator with an integrator-type circuit driving the y-actuator. The z-direction is controlled by software. A single CRT is used to display image data in a window as soon as it is available, an "oscilloscope window" (see text) on which the profile of a specific line can be inspected, and an interactive workbench for the control of all functions.

Sampling, analogue-to-digital conversion, and temporary data storage in our system are all carried out by a commercial "slow-scan" frame grabber (Data Translation model DT 2851, Marlborough, MA, USA), to which synchronization signals from the scan generator are sent. Through the use of a multiplexing technique, this module may also be used for simultaneous two-channel detection. This is particularly useful in a confo-

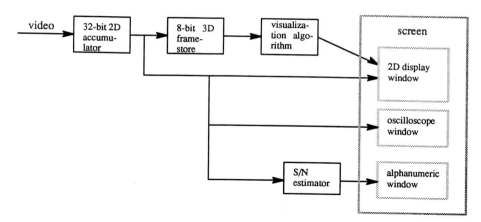

Fig. 2. Data flow in the workstation-based confocal microscope.

cal fluorescence microscope equipped with two or more detectors, each sensitive to a specific band in the fluorescence emission spectrum.

The sole function of the image memory on the frame grabber is to act as a buffer between the real-time world of the scanner and the non-real-time Unix system. This leaves two important tasks to the workstation: (1) reading a freshly scanned line out of the buffer as soon as possible, and storing it in an accumulator buffer located in (virtual) memory; and (2) updating a image display window and an "oscilloscope" window on the display monitor with the most recent image data. In the oscilloscope window, plots of the detected signal versus the x-position of the scanner are drawn with high repetition rate for use during adjustment of the detector sensitivity or alignment of the confocal optics. The accumulator buffer serves to provide an averaging function and a signal-to-noise (S/N) ratio estimator (see Fig. 2). The size of the image display window is fixed and independent of the image format. In that way, the magnification factor of the microscope, for the moment defined as the size of an imaged object in the display window divided by its actual size, is kept independent of the data size. Therefore, when scanning only a few lines for fast user interaction, pixels dupli-

cation is necessary to fill the image display window. The workstation we used proved capable of handling all the tasks mentioned above while maintaining normal user interaction by mouse-driven menus and scroll bars (a computer-drawn lever which can be adjusted by means of mouse interaction) on the interactive workbench window.

After completion of a 3D scan, the results are immediately available to other workstations (nodes) on the network. This creates the possibility of carrying out more elaborate and often time-consuming image processing on another node in the network, if need be simultaneously, or of distributing the computational load over all available nodes. This improves the throughput of the microscopic facility, as it frees the microscope for data acquisition by other users.

III. DATA ACQUISITION AND IMAGE RESTORATION

A. Data Acquisition

The image acquisition in the confocal microscope is dominated by two basic properties of the instrument: the bandwidth difference in the axial and lateral directions, and the limited axial size of the scanned volume. In the limiting optimal case of equal excitation and emission wavelength, the axial

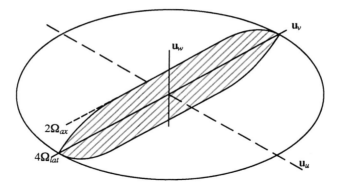

Fig. 3. Spatial frequency bandpass region of a fluorescence confocal microscope. The spatial frequency axis are indicated by $\mathbf{u}_w$ (axial direction), $\mathbf{u}_v$, and $\mathbf{u}_v$ (lateral directions). Ω_{lat} and Ω_{ax} are the lateral and axial bandwidths of a conventional microscope, respectively. The bandpass volume is cylinder-symmetric around the $\mathbf{u}_w$ axis. The shaded plane indicates a cross section through the bandpass volume for spatial frequencies with $\mathbf{u}_u = 0$. In the $\mathbf{u}_w = 0$ plane, the bandpass area is limited by a circle with radius $4\Omega_{lat}$.

and lateral bandwidth of the fluorescence confocal microscope [Sheppard, 1986a,b; van der Voort, 1989] can be found as

$$\Omega_{ax} = \frac{2n}{\lambda}(1 - \cos\alpha); \Omega_{lat} = \frac{4n}{\lambda} \sin \alpha, \tag{3}$$

where n is the refractive index of the medium, λ the wavelength of the excitation or emission light, and α the half-aperture angle of the objective(s) used (Fig. 3). The ratio of the axial and lateral bandwidths is

$$r_{band} = \frac{\Omega_{lat}}{\Omega_{ax}} = \frac{2(1 + \cos\alpha)}{\sin\alpha} \tag{4}$$

and consequently

$$r_{band} \geqslant 2, \alpha \leqslant \frac{\pi}{2}. \tag{5}$$

In the case of an oil-immersion objective with NA = 1.3 (n = 1.515; $\alpha \approx 1$ rad), as is often used in fluorescence work, this results in r_{band} = 3.66. The aspect ratio (axial dimension divided by lateral dimension) of the corresponding samples or voxels should reflect

this value, which is plotted in Figure 4. According to the Nyquist theorem, no information about the image is lost when sampling at a rate twice the bandwidth(s) (see Castleman, Chapter 3 of this volume).

In other words, confocal images consist of 3D arrays of numbers where each number (voxel value) represents a physical quantity measured in a small volume at a specific location in the specimen. The size and shape of this volume, and therefore the smallest significant distance between the centers of adjacent voxels, is determined by the optical properties of the microscope.

Although fluorescence confocal microscopy shows a bandwidth improvement of (in the limiting case of equal excitation and emission wavelengths) a factor of two in *all* directions [Cox et al., 1982; Wilson and Sheppard, 1984], the band is not well filled, resulting in an effective bandwidth which is significantly lower than the theoretical band limit (see Fig. 5).

In the case of a properly sampled cubic viewing volume, $r_{band} \geqslant 2$ leads to at least a two times lower number of samples in the axial direction. For many objects, the axial size of the viewing volume is limited by the maximum penetration depth of the micro-

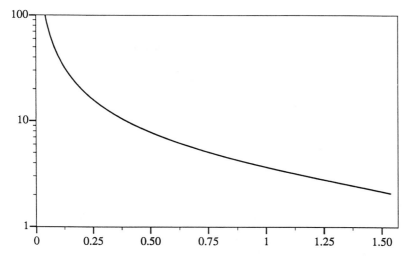

Fig. 4. Bandwidth ratio Ω_{lat}/Ω_{ax} (vertical axis) as a function of the half-aperture angle α (radians; horizontal axis).

scope. This penetration depth is object-dependent, and may be as low as 10 μm in fat emulsions, or up to 100 μm in some plant tissues. Therefore, for the aspect ratio of the viewing volume r_{vol} (axial size/lateral size) we usually have $r_{vol} < 1$. The aspect ratio M/N of the voxel array

$D(k,l,m)$, $(k,l = \{0,1, \ldots, N-1\}$, $m = \{0,1, \ldots, M-1\}$, m corresponding with the direction of the optical axis) can be expressed as

$$\frac{M}{N} = \frac{1}{r_{band}r_{vol}}. \tag{6}$$

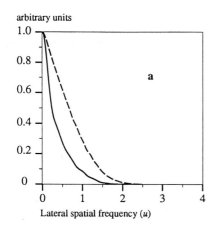

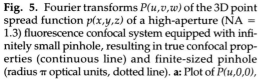
Lateral spatial frequency (u)

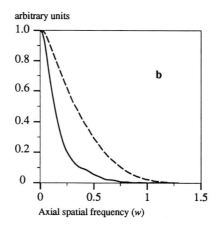
Axial spatial frequency (w)

Fig. 5. Fourier transforms $P(u,v,w)$ of the 3D point spread function $p(x,y,z)$ of a high-aperture (NA = 1.3) fluorescence confocal system equipped with infinitely small pinhole, resulting in true confocal properties (continuous line) and finite-sized pinhole (radius π optical units, dotted line). **a:** Plot of $P(u,0,0)$, horizontal axis; lateral spatial frequency relative to lateral bandwidth of a coherent conventional microscope; **b:** plot of $P(0,0,w)$, horizontal axis: axial spatial frequency, relative to axial conventional bandwidth. Calculation by M. Bertero.

In typical circumstances we have $r_{band} = 4$ and $r_{vol} = 2$, and therefore $M/N = \frac{1}{8}$. This large deviation from a cubic array poses a special problem for visualization: An effective method will have to recreate the illusion of a more or less cubic image volume.

B. Image Restoration

In the case of incoherent fluorescence confocal microscopy, the imaging process can assumed to be linear. By making use of linear inversion techniques, it is possible to restore the image to a considerable degree by off-line numerical processing, provided that the imaging characteristics of the microscope are known. In this section we briefly discuss the Tikhonov restoration method; more elaborate discussions on this topic can be found elsewhere [Bertero et al., 1989, 1990] (see also Shaw, Chapter 9 of this volume, for a discussion of restoration of conventional and confocal fluorescence images).

In the absence of noise, the intensity distribution g in the image is given by:

$$g(\mathbf{x}) = p(\mathbf{x}) \otimes f(\mathbf{x}), \qquad (7)$$

where $\mathbf{x}$ is the position vector (x,y,z), $p(\mathbf{x})$ the instrumental point spread function (PSF), $f(\mathbf{x})$ the distribution of fluorochromes, and $\otimes$ denotes the convolution operation. With the convolution theorem we have also:

$$G(\mathbf{u}) = P(\mathbf{u})F(\mathbf{u}), \qquad (8)$$

where the capitals denote Fourier transform of the corresponding functions, and $\mathbf{u}$ the spatial frequency vector. By making use of linear inversion techniques, it is possible to restore the image to a considerable degree by off-line numerical processing, provided that the imaging characteristics of the microscope are known. The PSF can be obtained either by experimental measurement or by calculations based on diffraction theory. Direct measurement of the fluorescence PSF with sufficient accuracy for use in restoration is difficult, due to the small signal

obtained from a fluorescent volume of an object small enough to be considered point-like [van der Voort et al., 1987a]. When using objects with known geometry such as line patterns [Janssen et al., 1987; van der Voort et al., 1987b] or latex spheres, care should be taken that the Fourier transform of these objects shows no zeros within the pass band of the imaging system. In the case of latex spheres, this demand results in an upper limit to the diameter of $1/\Omega_{lat}$. The small number of fluorescent molecules in such a particle, together with bleaching effects, result in a weak signal and consequently a poor signal-to-noise ratio in the image [Brakenhoff et al., 1986].

For these reasons, a theoretical model of the microscope was developed [van der Voort and Brakenhoff, 1990]. Based on electromagnetic diffraction theory, this model takes into account the high aperture of the objectives, the polarization states of the excitation and emission light, a finite-sized detector pinhole, and (where necessary) the spectrum of the emitted light. Depending on the experimental parameters, the model is capable of computing the actual PSF by numerical means.

Equation 8 suggests that it would be possible to reconstruct the spectral components of the object F within the pass band where P is nonzero by dividing the image G by P and restrict this operation to the pass-band area where P is nonzero. However, this is not feasible, due to the fact that P approaches zero in large areas within the pass band (see Fig. 5), and thus Equation 7 should in fact be replaced by

$$g(\mathbf{x}) = p(\mathbf{x}) \otimes f(\mathbf{x}) + n(\mathbf{x}), \qquad (9)$$

where n represents additive noise. Straightforward dividing G by P would therefore result in

$$\frac{G(\mathbf{u})}{P(\mathbf{u})} = F(\mathbf{u}) + \frac{N(\mathbf{u})}{P(\mathbf{u})}. \qquad (10)$$

Fig. 6. Vertical *(x–z)* section of chromosomes of a *Crepis cappilaris* root-tip cell. Vertical dimension of the section: 3.5 μm. Original **(a)** and reconstructed image **(b)** with reconstruction parameter α = 0.1. The reconstruction was based on the Tikhonov regularized inversion method and a theoretical point spread function as described in the text.

To obtain reliable results from this equation, the contribution of noise should be small with respect to the object spectrum:

$$|F(\mathbf{u})| \;\gg\; \left|\frac{N(\mathbf{u})}{P(\mathbf{u})}\right|. \tag{11}$$

However, because of the appreciable areas within the pass band where $[P(\mathbf{u})]^{-1}$ is large with respect to unity, application of this method would require an extremely (and, depending on the asymptotic behavior of $P(\mathbf{u})$, impossible) noise-free image. For these reasons the result $G(\mathbf{u})/P(\mathbf{u})$ of Equation 10 is extremely sensitive to noise.

A more robust method to solve these so-called "ill-posed" problems is to apply regularized inversion techniques. In the regularized inversion method, due to Tikhonov [Tikhonov and Arsenine, 1977] the reconstructed object function is defined as follows:

$$F(\mathbf{u}) \;=\; \frac{G(\mathbf{u})P^{*}(\mathbf{u})}{|P(\mathbf{u})|^{2} + \alpha^{2}}, \tag{12}$$

where the asterisk denotes the complex conjugate. It can be shown that the regularizing constant α should be chosen equal to the estimated signal-to-noise ratio of the image function $G(\mathbf{u})$ [Bertero et al., 1990].

Using this method in combination with a point spread function $p(\mathbf{x})$ obtained by the numerical model outlined above, promising results have been obtained, especially with respect to the performance of the overall system (microscope and regularized inversion) in the axial direction (see Fig. 6).

IV. IMAGE PROCESSING

A. Introduction

Once an image is acquired and is available as a digital data array in the computer, various image processing procedures may be applied.

First of all, image restoration aims at correcting the 3D data array, such that the numbers of the array more faithfully represent the fluorescent intensity of the corresponding local volume in the specimen. Image restoration is an important procedure, specifically when quantitative measures of the fluorophore amount are at stake. Advanced image restoration techniques incorporate specific information of the staining techniques (for example, the bleaching characteristics thereof), as well as the optical characteristics of the microscope, as discussed in Section IIIB.

Second, techniques for image enhancement aim at visualizing hardly perceptible image details. In contrast to restoration techniques, image enhancement procedures distort the original fluorescent intensity distribution in the process. In microscopical biology, image enhancement techniques are particularly important when visualizing very faint fluorescent images, for example, to show the spatial distribution of DNA-specific sequences by the in situ hybridization technique [Van der Ploeg et al., 1980]. By the nature of their effect, image enhancement techniques are often nonlinear, as discussed in Section IVB2.

A third processing procedure is image segmentation, in which the data array is divided into contiguous sets such that each set corresponds to a microscopical object. Image segmentation is mostly obvious to the human eye, but appears to be very complex in the digital processing of an image. Techniques of segmentation in 3D images are discussed in Section IVC. Such image segmentation is a precondition for advanced visualization and image analysis.

Image analysis techniques are applied to qualitatively describe an object to measure features of the object quantitatively [Vossepoel et al., 1979]. Quantitative measures of geometry and shape are discussed in Section IVD.

B. Basic Paradigms

Before we discuss specific applications of image processing, first two major paradigms of low-level image processing techniques are discussed: linear theory and the theory of mathematical morphology.

1. Linear theory. In linear theory, the image is considered to be the result of a (local) linear process. So, the rationale behind linear neighborhood operations is that a voxel value may be restored, recovered, or represented as the result of some function of the voxel values in a local neighborhood. Examples of the use of local linear operations are the removal of noise and other restoration purposes, or the enhancement of

edges, e.g., in a segmentation procedure, where the effect of the operation depends on the size of the neighborhood and the value of the weighting coefficients.

It is important to notice that such a local neighborhood of voxels should have an extent in three dimensions, as a confocal image essentially is three-dimensional. In the digital processing of confocal images it is simply insufficient to treat the image as if it were a stack of two-dimensional images to be processed one by one as separate two-dimensional entities. With such a sequential processing, the properties of the image in the third dimension are ignored and cannot be improved. This is particularly regrettable as the properties in the third dimension suffer most from the limitations in optical resolution intrinsic to the confocal microscopy. So, linear filters (as well as any other filter) should have an extent in three dimensions, $N_x.N_y.N_z$, in general (where mostly N_x equals N_y and N_z is smaller to compensate for the noncubical shape of voxels).

A commonly used 3D filter is the uniform filter with equal-weight coefficients in all dimensions (see Fig. 7). The uniform filter has low-pass characteristics and improves the perceived quality of the image (at the expense of losing detail). The uniform filter admits a statistical interpretation in that the average value is computed over the local neighborhood, decreasing the noise in the voxel values approximately by a factor $1/\sqrt{N_x.N_y.N_z}$. The Gaussian filter computes the Gaussian weighted local average and has similar smoothing characteristics, but with higher emphasis on the central value. This can be understood from the fact that the Gaussian filter can be approximated by repeated application of a uniform filter. One can prove that the Gaussian filter is the only linear filter which does not introduce new local minima and maxima in the image. In smoothing a noisy image this may be an important property, optimally safeguarding fidelity with the original.

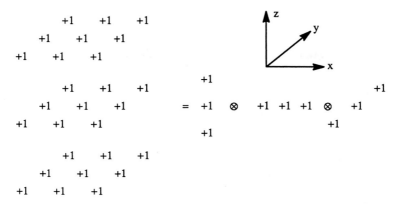

Fig. 7. Decomposition of the 3D uniform filter into three small 1D filters. Successive application of the three three-element filters yields the same result as direct application of the 27-element filter, though at significantly smaller computational cost.

The differential filter has positive- and negative-weight coefficients, an example of which is given in Figure 8 (the Söbel differential operator). It will yield an estimate of the derivative of the voxel values in the x-, y-, or z-direction, clearly showing the locations in the image with a ramp in the voxel value. The Laplace filter computes the second derivative of the voxel values.

All filters should be sized so that they make a compromise between reliable filter value response and local fidelity, taking the x-, y-, and z-resolution into account. A larger-sized neighborhood averages over more voxel values, resulting in a more reliable estimate, whereas a smaller-sized neighborhood results in a more faithful representation of the local situation at the expense of less noise reduction.

Some commonly used linear 3D filters can be computed more efficiently by decomposition into a series of 1D filters. Thus the $3 \times 3 \times 3$ uniform filter can be decomposed into three uniform 1D filters over three elements, as shown in the scheme in Figure 7. The net effect is an improve-

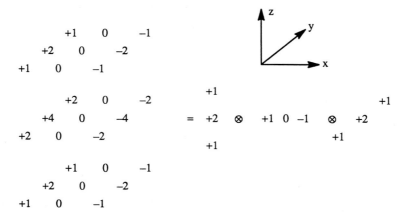

Fig. 8. Decomposition of the 3D Söbel filter into three small 1D filters. Successive application of the two smoothing filters in the y- and z-direction and a differentiating filter in the x-direction yields the same result as direct application of the Söbel filter, though at smaller computational cost.

ment of the efficiency by a factor of three, as in total a sum over 9 voxels is now taken rather than a sum over 27 voxels in the 3D filter case. A similar improvement in the computational efficiency can also be achieved for the 3D Gaussian filter, decomposing it into three 1D Gaussian filters, and for the 3D differential filter by replacing the filter by one 1D differential filter and two 1D Gaussian filters.

2. Nonlinear variations. In the case of image enhancement, the application of strictly linear filters often gives undesirable results, as the edges of objects are smeared out over the surroundings. In detecting or sharpening edges, it is therefore advisable to apply variations of linear theory such as the Kuwahara filter:

1. Partition the $N_x.N_y.N_z$, voxel neighborhood, where $N. \geqslant 3$, into subneighborhoods of size $(N_x - 1).(N_y - 1).(N_z - 1)$, such that the central voxel is the vertex of all eight of them.

2. Compute the mean m and standard deviation s of the voxel values in all $i = 1, 2, \ldots, 8$ subneighborhoods.

3. Assign as the new value for the central voxel that m for which s is minimum over i.

The Kuwahara filter acts as a smoothing filter in regions between edges, whereas it sharpens edges between the regions with different values.

The median filter is another variation of the linear filter, which may be very effective for smoothing, especially when at specific spots in the image the voxel value severely deviates from the true value [Gonzalez and Wintz, 1987]. In the median filter, the central value of the neighborhood is replaced by median value of the neighborhood, a statistic that is less sensitive to extreme values than is the average value used in the linear, uniform filter. From a statistical point of view, the uniform filter can be regarded as the best estimate for the mean

value, whereas the median value can be regarded as a robust estimate for the mean value.

3. Mathematical morphology. Next to linear theory, the other major paradigm in image processing is mathematical morphology [Serra, 1972]. In mathematical morphology, images are seen as sets, one set to each image entity. Thus, the image of each cell or chromosome forms a set. A set is denoted by Y, and in our case it is a 3D set of voxels. The elementary processing unit is also a set (of points), much the same as a neighborhood but now called a structuring element. The 3D structuring element is denoted by X, and its version mirrored through the origin by $\check{X}$. A set X shifted over a 3D displacement vector $\mathbf{t}$ is indicated by $X_\mathbf{t}$.

The mechanism of mathematical morphology is best explained for binary images, having only two different values (by convention "0" for "background" voxels, "1" for "object" voxels), but the theory also works for ordinary images with gray-level intensity values, say integers between 0 and 255. One of the basic operations is the Minkowski set addition (indicated by the symbol $\oplus$) defined by

$$Y \oplus X = \bigcup_{x \in x} Y_x, \qquad (13)$$

where $\cup$ indicates the union. In words, the second half of this definition reads: The union of all sets Y translated over vectors $\mathbf{x}$ contained in the set X (see Fig. 9a). Another basic operation is the dilation: When a set Y is dilated by the structuring element X, the resulting set Y_d contains all voxels $\mathbf{t}$ where the translated origin of the structuring element $X_\mathbf{t}$ *touches* or *overlaps* with the set Y (see Fig. 9c). This overlap/touch condition can be formally expressed as $\mathbf{t}|\epsilon \subset Y \cap X_\mathbf{t}$, where "$\mathbf{t}|$" may be read as "all $\mathbf{t}$ for which there is," ϵ is an arbitrary small set, $\subset$ denotes "subset of," and $\cap$ denotes the intersection. So for the dilated set Y_d we have:

$$Y_d = \{\mathbf{t} \mid \epsilon \subset Y \cap X_\mathbf{t}\}. \qquad (14)$$

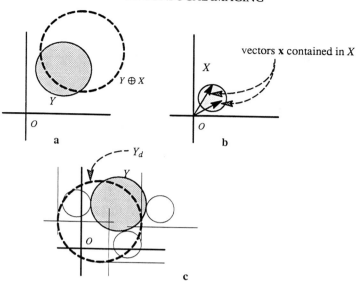

Fig. 9. a: Two-dimensional example of the Minkowski addition $Y \oplus X$ of a set Y with a set X (defined in **b**): The union (area within dotted circle) of sets Y_x (set Y translated over vector x) for all x contained in the set X. **c:** Dilation of Y with set X, now called structuring element: the collection of all positions t of the origin of the translated structuring element X_t where it touches or overlaps with Y (see text). The dilated set Y_d is indicated by the dotted circle. Three instances of just-touching, translated sets X are shown (thin lines). O denotes the origin of the coordinate system in which the sets X and Y are defined. Note that the result Y_d can also be obtained by the Minkowski addition of Y with X (see text).

This is equivalent with taking Y, shifting it over all positions contained in "X and taking the union of all these shifted versions of Y. We can therefore express the dilation in terms of a Minkowski addition, as follows:

$$Y_d = \{t \mid \epsilon \subset Y \cap X_t\} = Y \oplus {''}X. \quad (15)$$

Loosely speaking, the dilation makes an object fatter by the size and shape of the structuring element. When X is a circle with radius r, the dilation operation expands the object Y by r; that is, Y_d contains those points which are within a distance circle r away from Y (see Fig. 9c).

The dual operation of a dilation is the erosion, which makes the object slimmer in a similar way as before, now requiring that the structuring element totally fits within the object Y. One can easily see that taking the erosion on object Y is equal to taking the dilation of the background surrounding the object.

From the viewpoint of neighborhoods and neighborhood operations, the dilation is implemented as the replacement of the central voxel value of the neighborhood by the maximum voxel value in the neighborhood. The erosion is implemented as the minimum voxel value of the voxel values in the 3D neighborhood whose size and shape are given by the structuring element.

4. Mathematical morphology applications. The "closing" is a sequence of a dilation of an object Y by a structuring element "X followed by a erosion of the result by structuring element X. Such a closing operation applied on binary images has the effect of filling holes in the image no larger than the transection of X. It also fills all concavities in the contour of the 3D binary object no broader than the transection of the structuring element X.

The "opening" operation is an erosion followed by the dilation, i.e., the same sequence in reverse. The effect is that two touching

3D objects may be separated into two. The operation can also be used to seive out objects with a minimum transection dimension as indicated by the size of the transection of X.

Mathematical morphology applications are used in the topographical description of images, the automated filling in of holes in objects, the filling in of holes in the intensity value pattern, and similar applications. The effect of the closing is a less-detailed contour or gray-value pattern, whereas the effect of an opening is to break touching objects apart or to open up the gray-value pattern.

In a similar way, the very important skeleton operation is an erosion with limitations such that the topological structure of the object remains the same. Skeletonalization results in a 3D threadlike representation of the image, called the skeleton, useful for describing and visualizing the essential form of an object.

C. Image Segmentation

Image segmentation is the partitioning of the image into regions of contiguous voxels, each region representing a microscopical object. An object may be visible on the basis of a general homogeneous feature discriminating it from the background, or it may be visible on the basis of a transition of the intensity value near the edge of the object.

1. Edge detection. Detection and definition of an edge in the image intensity may be achieved by finding an increase in the first or second derivative. To that end, computing the derivative of voxel values in the x-direction, $\partial f/\partial x$ gives a definition of edges oriented in a plane perpendicular to the x-direction, as discussed above under linear filters.

Repeating the procedure in the other two dimensions, the edge is then located in the image where the edge magnitude, defined as:

$$\left(\frac{\partial f}{\partial x}\right)^2 + \left(\frac{\partial f}{\partial y}\right)^2 + \left(\frac{\partial f}{\partial z}\right)^2 \quad (16)$$

reaches a local maximum.

More advanced techniques for the definition of an edge include information on the local consistency in the direction of the gradient vector,

$$\left[\begin{array}{c} \dfrac{\partial f}{\partial x} \\[4pt] \dfrac{\partial f}{\partial y} \\[4pt] \dfrac{\partial f}{\partial z} \end{array} \right] \quad (17)$$

Yet other techniques are based on the observation that taking a linear sum across an edge is not realistic, as the inside and outside of the object have really very different voxel values characteristics. In these cases, techniques such as the Kuwahara filter technique, or a technique of computing the local difference between a closing and an opening, are used.

2. Region detection. The most simple technique to find an object is to threshold the voxel values in the 3D image with a threshold value higher than the values of the background and lower than the voxel values of the object in the image. The result of such an operation is a binary image such that in the ideal case the object pixels all have the value "1," and the background pixels all have the value "0."

In the presence of noise, a more robust method is required. Based on a model of the image properties formulated in terms of homogeneity criteria, the split, merge, and group (SMG) algorithm [Horowitz and Pavlidis, 1976] provides more reliable segmentation. The algorithm operates as follows:

1. The initialization phase: Split the image into a set of eight equal-sized subimages. In order to allow for convenient splitting of the image at the voxel level, the dimensions

of the image should be a power of two. Repeat this eightfold splitting procedure a number of times.

The resulting subimages can be thought of as nodes in a treelike data structure called an octree: from the "root" node (lowest), representing the original image, stem eight branches, each ending in a node representing the first-level subimage. From each of these stem eight further branches, and so on, until the appropriate initialization level is reached.

2. The split phase: Apply the homogeneity criterion to all end nodes. *Split* individually if the criterion is not met. Repeat until all new end nodes are homogeneous.

3. The merge phase: Apply the criterion to all groups of eight nodes, on the initialization level. *Merge* if the criterion is met. Repeat on the next (lower) level until no groups of eight nodes can be merged.

4. The grouping phase: Until now, only groups of eight equal-sized subimages have been merged. In the grouping phase, adjacent subimages on various levels in the tree are also examined: Apply the criterion to each pair of adjacent subimages. *Group* when together they satisfy the homogeneity criterion.

By careful selection of the number of splitting steps in the initialization phase, the number of operations in steps 2 and 3 can be minimized, and computational efficiency increased. Details of a 3D implementation of the SMG algorithm and application to confocal, noncubic images can be found elsewhere (K.C. Strasters and J.J. Gerbrands, personal communication).

D. Image Measurement

Many different features may be established from an image. Which ones are the best to measure depends on the purpose and problem at hand. We discuss only the geometrical features here.

1. Geometrical measurement. Geometrical features describe properties of place and shape, such as length, area, diameter, distance, and shape. Many aspects of designing unbiased and accurate geometrical measurement methods for direct application in 3D are still unknown or only recently solved, but the principles are now well known.

We discuss the problem of determining the "length of a digital line" [Dorst and Smeulders, 1987] to illustrate the principle of measuring reliably from a digital image.

The length of a line, or in case of a closed curve the perimeter, is a basic feature which seems trivial to compute from a digital (2D) image. Counting the number of pixels, N, along a digital line gives for the length $L_c = N$. A more reliable approximation for the length follows from assigning a weight of 1 for each step between points on a line along the main directions of the grid, total number N_e, and a weight of $\sqrt{2}$ for all diagonal displacements, total number N_o, thus: $L_f = N_e + \sqrt{2} N_o = 1.000 N_e + 1.414 N_o$.

However, what one really wants to know is not the length of the digital line but rather that of the original continuous line in the specimen prior to digitization. A number of continuous straight lines, each having a slightly different length, can all generate the same digital line. It can be shown [Smeulders and Dorst, 1985] that both the above expressions are *biased* (over)estimates for the length of the original line, the bias in L_f being 5.2%. This bias in L_f can therefore be removed by correction with this amount: $L_p = 0.948 (1.000 N_e + 1.414 N_o)$, L_p being recommended for practical use, as it is easy to compute and will give an average inaccuracy of only 2.3%. A similar length determination result holds for 3D (A.L.D. Beckers and A.N.M. Smeulders, personal communication), but the computations are much more complex. This example shows the problems of obtaining reliable digital measurements.

This concludes our quick review of current image processing techniques for 3D confocal imaging, which is still under rapid development.

V. 3D VISUALIZATION

Visualization techniques can be divided into two classes: those suitable for use during a microscopical session (on-line usage) and those suitable only for off-line purposes. The two basic requirements for on-line visualization techniques are that they should be capable (1) of operating on raw data sets without the necessity for user interventions, and (2) of acceptable rendering speed. For interactive purposes an acceptable rendering speed is 0.5 sec, while for inspection purposes times up to 10 sec are still acceptable. The voxel- or volume-rendering techniques discussed in this section all meet the requirements for on-line usage, and, depending on the equipment used, are also suitable for interactive purposes.

Standard graphics packages[4] for 3D visualization are designed for rendering surfaces or boundaries of known objects [Rogers and Adams, 1976]. Before these surface-rendering techniques can be used for volume rendering, as is necessary to visualize confocal images, the voxel-type representation must be converted to a description of surfaces. Therefore, before graphics techniques can be applied to confocal images, three problems have to be resolved: first, segmentation of the voxel representation image into objects; second, detection of the boundaries (if any) of these objects; and third, converting these into a data structure which can be accepted by the graphics package. As the first two problems are not solved in general, and except in very simple cases [van der Voort et al., 1989] require user interaction, surface-rendering techniques are not suitable for on-line visualization of objects.

However, it must be noted that in principle it is possible to model *all* voxels in the data set by a small structure, for example, an ellipsoid of the dimensions of a voxel inserted in the "scene" at the location of the original voxel, and to assign reflectivity and transparency attributes depending on the voxel value to this structure. In this way, surface-rendering techniques can be directly applied to voxel rendering.

For a visualization method to be effective, it should provide clear visual cues to enable the perception of depth. Visual cues for space perception fall into two categories: monocular and binocular [Graham, 1965]. To provide for binocular cues, special presentation techniques like stereoscopic pairs or holograms are necessary. Much more convenient are monocular cues, which can be conveyed by conventional 2D display means such as CRTs or photographs. These include the interposition of objects, light and shade, perspective, and motion parallax. The first requires the removal of hidden surfaces, the second requires variable shading of curved surfaces and the generation of shadows. Perspective only works well in the presence of familiar objects, and as in most cases these are not present in biological images, they must be added for the perspective to provide a cue. In Figure 12, for example, the familiar object "grid" was added to convey perspective. Motion parallax is a particularly powerful cue, especially if it is directly controlled by the observer. However, the rendering speeds needed to realize smooth motion require computational resources considerable higher than those necessary for static purposes, and can be achieved currently on standard equipment only in special cases (see Section VB).

A. Voxel Rendering

The most straightfoward method to inspect a 3D data set is to map all voxels as a series of planes to the screen. As this method provides the human visual system with none of the cues mentioned above, this method cannot be called a 3D visualization technique. Still, the method can be useful for purposes such as verifying whether an object is contained within the image, or whether the object was bleached during the scan, causing attenuation of the signal in the most recently scanned areas.

---

[4]Currently, the most widely accepted 3D graphics standard is PHIGS (Programmers Hierarchical Interactive Graphics Standard).

As explained in Section IIIA, the voxels in confocal data sets usually show an aspect ratio $r_{band} \approx 4$. As a consequence, a voxel-rendering technique must be capable of "magnifying" the z-direction with respect to the lateral directions to recreate the physical dimension of the original viewing volume.

1. Resampling along a line. A basic technique common in voxel rendering discussed in this chapter is the resampling of an image along an arbitrarily oriented line. Subsequently, the resulting one-dimensional array of samples is evaluated in a manner that is different for each specific rendering technique.

The coordinates of the sample points along a line can be expressed as

$$\mathbf{x}_n = \mathbf{p} + n.d.\mathbf{v}, \, n = \{0 \ldots n_{max}\}, \quad (18)$$

where $\mathbf{p}$ is a vector pointing to the first sample, $n_{max} + 1$ the number of samples to be taken, d the distance between two samples, and $\mathbf{v}$ a vector of unit length determining the orientation. Only in special cases will the coordinates $\mathbf{x}_n$ match the integer coordinates of the 3D array of sampling points D which represents the image. Although these special cases play an important role in the rendering techniques discussed below, in general the new sample points will lie in between the original sample points. Therefore, a suitable strategy must be found for the computation of the values at the new sample points. A simple method is to assign the value of the nearest neighbor to the sample point. An improvement over this method is to use trilinear interpolation according to:

$$\begin{aligned}
C_n = &D(k_n, l_n, m_n)(1 - x'_n)(1 - y'_n)(1 - z'_n) + \\
&D(k_n + 1, l_n, m_n)x'_n(1 - y'_n)(1 - z'_n) + \\
&+ \ldots + D(k_n + 1, l_n + 1, m_n + \\
&1)x'_n y'_n z'_n
\end{aligned}$$

with

$$k_n = \text{trunc}(x_n), l_n = \text{trunc}(y_n), m_n = \text{trunc}(z_n)$$

and

$$x'_n = x_n - k_n, y'_n = y_n - l_n, z'_n = z_n - m_n, \quad (19)$$

where k_n, l_n, m_n are indices in the array of original sampling points D, which are computed by truncation of the real-valued coordinates of a sample point $\mathbf{x}_n = (x_n, y_n, z_n)$ along the viewing line. The coordinates of the eight nearest sampling points of D to $\mathbf{x}_n$ are therefore $(k_n, l_n, m_n), (k_n + 1, l_n, m_n), \ldots, (k_n + 1, l_n + 1, m_n + 1)$. The real-valued symbols x'_n, y'_n, z'_n denote the distance of $\mathbf{x}_n$ in the x-, y-, and z-direction, respectively, to the original sample point (k_n, l_n, m_n).

2. Stereoscopic pairs. Stereoscopic pairs can be generated from 3D data sets by tracing "viewing lines," originating from two different viewpoints through a projection plane into the viewing volume corresponding to the data set. A discrete left image of the stereoscopic pair is formed by restricting the viewing lines to those that intersect the projection plane at the center of a pixel (located on its projection plane) of the left stereo image. In this way, a viewing line is assigned to each pixel in the stereoscopic pair. Subsequently, the data set is sampled along the viewing line using the method described above.

The value of the pixel in the stereoscopic pair corresponding to the viewing line can now be computed on the basis of the samples taken along the viewing line. Several methods to achieve this can be used: The samples can be averaged, exponentially weighted, or the largest value can be extracted [Shotton, 1989]. This last method, also known as the "maximum" projection method, is the one most frequently used by our group. To avoid interpolation, the values of $\mathbf{p}$, d, and $\mathbf{v}$ in Equation 18 are chosen as follows:

$$\begin{aligned}
\mathbf{p} &= (k, l, 0) \text{ with } k, l = \{0, 1, \ldots, N - 1\} \\
d.\mathbf{v}_L &= (v_{L;k}, v_{L;l}, 1) \text{ with} \\
v_{L;k}, v_{L;l} &= \{0, 1, 2 \ldots\}, \quad (20)
\end{aligned}$$

$\mathbf{v}_L$ determining the direction of the viewing lines corresponding to the left stereoscopic image. With a similar vector $\mathbf{v}_R$ for the right image, the stereoscopic pair can now be computed as:

$$L(k,l) = \text{Max}\{D(k + n.v_{L;k}, l + n.v_{L;l}, n)\}$$
and
$$R(k,l) = \text{Max}\{D(k + n.v_{R;k}, l + n.v_{R;l}, n)\},$$
$$(21)$$

with $n = \{0 \ldots M - 1\}$. Boundary problems are solved by substituting $D = 0$ for voxels outside the data array D.

Selecting the largest voxel value has three advantages: It is computationally simple, there is no need for scaling of the result, and, most important of all, there is no need for pre-processing. The restriction of the viewing directions makes the algorithm fast: fifteen standard instructions per voxel (see Section IIA3). This, together with the fact that the algorithm requires no pre-processing, makes it particularly suitable for on-line usage.

The resulting depth in the stereo image is determined by the angle between the vectors $\mathbf{v}_L$ and $\mathbf{v}_R$, and can be made sufficiently large to suggest a cubic image. The number of x–y planes M needed for a strong depth impression is small. We found that 16 of these planes were more than sufficient to create an image in which the individual planes can no longer be seen. Although a practical advantage, this points to two weaknesses in the approach. First, as is well known, it is much more difficult to estimate distances in depth than in lateral directions. Second, since the algorithm searches only for the largest pixel value along a viewing line, hidden lines or surfaces are not always removed. As a result, correct interposition cues are not reliably provided, leading to ambiguous or even confusing stereo images. More elaborate methods as mentioned above for assigning pixel values can eliminate this problem [Carlsson and Åslund, 1987; Shotton, 1989], but will negate the speed and simplicity advantages of this method over the more sophisticated visualization methods discussed below. Moreover, sophisticated methods can also be used to generate stereoscopic pairs (see Fig. 10). An additional but important practical problem is that 10% of the population is not able to perceive depth from stereoscopic pairs [Richards, 1970].

3. The simulated fluorescence process (SFP) algorithm.

The SFP voxel-rendering algorithm is based on simulation of a fluorescence process. It does so by "exciting" a 3D distribution of fluorophores (corresponding to the image), and subsequently calculates how much light will be emitted in the direction of the observer. Because of the basically 3D nature of the fluorescence process, the SFP algorithm does not depend on detecting gradients or boundaries in the image, and is therefore ideally suited for visualizing confocal images, which, especially at high magnification, often do not contain sharp boundaries. The algorithm consists of two steps: The first corresponds to the excitation phase in a fluorescence process; the second step corresponds to the emission phase. In the first step, the measured data set $D(k,l,m)$, as defined above, is "excited" by a light source placed at infinity. Consider a collection of voxels

$$C_{n;k,l} = D(k + s_k.n, l + s_l.n, n), \text{ with}$$
$$n = \{0, 1, \ldots, M - 1\}, \qquad (22)$$

encountered along an "exciting ray." The direction of this ray is determined by the vector $\mathbf{s} = (s_k, s_l, 1)$. The coefficients of $\mathbf{s}$ are usually chosen to be integers in order to avoid the need for interpolation between voxels as in Equation 19, and to allow for fast integer arithmetic voxel address calculation. From the C_n (the indices k,l are hereafter omitted for convenience) a new collection E_n is formed according to

$$E_n = I_n.C_n \text{ and } I_{n+1} = I_n(1 - c_{ex}.C_n) \quad (23)$$

with $n = \{0, 1 \ldots M - 1\}$, and I_0 set to some constant initial value. The absorption of the "exciting radiation" is controlled by the extinction factor c_{ex}, with $c_{ex}.D_n \leqslant 1$ for all values of D_n. The E_n are written back to form a new data set $E(k,l,m)$, which can be thought of as representing a distribution of emitting fluorochromes.

In the second step, the amount of "fluorescence light" emitted in the direction of

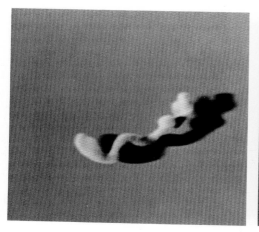 

Fig. 10. Stereoscopic pairs of a single chromosome of a *Crepis capillaris* root-tip cell. This pair was generated by application of the SFP algorithm from two different viewpoints. The same original image data were used in Figure 12.

the observer is calculated. Again, to simplify voxel address calculations, the observer is placed at infinity. The location of the observer is determined by the vector $\mathbf{v} = (v_k, v_l, 1)$, and is independent of the illumination vector $\mathbf{s}$. From each of the collections of voxels, one calculates

$$E'_n = E(k + v_k.n, l + v_l.n, n) \quad (24)$$

and

$$D'_n = D(k + v_k.n, l + v_l.n, n). \quad (25)$$

A new collection I_n, representing the emitted light traveling in the direction of the observer, is obtained by:

$$I_{M-1} = E'_{M-1}$$
$$I_{n-1} = I_n(1 - c_{em}.D'_n) + E'_n,$$
$$n = \{0, 1, \ldots, M - 1\}. \quad (26)$$

The absorption of the emitted radiation is controlled by the extinction factor c_{em}, with $c_{em}.D_n \leqslant 1$ for all values of D_n. The rays leaving the voxel array are the $I_{-1;k,l}$ and represent the final 2D output image.

The extinction factors c_{ex} and c_{em} may be chosen independently. In this way, different types of objects can be optimally visualized. For instance, high-magnification images are best rendered by the SFP algorithm with low values for c_{ex} and c_{em}; complicated low-magnification images generally require a low value of c_{ex}. In addition, a specific object can be visualized in different ways (see Fig. 11) to emphasize different aspects of its structure. With D_n in the usual range of $0 \ldots 2^8 - 1$, typical values for c_{ex} and c_{em} are in the range between $0.1 . 2^{-8}$ and 2^{-8}.

The algorithm can be implemented very efficiently due to the parallel projection and the simplicity of the required calculations, resulting in $K_{SFP} = 180$ standard instructions per voxel. By treating the image as a series of planes defined by the vectors $\mathbf{s}$ and $\mathbf{v}$, the algorithm can be parallelized as these planes, typically $x–z$ or $y–z$ planes, and may be processed independently by separate processing units. In that way, data rates of fast scanners can be matched.

B. Graphic Techniques

In many industrial design applications, surface-rendering techniques are used. The designer usually controls the shape of objects by specifying surfaces or intersections of these with a given plane. Because of the importance of these applications for the industry, many surface-rendering packages have been written and special rendering

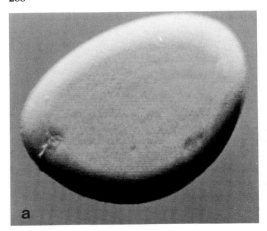
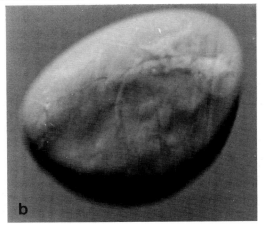

Fig. 11. Eggs of *Acari* (mite) rendered with the SFP algorithm. Smallest diameter of the egg: 70 μm. **a:** High values of both c_{ex} and c_{em} result in the egg being displayed as a solid. **b:** Low values of both c_{ex} and c_{em} render the egg transparent, with internal structures visible.

hardware has been developed. As design is typically an interactive process, many manufacturers strive toward, or have already realized, systems which are capable of rendering complex objects with sufficient speed to meet the requirements for interactive usage or even smooth motion. Apart from these speed advantages, modern surface-rendering techniques offer features such as complete freedom in the location of the viewpoint, the possibility of examining individual structural components separately, and the possibility of assigning attributes such as color, texture, or transparency separately to different objects in the image. However, before these features can be used, the voxel image from the microscope must be converted to a surface description. These surfaces can be specified to the package in the form of either an analytical description, or more explicitly, as a set of control points (3D vectors) through which the surface is fitted, or as a collection of graphic primitives such as spheres, cubes, or planar polygons.

The critical step in the conversion of a voxel representation to a description of boundary surfaces is segmentation. At present, a general solution to the problem is not available (see Section IVC) and automatic segmentation is feasible only in simple cases. In all other cases, segmentation of a voxel image has to be carried out fully interactively [Tuohy, 1987] or by human-aided automatic techniques. This makes surface-rendering graphic techniques currently unsuitable for on-line visualization. However, the advantages of these graphic techniques, as mentioned above, make them especially useful for visualizing the results of (off-line) image analysis.

A typical result of the latter strategy is shown in Figure 12. Here, the shape of a single prophase chromosome of a *Crepis capillaris* [Babcock, 1947] root-tip cell is visualized using graphics techniques. Before the object could be visualized in this way, the original image was edited interactively to remove all other components. Subsequently, along the chromosome, a chain of local centers of mass was determined using an interactive technique called the homing cursor [Houtsmuller et al., 1990; Oud et al., 1989].

This technique consists of moving a 3D cursor through the image, which is rendered (including the cursor) using the SFP algorithm. The cursor is first manually positioned onto an end of the chromosome, after which

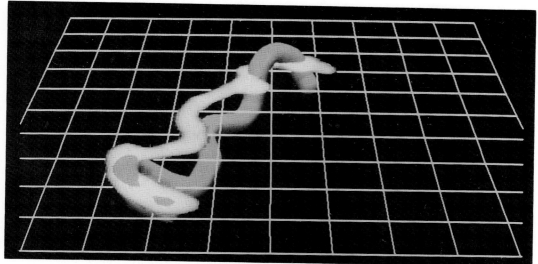

Fig. 12. A single chromomsome of *Crepis capillaris* as visualized after shape analysis (same data as in Fig. 10). In the analysis, the "spine" of the chromosome was determined by chaining a series of points, each corresponding to a local center of mass (see text). Using building elements resembling the cylindrical and elbow pieces, a stovepipe-like structure was constructed, centered around the spine of the chromosome **(red structure)**. To convey the perspective of the scene, a rectangular grid was added. To show the gain of the combination confocal imaging and 3D analysis, the 2D image of the chromosome as it would have been seen in a conventional microscope was projected, but without out-of-focus blur **(blue-gray structure)**. Figure appears in color in Color Figure Section.

it automatically homes in on the local center of mass. With the consent of the operator its location is recorded, and the cursor then moves automatically to the next candidate point. In this manner a gray-weighted medial axis ("spine") of the chromosome is determined. The medial axis can then be visualized using wire frames or by using a geometrical model of the chromosome, consisting of jointed cylinders and elbow pieces of a diameter corresponding with the typical actual chromosome diameter (the stovepipe model; see Fig. 12). In the wire-frame technique, the medial axis is displayed as a single line in space. As such "space-curves" can be rendered rapidly even on workstations not equipped with special hardware, *motion parallax* can be used to study the 3D structure of a chromosome or group of chromosomes rendered in this way.

VI. CONCLUDING REMARKS

In this chapter we have reviewed the most important topics in the information processing in confocal microscopy: 3D image processing (including restoration), image analysis and image visualization, and instrumentation. We have placed considerable emphasis on this last topic, as in our opinion the full potential of the confocal microscope can only be realized when the optics are completely integrated with a 3D image processing/visualization system. We found that the use of a standard Unix workstation for this particular instrumentation problem is an extremely efficient solution in terms of software development effort and total hardware cost.

In the section about rendering techniques for voxel images, we have discussed several algorithms which are suitable for the visualization function in such a "confocal workstation." The simple stereoscopic algorithm discussed in Section VA2 is sufficiently fast to allow real-time visualiation on standard equipment. The SFP algorithm offers more, and more effective, space perception cues and may in future turn out to be a key element in interactive image analysis.

Surface-rendering graphics techniques are less suitable for on-line visualization or interactive analysis, as they must rely on segmentation and analysis techniques which are not yet developed to the necessary degree. However, it may very well be possible that the performance increase of surface-rendering hardware expected in the next decade will allow interactive voxel rendering by the "brute force" method. To summarize, the image processing and voxel-rendering techniques described in this chapter together constitute an already sufficiently powerful toolbox for qualitative assessment of 3D confocal images. In quantitative image analysis, much work has still to be done, especially in the fields of image restoration, image segmentation, and measurement.

ACKNOWLEDGMENTS

We thank J. Heijmink, J.J. Krol, and B. Mosterd for their assistance in software development, and W. Takkenberg for his work in the darkroom. We are grateful to A. Houtsmuller and A. Mans for supplying raw image data and specimens. We are particularly indebted to Professor M. Bertero for many inspiring discussions on the topic of image restoration. This work has been supported by the Stichhting Technische Wetenschappen (STW), the Foundation for Fundamental Biological Research (BION/NWO), E.C contract BAP 0293-NL, and the SPIN research program for Three-Dimensional Image Analysis.

REFERENCES

Babcock FB (1947): The genus Crepis. University of California Publications (Botany) 21:22.

Bertero M, Boccacci P, Brakenhoff GJ, Malfanti F, van der Voort HTM (1989): Three-dimensional image restoration in fluorescent confocal scanning microscope. Opt. Storage Scanning Technology 1139:86–91.

Bertero M, Boccacci P, Brakenhoff GJ, Malfanti F, van der Voort HTM (1990): Three-dimensional image restoration and super-resolution in fluorescent confocal microscopy. J Microsc 157:3–20.

Brakenhoff GJ, van der Voort HTM, van Spronsen EA, Linnemans WAM, Nanninga N (1985): Three-dimensional chromatin distribution in neuroblastoma nuclei shown by confocal scanning laser microscopy. Nature 317:748–749.

Brakenhoff GJ, van der Voort HTM, van Spronsen EA, Nanninga N (1986): Three-dimensional imaging by confocal fluorescence microscopy. Ann NY Acad Sci 483:105–415.

Carlsson A, Åslund N (1987): Confocal imaging for 3-D digital microscopy. Appl Opt 26:3232–3238.

Cox IJ, Sheppard CJR, Wilson T (1982): Super-resolution by confocal scanning fluorescent microscopy. Optik 60:391–396.

Dorst L, Smeulders AWM (1987): Length estimators for digitized contours. Comp Graph Vis Im Proc 40:311–333.

Gonzalez RC, Wintz P (1987): "Digital Image Processing," 2nd ed. Reading, MA: Addison-Wesley.

Graham CH (1965) Visual space perception. In Graham CH (ed): "Vision and Visual Perception," pp 504–547. New York: J Wiley & Sons.

Groen FCA, Van der Ploeg M (1979): DNA cytophotometry of human chromosomes. J Histochem Cytochem 27:436–440.

Horowitz SL, Pavlidis T (1976): Picture segmentation by a tree traversal algorithm. J Assoc Comp Machinery 23:368–388.

Houtsmuller AB, Oud JL, van der Voort HTM, Baarslag MW, Krol JJ, Mosterd B, Mans A, Brakenhoff GJ, Nanninga N (1990): Image processing techniques for 3-D chromosome analysis. J Microsc 158:235–248.

Janssen GCAM, Rousseeuw BAC, van der Voort HTM (1987): Testpattern for fluorescent microscopy. Rev Sci Instr 58:598–599.

Marsman HJB, Stricker R, Wijnaendts van Resandt RW, Brakenhoff GJ, Blom P (1983): Mechanical scanning system for biological applications. Rev Sci Instru 54:1047–1052.

Oud JL, Mans A, van der Voort HTM, van Spronsen EA, Nanninga N (1989): 3-D chromosome arrangement of *Crepis capillaris* in mitotic prophase and anaphase as studied by confocal scanning laser microscopy. J Cell Sci 92:329–339.

Richards W (1970): Stereopsis and stereoblindness. Exp Brain Res 10:380–388.

Rogers DF, Adams JA (1976): "Mathematical Elements for Computer Graphics." New York: McGraw-Hill.

Serra J (1972): "Image Analysis and Mathematical Morphology." New York: Academic Press.

Sheppard CJR (1986a): The spatial frequency cut-off in three-dimensional imaging. Optik 72(4): 131–133.

Sheppard CJR (1986b): The spatial frequency cut-off in three-dimensional imaging II. Optik 74(3): 128–129.

Shotton DM (1989): Confocal scanning optical microscopy and its applications for biological specimens. J Cell Sci 94:175–206.

Smeulders AWM, Dorst L (1985): Measurement issues in morphometry. Anal Quant Cytol 7:242–249.

Tikhonov AN, Arsenine VY (1977): "Solutions of Ill-Posed Problems," Washington, DC: Winston/Wiley.

Tuohy M, Conchie C, Knox RB, Szarski L, Arkin A (1987): Computer-assisted three-dimensional reconstruction technology in plant cell image analysis: Applications of interactive computer graphics. J Microsc 147:83–88.

Van der Ploeg M, Bauman JGJ, Wiegant J, Borst P, van Duijn P (1980): A new method for fluorescence microscopical localization of specific DNA-sequences by in situ hybridization of fluorochrome labeled RNA. Exp Cell Res 138:485–490.

van der Voort HTM, Brakenhoff GJ, Valkenburg JAC, Nanninga N (1985): Design and use of a computer-controlled confocal microscope. Scanning 7:66–78.

van der Voort HTM, Brakenhoff GJ, Janssen GCAM, Valkenburg JAC, Nanninga N (1987): Confocal scanning fluorescence and reflection microscopy: Measurements of the 3-D image formation and application in biology. SPIE 809:138–143.

van der Voort HTM, Brakenhoff GJ (1987): Determination of the 3-dimensional optical properties of a Confocal Scanning Laser Microscope. Optik 78:48–53.

van der Voort HTM (1989): Image formation and processing in the confocal microscope. Thesis, University of Amsterdam.

van der Voort HTM, Brakenhoff GJ, Baarslag MW (1989): Three-dimensional visualization methods for confocal microscopy. J Microsc 153:123–132.

van der Voort HTM, Brakenhoff GJ (1990): 3D image formation in a high-aperture fluorescence confocal microscope: A numerical analysis. J Microsc 158:43–54.

Vossepoel AM, Smeulders AWM, Van den Broek K (1979): DIODA: Delineation and feature extraction of microscopical objects. Comp Prog Biomed 10:231–244.

Wilson T, Sheppard CJR (1984): "Theory and Practice of Scanning Optical Microscopy. London: Academic Press.

Wijnaends van Resandt RW, Marsman HJB, Kaplan R, Davoust J, Stelzer EHK, Sticker R (1985): Optical fluorescence microscopy in 3 dimensions: Microtomoscopy. J Microsc 138:29–34.

CHAPTER 12

High Scan-Rate Confocal Laser Scanning Microscopy

A. Draaijer and P.M. Houpt

I. INTRODUCTION

A. The Conventional Microscope

The light microscope, either laser scan or with a more conventional light source, sometimes supported with a video- or a cooled charge-coupled device (CCD) camera, is an invaluable tool for looking at features down to the submicron range in both living and dead specimens. It suffers from a major disadvantage, i.e., a small depth of focus, especially at high resolution. The depth of focus can be as low as a few tenths of a micron. This depends upon the total magnification and the numerical aperture (NA) of the microscope objective used. Images from a

Electronic Light Microscopy, pages 273–287 © 1993 Wiley-Liss, Inc.

Fig. 1. Micrograph of a Smithsonite sample recorded with a laser scanning microscope in a conventional setup. (Horizontal field of view = 80 μm, NA of the objective = 0.8.)

sample with a larger thickness than the depth of focus show a confusing mixture of blurred and sharp parts (Fig. 1).

If the microscope "records" information from the full field of view, the information in the resulting image can get lost because of an undesirably high influence of the background caused both by light coming from above and below the focal plane and also from stray light from the same focal plane. This is typically the case with fluorescence microscopy of large-volume samples having many fluorescent structures (Fig. 2).

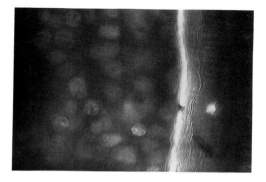

Fig. 2. Extremely thick sample of bone embedded in plastic (PMMA) and stained with brilliant sulfo-flavine, recorded with a laser scanning microscope working in the conventional fluorescence mode. (Horizontal field of view = 110 μm, NA objective = 1.4.)

B. The Confocal Concept

Minsky [1961] already recognized the advantage of the so-called confocal concept. In such a system, a point light source is focused in or on the object. Out-of-focus information is rejected by means of a spatial filter in the image plane of the projected (transmission) or back-projected (reflection) focal spot (Fig. 3). This, however, reduces the field of view to one spot, making it necessary to probe the sample not only in the z-direction as in a conventional light microscope but also in the x- and y-directions. The resulting scanning optical microscope (SOM) has strong depth-discriminating properties, a slightly improved x–y resolution, and a much better contrast caused by the rejection of flare light (stray light surrounding the focal point of the point light source) by the spatial filter.

The micrograph on the left in Figure 4 shows a typical confocal reflection image of the same sample, at the same position and under the same circumstances as the micrograph in Figure 1. The depth discrimination is remarkable. By adding all the consecutive optical sections, a so-called extended depth of focus image was obtained (Fig. 4, right-hand image). Figure 5 shows a confocal fluorescence image of the same sample at the same position as in Figure 2. The increase in contrast is overwhelming.

The depth discrimination can be used to obtain accurate height information from the sample. Moreover, a full 3D data set from a volume in a sample can be recorded and used to generate stereo pairs, holographic representations, topographic images, skeletal images, and so on, in order to obtain a better understanding of the sample under study.

C. The Confocal Microscope

There are numerous ways to realize a confocal microscope, of which the simplest is the scanning stage type, where the sample is probed by scanning it through the confocal point of the point light source and the

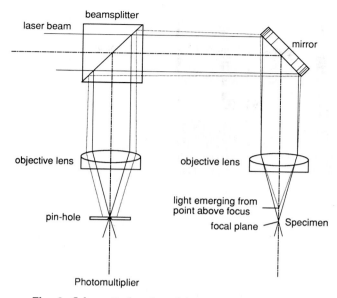

Fig. 3. Schematic drawing of the confocal concept.

point detector. Although from a theoretical point of view those systems have a number of advantages [Sheppard, 1986; Brakenhoff et al., 1988], scanning the sample severely limits the sample choice and the scan speed. Scanning beam systems do not have these disadvantages.

There are two fundamental types of scanning beam systems:

- Single beam systems, where the sample is probed with one beam at a time.
- Multiple beam systems, where more than one beam probes the sample.

An example of a multiple beam system is the spinning disk confocal microscope described in detail by Boyde (Chapter 13) and by Masters and Kino (Chapter 14, this

Fig. 4. Confocal reflection images at the same spot, under the same circumstances as in Figure 1. The **left-hand** image shows one confocal section, the **right-hand** image shows a so-called extended depth of focus image, formed by summing the in-focus information present in all the consecutive sections. (Horizontal field of view = 80 μm, NA of the objective = 0.8.)

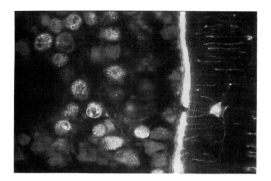

Fig. 5. Confocal fluorescence image of a PMMA-embedded, brilliant sulfoflavine-stained bone sample recorded at the same position as the conventional image in Figure 2. (Horizontal field of view = 110 μm, NA of the objective = 1.4.)

through an eyepiece or by a camera. Since the pinholes are rotating very fast, a live image is obtained. In 1968, Petrán et al. constructed the first confocal microscope with a spinning Nipkow disk and a conventional light source. Figure 6 outlines the principle of such a spinning disk system. Microscopes based on this principle are commercially available.

Single beam systems that have a laser as their light source are also commercially available, and most of them use scanning mirrors to scan the laser beam through the object. It takes about 2 to 4 sec to build up one full-resolution frame with these kinds of systems. Building up a full-resolution 3D data set takes proportionally longer. Moreover, it is very difficult to search the sample confocally with such a slow scan rate.

The advantages of fast scanning systems are obvious. Not only is it possible to search the sample rapidly in a confocal mode but a full 3D data set may be obtained much faster too. Even photobleaching can be prevented in certain cases. For instance, in the fluorescence mode a 60 ns (video rate) pixel dwell time prevents the build-up of fluorescein molecules in the triplet state. Tsien and

volume). In this system, a conventional light source illuminates a rotating disk containing many regularly spaced pinholes. These rotating point light sources are imaged in or on the object. The reflected light or the fluorescence coming from a point illuminated by one of the point sources is back-projected through the same pinhole or through a pinhole positioned diametrically opposite to the illumination pinhole which affords confocal operation. The image can be viewed

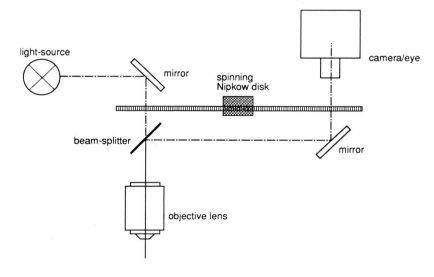

Fig. 6. Schematic drawing of the spinning disk microscope.

Wagonner [1990] calculated that for the typical case of 1 mW of laser light being focused by a 1.25 NA objective, a steady state of 80% of the fluorescein molecules in the triplet state is reached after 180 ns.

D. High-Speed Laser Scanning Microscope Scenarios

A high-speed laser scanning microscope can be realized by either scanning slowly with more than one laser beam (multiple beam principle) or by scanning a single beam faster. The multiple beam approach necessitates more complicated electronics. One needs multiple detectors and special frame grabbers to capture the image [Electronics, 1985]. The single beam approach can be realized by simply taking fast nonmechanical laser beam deflectors. Recently, Goldstein et al. [1990] realized a complete solid-state confocal laser scanning microscope by making use of acousto-optical deflectors for beam scanning and an image dissector tube as a steerable point detector, scanning synchronously with the laser beam.

Video-rate scanning necessitates the use of at least one nonmechanical laser beam deflector to achieve a line-scan rate of 20 kHz. Scanners like the Hologon scanner, the acousto-optical deflector (AOD), and the electro-optical deflector (EOD) can reach those scan rates. However, the EOD is not readily available and the Hologon scanner does not have the flexibility to vary the scan angle, which is necessary for zooming the field of view.

This chapter provides a description of a real-time confocal laser scanning microscope incorporating an AOD and a scanning mirror as laser beam deflectors. It appeared to be able to work efficiently in the fluorescence as well as in the reflection mode. Furthermore, the spinning disk confocal microscope, which is also a fast scanning confocal system, is compared with a confocal laser scanning microscope from the point of view of illuminance (photon flux per unit area) and fluorescence sensitivity.

II. THE ACOUSTO-OPTICAL DEFLECTOR (AOD)

A. General

When sound waves are generated in a transparent slab of optical material by means of a piezo-transducer, these traveling waves result in regions of compression and rarefaction that behave like a transmission diffrac-

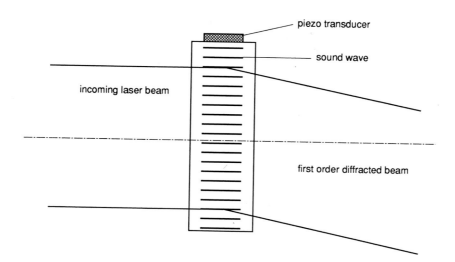

Fig. 7. Schematic drawing of an acousto-optical deflector.

tion grating. If a monochromatic laser beam enters this grating, it is diffracted proportional to the distance between the regions of equal density. Scanning the diffracted beam is possible by changing the acoustic frequency of the sound waves. This change will cause a proportional change in distance between the waves. Such a device is called an acousto-optical deflector. Figure 7 outlines its principle.

Depending upon the thickness of the crystal and the acoustic and optical wavelengths, the AOD can operate in the Raman-Nath regime or in the Bragg regime. Contrary to the Raman-Nath diffraction, the Bragg diffraction couples all the light into a specific diffraction order with higher efficiency. It goes beyond the scope of this article to go deeper into this matter. An extensive treatment on the theory of AODs and EODs is given by Gottlieb et al. [1983].

B. Application of Acousto-Optical Deflectors in a Confocal Laser Scanning Microscope

Specific features are important when an AOD is used as a laser beam deflector in a confocal laser scanning microscope.

1. Resolution. The AOD can resolve a specific number of points, dependent upon the acoustic frequency range and the time it takes for an acoustic wave to travel from one end of the aperture to the other (aperture time). The high-resolution devices which are commercially available can typically attain 1,000–2,000 resolvable points.

2. Scan-angle range. The scan-angle range (typically between 10 and 30 mrad for commercial devices) has to be adjusted to the field angle of the microscope in such a way that with the proper sampling frequency (the number of resolvable points of the AOD) the desired field of view is obtained. The scan angle can be "amplified" optically.

3. Diffraction efficiency as a function of the scan angle. The diffraction efficiency of the AOD is not constant over the full scan-angle range. This means that a certain amount of shading is inevitable unless the intensity of the diffracted laser light is measured and used in a feedback loop to maintain the diffracted light intensity at a constant value. All optical microscopes, except the scanning-object types mentioned earlier, suffer from shading. Figure 8 shows the diffraction efficiency as a function of scan angle for two different types of AODs.

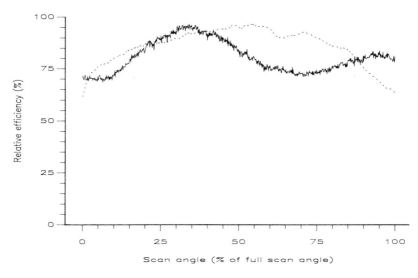

Fig. 8. The diffraction efficiency versus scan angle for two different AODs.

4. Dispersion. The AOD acts as a diffraction grating. This causes some special problems if fluorescence light has to be detected confocally. The fluorescence light has an emission wavelength different from the excitation wavelength (Stokes shift). Moreover, it is emitted over a broad spectral range. To detect the fluorescence confocally it should be back-projected along the same optical path as the illuminating laser beam so that it can be projected on a pinhole. Because the AOD is dispersive, the fluorescent light deviates from the optical axis when it is back-projected through the AOD because the excitation wavelength and the emission wavelength are different. This can be compensated for by shifting the pinhole proportional to the Stokes shift. The dispersion of the AOD also causes the fluorescent light, being emitted over a broad wavelength band, to be dispersed over the pinhole. Both those effects cause a significant reduction of the efficiency of the AOD system for fluorescent light. The fluorescence sensitivity is of the utmost importance, so this is not desirable. Section III gives a solution for this problem.

5. Optical setup. To get a good resolution from a diffraction grating, it has to be filled completely over the aperture so that enough lines of the grating are used. Apertures of AODs are rectangular (2×40 mm), which necessitates the use of cylindrical lenses. Figure 9 shows a typical optical setup used for AODs. Although it looks rather complicated, alignment of this system is not difficult to achieve.

6. Cylindrical lens effect. As mentioned, the laser beam is scanned by changing the frequency of the acoustic waves traveling in the AOD crystal. It takes a specific time—the aperture time—to fill the entrance aperture of the AOD with acoustic waves of equal frequency. When the pixel time approaches this aperture time, the so-called cylindrical lens effect will appear. There will be a frequency gradient over the aperture of the AOD during one pixel, causing the top part of the laser beam to be deflected differently compared to the bottom part. Figure 10 shows this schematically.

It is obvious that the cylindrical power of the AOD depends upon the existing frequency difference during one pixel. A change in scan speed and scan amplitude causes a proportional change in the frequency gradient and consequently in the cylindrical power. So, changing the scan speed neces-

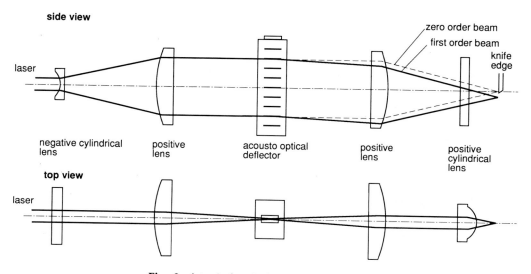

Fig. 9. A typical optical system for AODs.

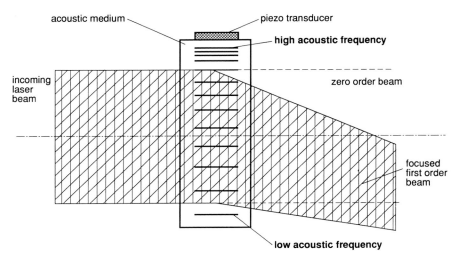

Fig. 10. The cylindrical lens effect in an AOD.

sitates a change in the optical system to compensate for this effect.

III. THE VIDEO-RATE CONFOCAL LASER SCANNING MICROSCOPE

To realize a fast confocal laser scanning microscope (CLSM), it is necessary for one of the two laser beam deflectors to be capable of scanning the laser at a line-scan frequency of about 20 kHz. This is possible with an AOD. The second deflector need only be able to scan at 50–60 Hz (frame-scanning frequency), which is possible with a scanning mirror. The scan motion should preferably be linear to prevent image distortion.

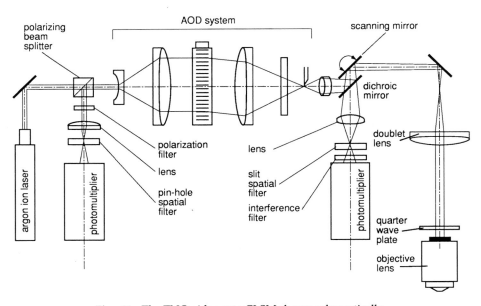

Fig. 11. The TNO video-rate CLSM shown schematically.

Figure 11 outlines a video-rate CLSM which has been developed in our institute. The laser beam coming from a 25 mW water-cooled argon ion laser (λ = 488 nm, TEM 00, Gaussian beam profile, linearly polarized) is expanded (5 times) to eventually fill the back pupil of the objective completely. After that it passes a polarizing beam splitter oriented in such a way that it transmits the laser beam. A combination of a negative cylindrical lens and a positive doublet lens expands the laser beam in one direction (8 times) and focuses it in the other direction (as shown in Fig. 9). The first-order diffracted laser beam is restored to a circular shape by a combination of a positive doublet lens and a positive cylindrical lens, both having the same focal distances as the first combination of the negative cylindrical lens and positive doublet. Behind the positive cylindrical lens is a third positive doublet lens (f = 50 mm) with a scanning mirror at the focal distance behind it. The third doublet is at such a distance from the positive cylindrical lens that the focal point of the laser beam emerging from this doublet coincides with the focal point of a fourth doublet in front of the microscope. This causes the laser beam to enter the (infinity-corrected) microscope objective in a parallel manner. To prevent reflections from the optical parts reaching the reflected light detector, antireflection optics is incorporated by means of a quarter-wave plate in front of the objective (rotating the reflected light coming from the sample by 90 degrees) and a polarizing beam splitter in front of the detector. The reflected light is subsequently back-projected through the system and so descanned. The polarizing beam splitter then selectively reflects the descanned laser light reflected by the specimen. An objective lens (NA = 0.1) focuses it on a spatial filter pinhole (ø 5 μm) in front of a photomultiplier tube. This combination serves as a confocal point detector.

Because of the dispersion of the AOD, it is not possible to descan the line-scan movement of the fluorescent light by back-projecting it through the AOD without a serious loss in intensity. Instead the fluorescent light coming from the specimen is coupled out of the optical path by a dichroic mirror between the scanning mirror and the AOD. So the frame scan is canceled but the line scan is not. This makes it necessary to use a slit-shaped spatial filter (5 μm wide) in front of the photomultiplier tube. An objective lens (NA = 0.1) focuses the light on a slit. The consequences of this type of confocal imaging are summarized in Section IV.

A new version of the video-rate CLSM which has been constructed in our institute (Netherlands Organization for Applied Scientific Research, TNO) has shunts to circumvent the confocal spatial filters, allowing an alternative nonconfocal view of the image.

IV. SLIT AND PINHOLE DETECTORS IN CONFOCAL MICROSCOPY

It was already pointed out by Sheppard and Mao [1988] and Wilson [1990] that confocal properties can be obtained with detector geometries other than circular ones. As discussed in Section III, the video-rate CLSM makes use of a circular aperture for the reflection mode and a slit aperture for the fluorescence mode. The results of the calculations made by Sheppard and Mao [1988] and Wilson [1990] show that the axial response of the slit detector is about 18% broader at half-height and falls off with $1/x$, while a pinhole detector falls off with $1/x^2$ (x is the distance from the focal plane). The lateral response differs for the two directions. The lateral response is roughly confocal in the direction perpendicular to the slit and conventional in the direction parallel to it. However, this will only be noticeable for objects of the same size as the resolution.

Wilson [1990] concludes that the difference between slit and pinhole detectors will be less dramatic when practical requirements of signal strength are considered. Sheppard and Mao [1988] even conclude that "for weakly scattering objects for which the signal level is low, it is possible to use a narrow slit whilst still maintaining adequate signal levels, which results in an improved depth discrimination as compared with a system with the usual circular pin hole."

In the case of the video-rate CLSM, a 5 μm pinhole is used in the reflection channel. Together with a 0.1 NA objective (effectively used at NA ~ 0.05), the normalized detector radius (being the ratio between the pinhole diameter and the Airy disk diameter) is approximately 1. This means that the depth profiles will be as small as possible and the gain in lateral resolution will be somewhat less than the factor 1.4 that can be reached in the limiting case of an infinitely small pinhole [Wilson, 1990]. As explained, the fluorescence channel is equipped with a slit aperture. The width of the slit is 5 μm and the objective in front of it has an NA of 0.1. Depending upon the wavelength, the slit width is approximately between 1 and 1.5 times the diameter of the Airy disk, giving optimal sectioning properties for a slit detector.

V. CONFOCAL LASER SCANNING MICROSCOPES VERSUS SPINNING DISK MICROSCOPES

Since both the spinning disk microscope and the system presented in this chapter are scanning optical microscopes with a very fast frame rate, it is useful to compare both systems. From a theoretical point of view, the spinning disk microscope and the laser scan systems are comparable. They are different with respect to the following:

1. **Frame time.** The spinning disk system is able to refresh the image faster than video-rate because of its parallel character.

2. **Availability of color information.** Because in the spinning disk systems a conventional white light source is used, the images are presented in real color. Although possible, this is very difficult to realize in a confocal laser scanning microscope. A system working with a red, green, and blue laser line is commercially available but is not designed for fluorescence applications.

3. **Illuminance and pixel time.** Because spinning disk systems use conventional light sources, the illuminance (photon flux per unit area) at any resolvable object-point is comparable to that in a conventional microscope.

If a 100 W tungsten halogen lamp with an $f/0.7$ condenser is used to illuminate the sample, and the image consists of 10^6 resolvable object-points, one point will receive approximately 0.9 μW of the lamp energy in the 450–700 nm wavelength region [Oriel catalogue, 1989] (not taking into account the losses in the rest of the optical system).

In the fluorescence mode, the situation is even worse since only a small spectral band (ca. 30 nm) is used to generate fluorescence. For a 450 W xenon lamp with an $f/0.7$ condenser, 0.5 W of the lamp energy between 460 and 490 nm has to be shared by 1 million pixels, so each pixel will get 0.5 μW [Oriel catalogue, 1989] (again without considering the losses in the rest of the optical system). A laser scanning microscope, in contrast, uses all the light to illuminate one single pixel. Usually a 10 mW laser is used.

The time available for one pixel also has to be taken into account. With 2% of the spinning disk filled with holes, 2% of the pixels are viewed simultaneously. This means 2×10^4 out of 10^6 as compared to 1 out of 10^6 in a laser scanning microscope. So, to achieve the same frame-refresh rate, the time available to collect one pixel is 2×10^4 longer for a spinning disk system. To summarize, it can be said that the illuminance of one pixel is approximately four orders of magnitude higher in a laser scanning microscope but, with the same frame-refresh rate, the time available for one pixel is approximately four orders of magnitude shorter than in a spinning disk microscope. This implies that if the fluorophore in the sample is not saturated [Tsien and Wagonner, 1990], a laser scanning microscope will generate the same amount of fluorescence photons in a time period four orders of mag-

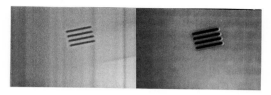

Fig. 12. Micrograph of a resolution sample fabricated with submicron techniques recorded both in the confocal reflection and confocal fluorescence mode with a 0.8 NA objective. The image shows four 0.5 μm-wide bars which are separated 0.5 μm. The bars are 650 nm-deep wells in a rhodamine-labeled photoresist.

nance per pixel as in the laser-scan microscope, and if both were equipped with the same detector, the spinning disk system would be much faster than a single beam laser scan microscope because of its parallel character, or if operated at the same speed would be much more sensitive (since the time available for one pixel is much longer). When a laser-scan microscope equipped with a 10 mW laser and a spinning disk microscope with a 450 W xenon lamp are compared, both systems being equipped with detectors with the same finite dark-count rate and quantum efficiency, the laser-scan system will have a better sensitivity in low light level applications (better signal-to-noise ratio). To reach the same sensitivity, the detection system of the spinning disk microscope would have to have a 10,000-fold lower dark-count rate.

nitude shorter. Consequently, both the dark-count rate and the quantum efficiency of the detection system used to generate the image determine the sensitivity of the two systems.

In this example (a spinning disk system with a 450 W xenon lamp and 2% of the spinning disk filled with holes, compared with a 10 mW laser in a laser-scan system), if the detection system used to generate the image has no dark counts at all, the two systems are approximately equal with respect to sensitivity. If it were possible to increase the intensity of the light source of a spinning disk microscope to attain the same illumi-

VI. RESULTS

A. Performance Data

Table I summarizes the performance data of the video-rate CLSM. The performance of this CLSM is further shown in Figures 12–14.

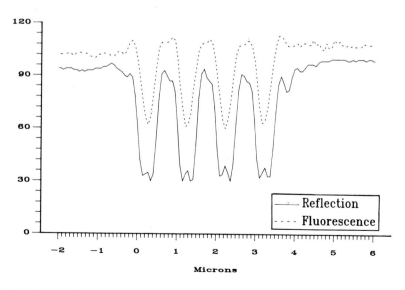

Fig. 13. Line profiles from 0.5 μm-wide structures in fluorescence and reflection mode.

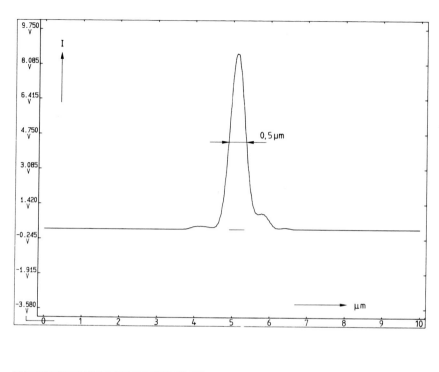

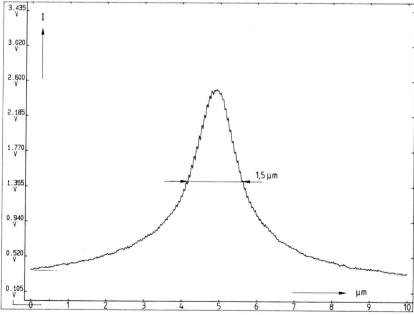

Fig. 14. Curves obtained while scanning a plane mirror **(upper curve)** and a wafer with a 650 nm layer of fluorescent material **(lower curve)** through the focal plane and recording the reflection and fluorescence intensities (NA objective is 0.8).

**TABLE I. Performance Data of the
Video-Rate CLSM**

Imaging modes	Confocal or nonconfocal reflection
	Confocal or nonconfocal fluorescence
	Nonconfocal transmission
Field of view	2×2 mm^2 ($5 \times$ objective)
Zooming	Maximum zoom factor
$\times$ 10	
Lateral resolution	0.27 μm (0.8 NA reflection)
Axial resolution	0.60 μm (0.8 NA reflection)
Lateral resolution	0.36 μm (0.8 NA fluorescence, average of x/y directions)
Axial resolution	0.85 μm (0.8 NA fluorescence)
Sensitivity	6,000 molecules of FITC spread over the surface of a 4.3 μm φ latex sphere are detectable

The micrographs in Figure 12 show that the 0.5 μm separated edges are well resolved in both fluorescence and reflection modes. The curves in Figure 13 show this again in a graph with line profiles taken perpendicular to the bars. Because the wells in the photoresist are probably not etched out completely, the fluorescence curve shows only good resolution when the microscope is focused at the top of the wells (as shown in Fig. 12). The depth profiles in Figure 14 show the depth discrimination for both the fluorescence and reflection modes. The micrographs in Figure 15 show the sensitivity of the system.

B. Other Micrographs

Additional micrographs obtained using the video-rate CLSM are shown in Figures 16–18.

VII. CONCLUSIONS

There are several ways to obtain the fast scan rates which are necessary for a real-time confocal microscope. The spinning disk system with a white light source is a very elegant solution, especially advantageous in specific applications since the reflection image is in color. Acousto-optical deflectors can replace mirror scanners to scan the laser beam. The only disadvantage of those systems is that the fluorescent light cannot be descanned by sending it back through the AOD because the efficiency is too low. This makes it necessary to use a slit as a spatial

Fig. 15. Micrographs of 4.3 μm diameter latex spheres with 6,000 ± 10% FITC molecules on the surface with **(upper micrograph)** and without **(lower micrograph)** frame averaging (32 frames).

Fig. 16. Stereo picture of an integrated circuit. (0.8 NA, field of view 80 μm, reflection mode, extended depth of focus, 5 μm.)

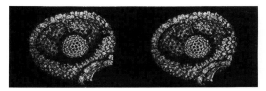

Fig. 17. Stereo picture of *Radiolaria*. (0.45 NA, field of view 400 μm, reflection mode, extended depth of focus, 200 μm.)

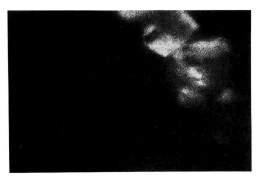

Fig. 19. Confocal Raman image of MoO_3 crystals. (Horizontal field width = 30 μm, NA = 0.8.)

filter instead of a pinhole. Compared to the pinhole systems with a $1/x^2$ fall-off of the depth curve, the depth curve of a slit system falls off with $1/x$. However, in reality pinholes are not infinitely small, therefore the differences between images obtained using pinholes and slits will not be extreme. Indeed, the combination of an AOD with a scanning mirror forms a video-rate laser beam scanning system which can be used very adequately in a confocal laser scanning microscope.

VIII. FUTURE DEVELOPMENTS

The prospects of the confocal laser scanning microscope depend largely on the developments in electro-optics. A matrix of random access surface emitting diode lasers [Jewell, 1990] as point sources and random access photodiodes [Walmsley, 1985] could replace the light source and the detector of the confocal microscope. But not only those kinds of developments are exciting. Recently a research project on 3D Raman mapping in our TNO institute resulted in a number of chemical substance-specific images (Fig. 19) [Draaijer et al., 1990]. There are many possibilities for obtaining contrast from microscopic samples. One of the areas in which much research is needed is the display of 3D images. A holographic monitor hooked up to a computer with 3D image analysis software would be an interesting and very useful device in confocal microscopy.

ACKNOWLEDGMENTS

The authors thank Alan Boyde for providing a fluorescently stained bone sample and the colleges from the MBL TNO for providing the fluorescently labeled DNA sample. They also wish to thank Noran Instruments (formerly Tracor Northern) for their financial support for parts of the research project.

Fig. 18. Micrograph of DNA damage sites in a nucleus of human white blood cells treated with pyrene-diol-epoxide and then successively with a monoclonal antibody against BP-DNA and with fluorescent monoclonal antibody against the latter. Three optical sections separated by 1 μm. (Horizontal field of view = 80 μm, NA = 1.25.)

REFERENCES

Brakenhoff GJ, van der Voort HTM, van Spronsen EA, Nanninga N (1988): 3-Dimensional imaging of biological structures by high resolution confocal scanning laser microscopy. Scanning Microscopy 2:33–40.

Draaijer A, van Dijk MH, Houpt PM (1990): Confocal laser scanning microscopy in combination with Raman spectrometry for 3D-Raman mapping. Trans R Microsc Soc 1:455–458.

Electronics (Author unknown) (1985): Laser beams speed up reticle writing. Electronics (October 7):40–42.

Goldstein SR, Hubin T, Rosenthal S, Washburn C (1990): A confocal video-rate laser-beam scanning reflected-light microscope with no moving parts. J Microsc 157:29–38.

Gottlieb M, Ireland CLM, Ley JM (1983): "Electro-Optic and Acousto-Optic Scanning and Deflection." New York: Marcel Dekker.

Jewell JL (1990): Surface emitting micro-lasers for photonic switching and interchip connections. Opt Eng 29:210–215.

Minsky M (1961): U.S. Patent 3013467. Microscopy apparatus, filed Dec. 1961.

Oriel (1989): Catalogue from Oriel Corporation. Light sources; Monochromators; Detection systems, 2:80–85.

Petrán M, Hadravsky M, Egger MD, Galambos R (1968): Tandem-scanning reflected-light microscope. J Opt Soc Am 58:661–664.

Sheppard CJR (1986): Scanning methods in optical microscopy. Endeavour (new ser) 10:17–19.

Sheppard CJR, Mao XQ (1988): Confocal microscopes with slit apertures. J Modern Opt 35:1169–1185.

Tsien RY, Wagonner A (1990): Fluorophores for confocal microscopy photophysics and photochemistry. In Pawley J (ed): "The Handbook of Biological Confocal Microscopy," pp 170–178. New York: Plenum Press

Walmsley C (1985): Linear photodiode array with serial and addressable real-time output capability. IEEE J Solid State Circuits 20:724–729.

Wilson T (1990): The role of the pinhole in confocal imaging systems. In Pawley J (ed): "The Handbook of Biological Confocal Microscopy," pp 113–126. New York: Plenum Press.

CHAPTER 13

Real-Time Direct-View Confocal Light Miscroscopy

A. Boyde

Electronic Light Microscopy, pages 289–314 ©1993 Wiley-Liss, Inc.

I. HISTORICAL INTRODUCTION

The tandem scanning microscopes (TSMs)[1] were invented by M. Petran and M. Hadravsky to solve the problem of imaging neurons in live brain tissue [Petran and Hadravsky, 1966a,b; Egger and Petran, 1967; Petran et al. 1968]. Because the TSMs are confocal, and, as such, optical sectioning microscopes, and since the interest comes from examining a large number of focal planes in sequence by focusing up and down rapidly (in real time), it was, and is, difficult to communicate the excitement and interest that comes from this sort of microscope in static images. The first published illustration deriving from such a microscope was a drawing [Egger and Petran, 1967], somewhat atavistic in the 1960s, which probably did not help to publicize the revolutionary achievement.

It is interesting to note that Minsky's confocal microscope, generally only known from the patent description, also used a conventional light source. Davidovits and Egger [1971] were to use a laser source in what appears to be the third type of confocal microscope built in 1970. Minsky [1961] designed his microscope to be able to exam-ine the three-dimensional architecture of neurons in prepared thick slices of brain stained by Golgi-related techniques.

The TSMs were—and still are—the only real-time scanning optical microscopes, if we accept the definition of real time to mean significantly faster than video or cinematographic frame repetition rates. All practical TSMs known to date are reflection or "epi-illumination" devices. However, a design for a transmission TSM using both sides of one aperture disk was published by Petran and Hadravsky [1966b], and Sheppard and Wilson [1981] use a concept diagram to explain the theory of the "direct-view" confocal microscope which shows two disks ganged together in a transmission mode. TSMs originally carried the acronym TSRLM—for tandem scanning reflected light microscopes—but they also work in fluorescence, as has been demonstrated by Boyde and Reid [1986], Boyde and Wolfe [1986], Boyde et al. [1986, 1990], Watson and Boyde [1985], and Wright et al. [1989], among others.

II. COMPARISON OF TSMs WITH CONFOCAL LASER SCANNING MICROSCOPES (CSLMs)

The reader of this volume will probably wish to see a comparison of the functional properties of TSMs with CSLMs before reading on. TSMs are optical sectioning microscopes which in their simplest form may be configured without any electronic aids. The microscope may look as simple as any conventional light microscope. A constant image is provided to the viewer looking through a single eyepiece or a binocular head. This image is in every way familiar to the light

---

[1]**Synonyms**: Tandem scanning reflected light microscopy (TSRLM); tandem scanning microscopy (TSM; real-time scanning optical microscopy (RSOM); real-time confocal microscopy; direct-view confocal microscopy; tandem scanning confocal microscopy; disk scanning microscopy.

Present variants: One-sided disk scanning; two-sided disk scanning.

Principal components: Conventional light source: e.g., tungsten filament or mercury or other arc; collector lens; field diaphragm; aperture diaphragm; beam steering mirror(s) (45°); field lens; multiple-aperture disk; beam splitter (dichroic?); objective lens; quarter-wave plate; eyepiece or other system for imaging the intermediate image plane.

microscope user, except that this is a reflected light or epifluorescence image in all TSM devices constructed to date.

Samples of zoological, botanical, or microbiological origin give reflection-mode images which are bright enough for comfortable viewing with the naked eye, if necessary in a somewhat darkened room. A great variety of natural and synthetic materials, ceramics, rocks, metals, semiconductor substrates, etc., may give images which are too bright for viewing without some reduction in intensity. Thus it is not strictly necessary to have any television or computer-related additional equipment. A considerable beauty of such a microscope is that it permits the examination of unstained, unfixed, unlabeled living biological tissue, such live tissue giving much better reflection contrast than tissue which has been fixed in any way [Egger and Petran, 1967]. Living tissues move or contain moving parts. The image of rapidly moving structures which can be seen in the TSM may not be bright enough for *recording* with conventional photographic materials, and this would be one reason for wishing to use a sensitive recording medium—perhaps an image-intensifying TV camera with videotape [Watson, 1990], videodisk or high-speed frame store [Jester et al., 1990c; Petroll et al., 1990]—to be able to capture and transport images over time and space. Such recording is, however, only necessary for the purposes of convincing third parties that a particular phenomenon had been observed. The quality of the *reflected* light image is, in most cases the author has had the opportunity to observe, considerably superior to that which can be obtained with any of the confocal laser scanning microscopes.

TSMs also work in the fluorescence mode, and will be found to be more convenient than CSLMs for the brighter classes of fluorescent objects such as tetracycline lines (vital labels) in bone and dentine, brilliant sulphaflavine-stained tissue embedded in PMMA and fluorescein-, rhodamine-, etc., labeled synthetic filling and bonding materials [see Boyde, 1985a; Reid et al., 1985;

Boyde and Reid, 1986; Boyde and Wolfe, 1986; Boyde et al., 1986, 1990c; Watson and Boyde, 1985, 1986, 1987b,c, 1989; Wolfe, 1988; Watson, 1989a]. With a properly designed TSM (i.e., with adequate disk transmission, suitable aperture size, a dichroic beam splitter, and good intermediate optics), even the lower fluorescence levels experienced in immunocytochemical labeling can be seen comfortably with the naked eye in a darkened room. Since most such preparations will be of fixed material, photographic film may be used for recording the image with longist exposure periods. However, in every case, the low light level fluorescent image can be captured from the TSM by the simple addition of an image-intensifying camera of one sort or another [Boyde and Watson, 1989; Wright et al., 1989; Boyde et al., 1990c]. These microscopes are then almost equivalent to the confocal laser scanning microscopes. The principal difference will be found in the size of the field of view, which is greater in the TSM with any given objective. There is considerable convenience in the high empty magnification available in the confocal laser scanning microscopes. According to any theory of resolution, this extra "zoom factor" should not be necessary to resolve what can be resolved, but in practice empty magnification may make it easier.

It is frequently claimed by those users of CSLMs who have had no experience with TSMs that dual- or multiple-wavelength fluorescence microscopy is not possible in the TSM. Given that the TSM uses an appropriately corrected objective with no significant longitudinal chromatic aberration, fluorescence excited at one or more shorter wavelengths will give rise to a colored image in which all wavelengths are imaged simultaneously, from the ultraviolet through to the near-infrared. The TSM is good for multiple-wavelength fluorescence microscopy [Boyde et al., 1990c]. Good objectives are very difficult to find.

A comparison of depth of field between CSLM and TSM depends strictly upon the confocal aperture dimensions and the numerical aperture and magnification of the objective lens being used. In the best case the TSM can be as good as the best laser confocal microscope. However, as it is less easy to change the aperture size in the disk scanning microscope, the usual compromise is to go for an acceptable degree of "confocality." The majority of users of CSLMs in practice, however, do not use their microscopes at maximum confocality (if they are confocal or type 2 rather than type 1 at all) and the instantaneous depth of field would be no less than that which we would normally see in the TSM.

In terms of the utilization of the possible field of view with any given objective lens, present (two-sided) TSMs give the broadest field of view. Some of the beam scanning confocal laser microscopes utilize only one-third of the field width. It should be noted, however, that object scanning confocal laser microscopes provide an essentially unlimited field of view. Such scanning is, of course, slow and not suitable for live, bulky, or moving objects.

III. CONSTRUCTION OF A TANDEM SCANNING MICROSCOPE

A. Basic Principle

A TSM is an epi-illumination light microscope (LM) to which is added an arrangement whereby an array of apertures lying at the intermediate image plane of the objective lens, on the illuminating side, can be scanned (Fig. 1). This aperture array will be imaged by the objective lens at a demagnification identical to the actual magnification of the objective, brightly illuminating discrete spots in the focal plane in the specimen. Returning reflected or fluorescence light can only pass a second, identical aperture array at the intermediate image plane of the objective on the image-forming side if it arises from these illuminated points. All other reflected or fluorescent light arising from other portions of the specimen, or scat-

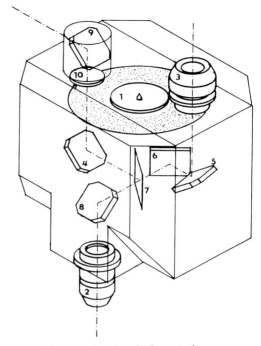

Fig. 1. Diagram showing the layout of components in a recent configuration of a two-sided tandem scanning microscope head design. (Courtesy of M. Petran and M. Hadravsky.) Illuminating light enters the microscope from the top left, is reflected via a front surface mirror (9) toward a field lens (10) located immediately above the disk (1). Passing the disk, it is reflected from mirror (4), off the front surface of a mica beam splitter (7) to another beam steering mirror (8), which directs the rays downwards to the standard RMS-threaded objective lens (2). Returning light passes through the beam splitter (7), suffers two more reflections from inverting mirrors (5,6) before arriving at the Nipkow (aperture) disk (1) on the observation side. In this microscope the image at the disk is viewed via a Ramsden-type eyepiece (3).

tered from other optical surfaces in the system, will be intercepted by the solid portions of the device containing the aperture array, and will be thereby prevented from contributing to the image.

B. "Nipkow" Disk Designs

In practice, the aperture array is a spinning disk with round or square holes (Fig. 2). Each aperture in the array lies at a unique

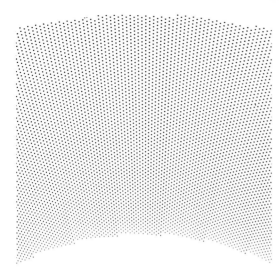

Fig. 2. Pattern of holes in a 1984 vintage Nipkow disk made by Petran and Hadravsky in Plzen.

radial distance from the center of the disk and so scans a single curved scanning line in illumination—the same or a corresponding aperture at the same radial distance in the detection array serving as the detector. If light goes in through the disk via a beam splitter, one and the same hole can serve simultaneously as illumination and detection aperture, the whole device being self-aligning. If, however, the waste illuminating light which will otherwise be reflected from the lit side of the disk back into the viewing path is to be physically stopped by the disk, then a separate set of apertures must be scanned in geometrical congruence with the first. The simplest way to achieve this is to use holes on the opposite side of the scanning disk for the detection apertures. In this case, identical twins of holes are situated at the same radial distance from the center of the disk on the same diameter. At any one instant, one such hole in a pair serves as the illuminating aperture and the opposite one as the detector. The functions of the two members of the hole pair are reversed 180° later in the rotation of the disk.

C. Etymology of "Tandem" Scanning

Some recent authors in this field have made the mistake of imagining that the name tandem scanning was married to the use of the two-sided disk. This was untrue from the outset. "Tandem" scanning means scanning illumination and detection simultaneously, and is the priority term (of those in present use) for all microscopes which are "confocal." We shall use the cumbersome terms one-sided and two-sided TSMs to make the distinction between the two sorts of device.

D. One-Sided Use of the Aperture Disk: 1sTSM

As already intimated, the elegant simplicity of the one-sided design is severely complicated by the fact that a very high proportion of the illuminating light may be reflected from the lit side of the disk back into the observation channel. This can be minimized in several ways: First, of course, the disk could be made of a minimally reflective material. Second, the disk could be highly polished to ensure specular reflection, and tilted so that the reflected rays are directed to a light trap where they can be removed [Xiao et al., 1990]. Third, it is possible to remove rays which were reflected on the top surface of the disk by the antiflex principle employing polarizing filters and a quarter-wave plate [Francon, 1961]. Polarized light is used for illumination: A quarter-wave plate is placed somewhere in the optical system—ideally between the objective lens and the specimen, but in practice often between the disk and the objective—and a second, crossed, polarizing filter is placed in the observation channel. Thus polarized light reflected from the disk will be largely intercepted by the second polarizing filter. The combination of all these means of eliminating this light still leaves an unwanted DC background signal of a much greater strength than the low level of light reflected from intracellular organelles in live biological samples. A further means of overcoming the difficulty is to use a good dichroic beam splitter and a fluorescence pass filter which will not transmit any of the exciting wavelength. It will thus be possible to build a good fluorescence microscope based upon the one-sided layout [Boyde et al., 1990d].

E. Two-Sided Use of the Aperture Disk: 2sTSM

Difficulties with the light reflected from the lit side of the disk are eliminated by using the disk in transmission only and a separate disk, or as already mentioned, matching holes on the opposite side of the disk, in the detection channel (Fig. 1). In this, the most favored design built so far, it is necessary to use other mirrors (4, 5, and 6 in Fig. 1) to produce top–bottom and left–right inversions of the image of one side of the disk so that it will be perfectly overlapped on the other side of the disk. The beam splitter (7 in Fig. 1) is then placed in the optical geometrical center of the TSM head with an equal path length from the disk to the beam splitter on both sides of the device. There will be one mirror between the disk and the beam splitter on one side of the head and two on the opposite side of the head to achieve the necessary alignments. The beam splitter is then between the disk and the objective lens instead of above the disk as in the one-sided TSM configuration. The three coated front surface, 45° mirrors (4, 5, and 6 in Fig. 1) necessary to produce the congruence of the two sides of the disk must be carefully aligned. In commercial instruments this would be a factory-based operation and the alignment would be permanent. In the earliest, near-prototype TSMs, such alignment was difficult to achieve for the amateur and this difficulty was one reason for the poor acceptance of these microscopes in the early years. It has been overcome in the currently released commercial devices.

Finally, in the 2sTSM design it is desirable in many instances to have a final 45° beam steering mirror (8 in Fig. 1) in order to permit the objective lens to be upright with the disk spinning in the horizontal plane. However, this final mirror is not an essential part of the microscope, and has been omitted in new designs specifically built for clinical use in human opththalmology, where it is highly desirable for the objective lens to lie in the horizontal plane so that the patient's eye can be adapted to the contacting (or *applanating*) objective [Cavanagh et al., 1988; Jester et al., 1988a–c, 1990a–c; Petroll et al., 1990].

In the rather more cumbersome original TSM designs [e.g., Petran et al., 1968; Egger and Petran, 1967], prisms or inverting devices with many more mirror surfaces were used. Several authors have overlooked this very important difference between Petran's 1968 published version and the more elegant design constructed in 1970 and used in the early 1970s [for example, Petran and Hadravsky, 1974; Petran and Sallam-Sattar, 1974; Sallam-Sattar, 1974; Sallam-Sattar and Petran, 1974]. This layout remained essentially the same for a long period but was only published much later [Petran et al., 1985a,b].

Regarding the asymmetry of the TSM head in the 2sTSM case: In the 1970 design [Petran et al., 1985a,b], light was reflected from two mirrors before reaching the beam splitter on the illuminating side and one mirror on the return half, after the beam splitter and before the eyepiece side of the disk. Petran and collaborators now use the TSM head in the opposite sense, the light bouncing off one mirror to the beam splitter on the illuminating side and two mirrors before the disk on the observation side, as in Figure 1. The choice of arrangement will be influenced by the design of dichroic beam splitter in a TSM optimized for fluorescence work.

F. The Merits and Demerits of the 1sTSM and the 2sTSM Cases

It cannot be argued that there is a greater simplicity in one design that the other [Boyde and Petran, 1990]. The 1sTSM is self-aligning,

Fig. 3. Longitudinal cut through a human tooth showing the internal structure of the dental enamel. Reflection image using a Nikon ×100/1.4 oil-immersion objective, Tracor TSM, 1988 vintage disk. Field is 95 μm high and 108 μm wide.

so that it is simpler to use smaller holes and thus to come close to the use of apertures sizes which would give maximal confocal performance [Kino, 1989; Kino et al., 1989; Xiao and Kino, 1987; Xiao et al., 1990]. On the other hand, the 2sTSM design makes it much simpler to view the intermediate image since it is now only necessary to use a Ramsden-type eyepiece (i.e., one whose focal plane lies in front of the eyepiece rather than within it) to view the observation side of the disk. The 2sTSM has the smaller number of optical components and optical surfaces (Boyde and Petran, 1990).

G. Intermediate (Relay) Optics

The need to have the beam splitter "above" the disk in the 1sTSM prevents the use of intermediate optics as simple as the Rams-den-type eyepiece. The intermediate optics in devices produced so far give a higher postdisk magnification than may be desirable. Although TSMs are capable of giving a lateral resolution slightly improved beyond the conventional LM limit and, therefore, merit some additional magnification to make this easier to demonstrate, the disadvantage of postdisk magnification is clearly that it reduces the light intensity at the viewing or recording device, be it an eyeball or a photographic or TV camera. With strong reflectors such as semiconductor devices, this will be no disadvantage, but for live biological tissues it may just make the one-sided TSM too difficult to use.

H. Disk Materials

As regards the materials of which the disk are made: Petran's original designs used

Fig. 4. Silver crystal grown by immersing copper wire in AgNO$_3$ solution, dried on glass slide. Reflection image using a Nikon ×200/0.95 210 mm tube-length objective. Tracor TSM with 50 mm extension tube, 1988 disk. Field, 38 μm high, 40 μm wide.

etched copper foils, 10 μm or 20 μm thick, which necessarily had to be supported centrally and peripherally. In the 2sTSM design by Tracor Northern (Middleton, WI, USA), the disks are silicon wafers, but again with real holes through the disk. In the 2sTSM design by Tandem Scanning Corporation (Reston, VA, USA), as well as in the one-sided disks produced by Technical Instruments (San Francisco, CA, USA) and others, the holes are in a layer of black chrome on glass. A glass disk may be very robust and safely spun at high speeds, but the chrome is relatively delicate and easily scratched. The fact that the glass disk is solid has disadvantages, not the least of which is that there are four reflecting surfaces for the two passes of the light through the disk. The light losses due to normal incidence reflections at these glass surfaces may alone be greater than the

losses from the necessary Nachet assembly of 45° mirrors in the two-sided design.

IV. THE USE OF CONFOCAL SLITS INSTEAD OF HOLES

To overcome early difficulties in the precise alignment of a 2sTSM with holes in the aperture disk, Egger et al. [1969] suggested and employed a disk with slit apertures. Slits are not as good as holes, but nevertheless provide excellent elimination of glare or halo from out-of-focus planes—which is one of the main advantages of the confocal instrument. They also provide some improvement in lateral resolution [Sheppard and Mao, 1988] and will be found in the fluorescence detector channel of recent video-rate laser scanning confocal microscopes (Draaijer and Houpt, Chapter 12, this volume). We should

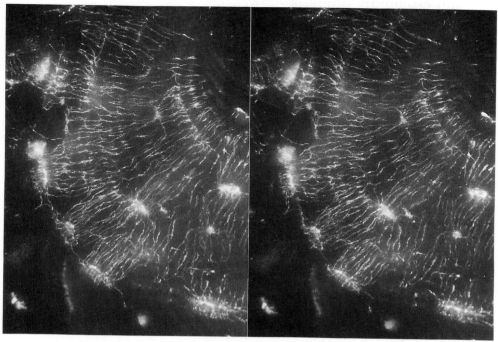

Fig. 5. Stereo pair. Cut compact bone tissue showing osteocyte lacunae and canaliculae as well as lamellae (see **left** image in top right part of field). It is clear that the strong reflection from the air-filled cellular spaces in the bone would account for the brightest features (lacunae and canaliculae). The differential reflection from the differently oriented collagen layers in the lamellae is rather more surprising. Reflection image using a Nikon ×60/1.4 NA oil-immersion objective. Field 183 μm high, 147 μm wide. Depth of field vertically (perpendicular to picture plane) is 9 μm, tilt angle difference is 12°. Tracor TSM, 1990 disk.

therefore look forward to more developmental work and trials with the use of slits, which could produce an extremely cheap version of the 2sTSM.

V. OBJECTIVE LENSES FOR TSMs

A. Tube length

The two-sided TSMs are necessarily designed for use with finite tube-length objectives. They can use all standard 160-mm tube-length objectives, while objectives with longer tube lengths, for example 170 mm or 210 mm (see Fig. 4), may be accommodated by adding an adaptor after the standard TSM head. Thus, the TSMs will accommodate all the good objectives designed for biological microscopy. Practically all such objectives were designed for use with 170 μm-thick coverslips, whether for use dry, or with oil, water, or glycerin immersion.

B. Coverslips and Immersion Media

In practice, one finds that the biological microscope user may not be aware of the imperative necessity to use a coverslip of the ideal thickness and with the correct immersion medium both above and below the coverslip. A biological object, which necessarily can only be alive in a watery fluid, must be examined with a water-immersion objective if one is to focus to a significant distance below the coverslip into the object. Apart from the image deformation which will result from spherical aberration, one will certainly

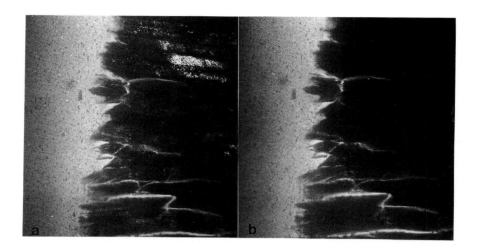

Fig. 6. Bond between a porcelain dental restoration and enamel, bonding agent labeled with rhodamine. (Sample courtesy of T.F. Watson.) **a:** Reflection image and fluorescence with green exciting light and no output filter. **b:** Rhodamine (red) fluorescence excited by green light, red output filter. Reflections from enamel prism boundaries are best seen in the top right part of the field in a. Bright, partly sinuous features in right part of both fields are cracked enamel prism boundary spaces filled with the fluorescently labeled polymer component of the bonding agent, which also outlines the filler particles in the left part of the field. Nikon ×100/1.4 NA oil-immersion objective. Field, 90 μm wide, 97 μm high. Tracor TSM, 1990 disk.

not "focus" the distance that one has "measured": The distance will be in proportion to the ratio of the refractive indices of the two media. This problem is just as important for confocal laser scanning microscopy as anywhere else in light microscopy.

A fluorescent object mounted in a glycerin-based mountant should be examined with a glycerin-immersion objective. Objects to be examined with oil-immersion lenses should be mounted in a medium having a refractive index close to that of the coverslip glass and the immersion oil, not in water or glycerin, for example. These caveats may be obvious, but it is also obvious from reading recent confocal applications literature that they are widely ignored. One will find many published examples of through-focus series with steps of a precise fraction of a micron specified, but made through a sample of a variable, undefined, lower refractive index imaged by an oil-immersion lens. The information accompanying such through-focus series must itself be examined with a magnifying glass! (The reader may like to calculate the true range of the series of three images in Fig. 12).

C. Applanating (Tissue Contact) Objectives

Special objectives will be required for use in the direct in vivo study of human tissues—in particular of the eye—in which it will be expected to have a conscious and cooperative "patient." Human eyes can be examined with conventional vertically mounted water-immersion objectives and several notable pioneers of TSM have been such horizontal specimens (Petran, ca. 1972, personal communication) [Lemp et al., 1985; Dilly, 1988]. However, the "patient's" position is then precarious. It is certainly much better to use an objective designed for optical coupling via a viscous, hydrated immersion

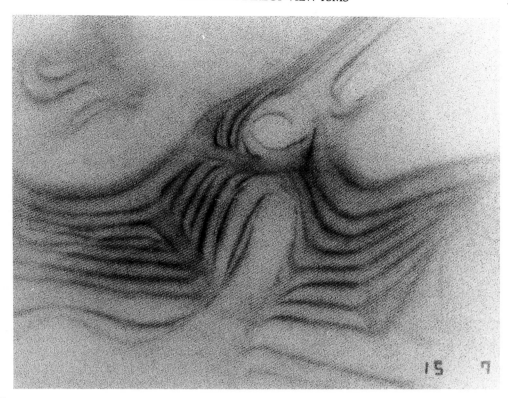

Fig. 7. Print from Kodachrome transparency, hence negative contrast, showing fluorescence from tetracycline lines in transverse section of tibia of a rat which received weekly tetracycline antibiotic injections: Note one interval of 2 weeks. Yellow fluorescence excited by violet light. Nikon ×20/0.8 NA glycerin-immersion objective. PMMA-embedded sample. Plzen TSM, 1983 vintage disk. Width of field is 600 μm.

medium which is stabilized by contact with the tissue (e.g., cornea) surface, and held at a fixed distance. Such objectives are called applanating objectives. Focusing is accomplished by changing the distance between lens components within the objective lens itself.

D. Resolution, Magnification, and Numerical Aperture

The resolution of the confocal microscope will depend upon the numerical aperture (NA) of the objective lens. All confocal microscopes are less confocal (i.e., have a reduced optical sectioning capability) with lower magnification objectives [Kino, 1989]. Nevertheless, lower-magnification objectives could be made with higher apertures, and it is thus necessary to campaign with the lens manufacturers to secure the construction of new low-magnification, high-aperture lenses. For some of the lowest magnification lenses which we require, it would clearly not be possible to construct them on the limited basis of the RMS thread. Indeed, conventional cine or video camera macro-lenses of high NA and low magnification have been used in electrophysiological applications of confocal laser microscopy, where a long working distance is required.

E. Dry Lenses for Dry Outer Surfaces

Many fields of application of TSMs involve the surface characterization of the outer

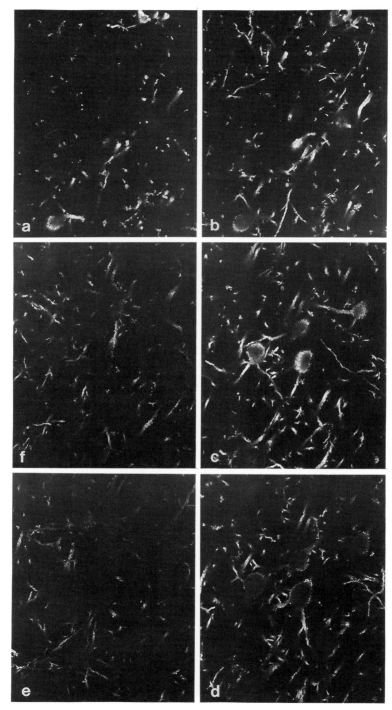

Fig. 8. a–f: Through-focus set of views at 4 μm focus plane intervals of a Golgi-stained preparation of hamster cortex showing pyramidal cell bodies, axons, and dendrites. (Preparation courtesy of Dr. M. Freire.) Reflection images obtained using a Nikon × 60/1.4 NA oil. Field is 177 μm high by 139 μm wide.

reflective surface of the object. TSMs have been particularly successful in the quantitative study of the resorption of test substrates, such as slices of bone and dentine, in assaying osteoclast (bone-resorbing cell) function [Boyde et al., 1985, 1990a; Jones et al., 1985]. In such cases, it may be desirable to use noncoverslip lenses, because if a coverslip is placed very close to a nearly flat object surface, intense reflection from the lower surface of the coverslip may interfere with the weaker reflection from the top surface of the real object. It is extremely difficult to procure 160 or 170 mm tube-length objectives which are not designed for use with a coverslip. Although restricted in maximal numerical aperture to 0.95, dry lenses would be optimal for most such applications because of the higher refractive index difference between air and any solid matter. There are some excellent dry, no-coverslip objectives available for 210 mm tube length, but these require the use of a 50 mm extension piece in a TSM designed for use with 160 mm tube-length lenses (see Fig. 4).

F. Turrets?

A question related to mechanical tube length is whether or not an objective turret should be used. Conventional LM users working with contrived specimens situated below a coverslip are practiced with this convenience. It is, however, dangerous to use an objective turret if one is looking at rough-surfaced, craggy objects as in many applications of the TSM. If objective lenses are not strictly parfocal, considerable damage can be done to the objective when rotating it against a rocky surface. Other than this, there are no reasons why turrets should not be used with TSMs, except that the extra tube length due to the turret should be designed into the dimensions of the TSM head. If this is not the case, then the turret may incorporate a tube-length-correcting lens, but at the disadvantage of adding four more unnecessary and undesirable glass–air interfaces at 90° to the optical path.

G. Chromatic Aberration

Both 1sTSMs and 2sTSMs, because they work with broad-band light and conventional objective lenses, can provide an interesting sideline on the confocal operating mode. Only absolutely exceptional refractive objectives have a negligible longitudinal chromatic aberration. This gives rise to the property of the TSMs to be able to produce image which appear to be color-coded depth "maps" of reflecting surfaces [Boyde, 1985a]. At present, objectives can be selected to give a particular color range for a range of depths which might be at best confocal focus (at different wavelengths) at the same time, but it is unlikely that lenses will have the ideal range and a satisfactory degree of flatness of field. In the future, one would hope for some new objective designs to be produced to exploit the maximum potential of this ultrafast surface-contouring and -mapping method.

VI. IMAGING MODES AND LIGHT SOURCES

A. Simultaneous Confocal Epi-illumination and Nonconfocal Transmission Imaging

The conventional biological microscopy object, thin enough to be viewed in transmitted light, can be viewed simultaneously in transmission and reflection. This much is obvious. Viewing the TSM as a complicated epi-illumination modification to the basic LM, it is therefore obvious that the object under the TSM can be illuminated by transmitted light—in any contrast-generation mode—synchronously with confocal epi-illumination. In this case [unless the disk is removed, as is very simply done with the 1sTSM: Kino, 1989; Xiao and Kino, 1987; Xiao et al., 1990] the disk will merely act as a neutral density filter in the observation side channel. Thus the live, real-time, direct-view confocal reflected or fluorescent TSM image can be viewed simultaneously with the nonconfocal transmitted light image (Fig. 10).

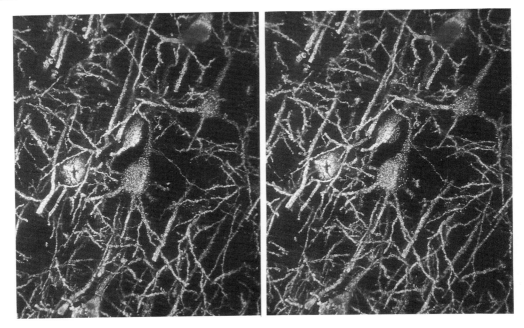

Fig. 9. Same field of view as Figure 8 presented as a stereo pair recorded by double, slightly oblique through-focusing over a range of 20 μm. Field is 177 μm high by 139 μm wide. Tilt angle difference is 10°.

The depth of field of the latter is inversely proportional to the sum of the numerical apertures of the condenser and the objective lens, and so can be influenced separately from the depth of field of the confocal channel. It is therefore possible to have a large depth of field transmitted light image upon which is superimposed the very shallow depth of field reflected or fluorescent confocal image. This advantage will have many fields of application [Boyde, 1989a].

When examining autoradiographs of tissue sections, it is possible to have the large depth of field image in transmitted light (showing the distribution of cell and tissue types) superimposed on the brilliantly lit image of the silver grains on the surface of the section. When examining Golgi-stained nervous tissue, it is possible to see the large depth of field transmission (shadow) image at the same time as the confocal image of a very shallow plane in which out-of-focus features are invisible (see Figs. 8–10). This makes it much simpler to search in three

dimensions. Those who have not used confocal microscopes do not really know what is the meaning of looking for a microscope needle in a haystack.

B. The Brightness of the TSM Image

The brightness of the TSM image is determined by the light source, the collector, and the (percent) transmittance of the disk. Typical early TSMs used 1% transmission disks, in other words throwing away 99% of the available light. At present, commercial TSMs for general use with the naked eye tend to have slightly higher transmission, 2% being common. On the other hand, a TSM has recently been designed solely for ophthalmological use (Tandem Scanning Corporation, Reston, VA 22090, USA) in which it is accepted that the movement due to the pulse and the patient's impatience combined with the movement necessary to focus through the tissue layers all constrain to make it impossible to follow what is happening in

Fig. 10. Same field of view as in Figures 8 and 9, viewed by simultaneous bright-field transmission (darkest shadowlike features) and confocal reflection (brightest features). Field is 177 μm high by 139 μm wide. Tracor TSM, 1990 disk.

real time. The designers have accepted that the image must be relayed via a low light level TV camera to a video recording system and played back in slow motion. In this case, it is possible to use a much lower transmission disk and 0.25% and 0.33% disks have been successfully employed [Petroll et al., 1990].

C. Light Sources for TSMs

The illumination sources used in TSMs have to date been conventional tungsten or mercury arc lamps, the latter obviously giving much higher brightness. In practice, however, nearly all reflection microscopy with the TSM, including a large proportion of work with live cell and tissue samples, is done with a 50 W tungsten lamp, which provides quite enough light. Mercury arc lamps give too much light, but have the advantage that selected wavelengths can be chosen for fluorescence work.

Given a particular source of illumination, there are only two things which can be done to improve the light flux to the sample. First, we can choose an objective with a larger exit pupil diameter (this is the hole that you can see when you look at an objective from the back). Second, one can improve the collection of the light from the source. Needless to say, it is important to align light sources correctly and this is nowhere more the case than with arc lamps.

Disks with very small holes may cause significant losses due to the diffractive spreading of the light as it passes through the disk [Kino, 1989].

D. Light Losses and Sources of Contaminating Light

The principal bugbear in any confocal microscope, the TSMs in particular, is light reflected from component surfaces perpendicular to the optical path. The only such components necessary in a 2sTSM with a disk with real holes are the lens surfaces. The surfaces of the collector and relay lenses in illumination are of some concern, but the main problem is reflection at the central portion of internal optical surfaces in the objective lens itself. In this respect the more complex the objective design, for example, to eliminate chromatic aberration, the greater the number of components and the greater the amount of internal reflection. We will be more concerned with eliminating this scattered light (because of its contribution to a high background signal) than the physical light loss. In the 2sTSM, it is possible to use a central stop in the aperture diaphragm to remove those rays which intercept the curved lens surfaces perpendicularly. This both improves contrast by reducing reflection in the objective, and also enhances the relative proportion of light reflected from more sloped portions of a reflective sample surface. Light losses can also occur in relay optics and careful choice of lenses should be made to minimize the number of optical surfaces in the train.

VII. RESOLUTION IN THE TSM

The subject of resolution in confocal scanning microscopes is somewhat confused. Nevertheless, it is agreed that, with suitably small confocal apertures, there is an improvement in lateral resolution in the TSMs which can just be demonstrated, and a dramatic improvement in resolution along the optical axis which can be demonstrated in every single case. In the TSM, the optical sectioning occurs due to the point sources and detectors in the disk. If one examines a plane mirror sample with the TSM, the image brightness peaks when the sample is in focus and falls off sharply when the sample is focused up or down. If one measures the intensity of the reflected light with some appropriate device, a characteristic through-focus intensity curve can be obtained. A commonly used measure of the vertical resolution is the full width at half-maximum of this curve. This author, however, doubts the practical value of this criterion.

If one examines real objects, one can always demonstrate an axial resolving power better than this theoretical value. If one measures the intensity of reflection at each pixel in an image of a rough-surfaced object as a function of change in focus, it is possible to determine the position of a point in the surface to a vertical resolution ≤ 0.1 μm if the light intensity is measured at each pixel, having relayed the image via a TV camera to a computer [Boyde et al., 1990a]. Nevertheless, with naked-eye inspection and an object with appropriate details, it is always possible to refocus on the same minute feature to better than 0.5 μm using a high-magnification, high-aperture immersion objective (see Fig. 4 for the highest magnification objective so far used). Theoretical computations of the axial resolution values for a range of objective numerical apertures and pinhole diameters have been derived by Kino [1989] and Kino et al. [1989].

In discussing the question of optimal axial resolution, it is not necessarily the case that the best microscope is the one with the smallest number for this value! Confocality is desirable. Optical sectioning is the single most important advantage of confocal microscopes in biology. However, optical sectioning can be carried too far. Our ability to recognize where we are depends upon a summation of a certain amount of information in the third dimension. Practical evaluation of given combinations of pinhole size, objective magnification and numerical aperture, and wavelength of light will be more important than theoretical calculations.

A. "Scanning Lines"

TSMs have the capability for an improvement in lateral resolution. In practice this has, as in all branches of confocal microscopy, not been nearly as important as the vertical sectioning ability. In earlier 2sTSMs, it was frequently difficult to demonstrate the inherent improvement in lateral resolution due to the interference caused by an image of scanning lines. Curved scanning lines can be recognized in all the earliest published TSM micrographs (see Fig. 7 for an example of an image recorded in 1984). It is an interesting fact that they are not noticed by the observer using the microscope when looking at the real sample, which will be moving laterally or vertically as the microscopist collects three-dimensional information in his brain. The pattern only becomes obvious in photographs. It is only significant if the periodicity in this curved line pattern interferes with the periodicity of real structure in the specimen. The curved line pattern can be removed entirely by correct disk design, and disks in which it is next to impossible to see such lines are now available from commercial sources for both one- and two-sided designs.

It is also possible to remove the curved disk line pattern from recorded images by image processing. Given that the image can be transferred via a TV camera to a framestore-based computer system, one may simply record an image of a featureless object

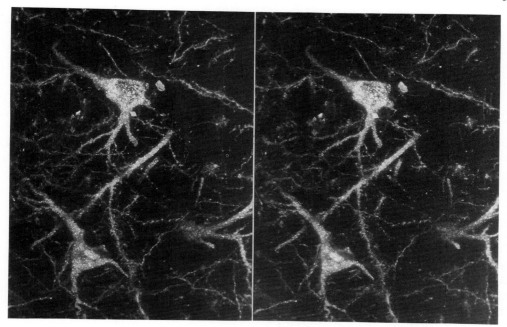

Fig. 11. Stereoscopic pair from a historic Golgi preparation made by the famous Spanish neuroanatomist Ramon Y Cajal. (Sample loanded by courtesy of the director of the Cajal Museum, Madrid.) Reflected-light image obtained with a Nikon × 60/1.4 NA oil-immersion objective. Field 183 μm high, 147 μm wide. Depth of field vertically (perpendicular to picture plane) is 17 μm, and the tilt angle difference is 12°. Tracor TSM, 1990 disk.

or an out-of-focus object and subtract this from the image of the real object. Alternatively, one can do a two-stage 2D fast Fourier transform operation, using a mask to remove the spatial frequencies corresponding to the disk lines in the reverse transformation. We have used such procedures in the past, but they are strictly unnecessary with the good-quality disks available at present.

B. Space-Filling Aspect of Disk Design

It is simpler to design the layout of pinholes in a disk for one-sided use. In this case, it is only necessary to ensure that there is the same density of holes across the diameter of the disk, so that a single simple Archimedian spiral would be employable, for example. More constraints are introduced if the disk is to be two-sided, since now holes have to be in identical positions on opposite sides of the disk (Fig. 2). The constraints are both in design and construction and, because of this, successful two-sided disks have been introduced later than successful one-sided disks, even though the two-sided design had been realized in practice at a much earlier date [Petran et al., 1990].

VIII. WHAT DOES ONE DO WITH THE IMAGE FROM THE TSM?

One looks, and assimilates 3D information by scanning through large volumes in three dimensions. One builds up a mental image of structures and a typical incidence of events. The standard TSM is a workhorse which can be used without any additional devices to acquire information in a way which was not previously possible. Eventually, however, if one cannot bring a colleague or an audience to the TSM to demonstrate a typical structure, an image may have to be recorded.

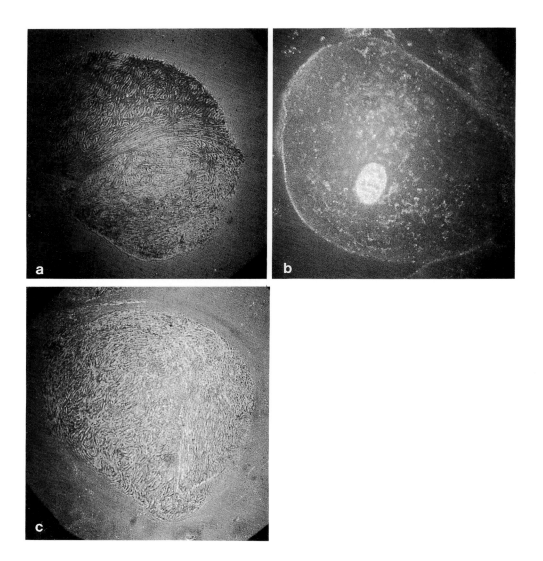

Fig. 12. Human buccal epithelial cell in water between glass coverslip and slide. **a:** Focused on the cell surface ridge pattern where the cell is attached to the coverslip. **b:** Focused 4.15 μm below this near the center of the cell body. **c:** Focused 9.5 μm below a at the contact of the cell with the slide. Note the interference reflection contrast in images a and c. Reflected-light images using a Nikon × 200/1.4 NA oil-immersion lens. Field is 88 μm high and 82 μm wide in each case. Tracor TSM, 1990 disk.

A. Recording the Image

The simplest means of recording the TSM image is to photograph it. The simplest method is to place a 35 mm photographic camera, with a standard lens focused at infinity, above the eyepiece of the microscope. If the camera back contains an automatic exposure meter, success will be consistently obtained from the first trial. Such is the means which we have used to record tens of thousands of TSM images. Images obtained in this manner are shown in Figures 3 to 12. The sole disadvantage of such automatic metering is that out-of-focus images (which are "out-of-focus" images in all confocal microscopes) are overexposed, and thus tend to show more out-of-focus plane features. This is cured by setting the exposure to that required for the brightest plane.

The TSM image is frequently attractively colored, and there will be a temptation, not a necessity, to use color recording materials. For economical reasons one would choose color transparency material rather than color negative material, and this is further advantageous because, for example, 400 ASA-speed slide film will give conveniently short exposures of a few seconds for a wide variety of reflecting-type biological objects when using a TSM with a tungsten lamp. Even with a mercury arc lamp, however, it may be necessary to increase the exposure time to 1–4 min for fluorescence applications (with a simple nondichroic beam splitter). A great many objects which we examine will be successfully recorded with ASA 125 black-and-white film.

B. Depth of Field and Extended-Focus or -Range Images

The depth of field in the TSM image will be determined by a number of factors, but it will be limited to a particular depth for a given optical setup. This depth of field can, however, be increased at will by through-focusing while recording an image (Figs. 5, 9, 11, and 13). If the camera shutter is opened and the objective moved vertically with respect to the specimen, the depth of field can be increased proportionately [Boyde, 1985a,b]. Such extended-focus or -range images recorded with the TSM have great beauty and contain a density of information which would be impossible to acquire with television- and computer-based image recording. The silver photographic emulsion is still the most economical high-density information-recording medium.

C. Stereoscopic Images

Three-dimensional stereoscopic views can be obtained by through-focusing along slightly inclined vertical axes [Boyde, 1985b]. The simplest of devices are all that is necessary to achieve this, but in the latest realization we have the objective lens of our TSM controlled by piezoelectric translators with the finest detail of positional movement controlled by computer (Figs. 5, 9, 11). We can now choose the tilt-angle difference and the range of a stereoscopic pair at will, and 3D imaging at high resolution is routine and economical [Boyde, 1989b]. Large series of images recorded by focusing along incremented tilted axes can be processed to produce integrams or pseudoholograms.

D. Pseudocolor Coding for Depth

In the case of objects which do not have an interesting range of natural colors, it is possible to use color coding to indicate relative depth of features within an object [Boyde et al., 1990a; Freire and Boyde, 1990]. This can be done in an "analogue" fashion in the case of a TSM equipped with a camera loaded with color film. One simply changes the color of filters in the illuminating channel as the focus of the microscope is changed. This can also be combined with the stereo imaging procedure just mentioned [Boyde, 1987b].

E. Pseudocolor Coding for Slope

The slopes of surfaces may also be color coded by the so-called Rheinberg method of illumination in a TSM, wherein a segmented color filter is placed in the plane of the aperture diaphragm in the illuminating channel.

A further method of introducing (real) color coding for slope is to evaporate colored metals (gold, copper, aluminum) onto the specimen surface from contrasting directions.

IX. USES OF TV CAMERAS

A. For Fast Motion

The addition of a TV camera to the TSM has many advantages, including the fact that

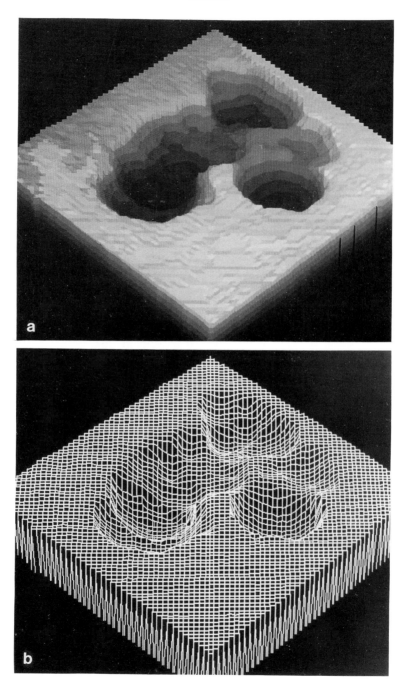

Fig. 13, Plate 1. Complex resorption pit excavated by an osteoclast which differentiated from chick bone marrow in 19-day culture on a slice of sperm whale dentine:
Sample prepared as for SEM. A map of the surface was generated by through-focusing
over a range of 24 μm—slightly more than the depth of the pit—to find the focal plane
at which the brightest signal was recorded for each pixel in a 512 × 512 field. This map
was then reprojected to produce the solid model display: Eight contour levels are shown

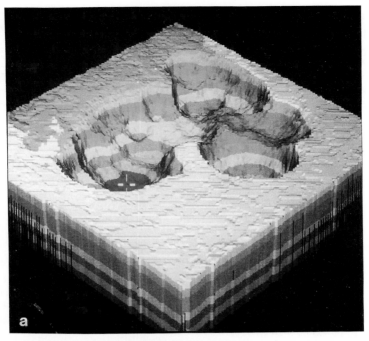

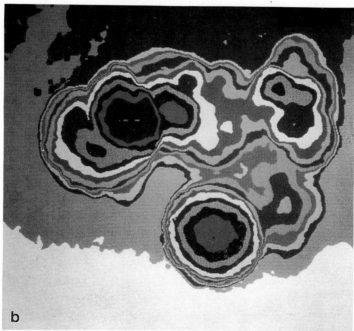

Plate 1.
in monochrome in Figure 13a and in pseudocolor in Plate 1a. Plate 1b shows an orthogonal pseudocolor height-coded image. A "wire frame" at 8-pixel interval intersections is shown in Figure 13b. Photograph of TV monitor on Tracor 8502 image-analyzing computer. Raw data collected by confocal reflection imaging using a Nikon ×60/1.4 NA oil-immersion objective. Figure on p. 309 appears in color in Color Figure Section.

several observers may see an image in real time simultaneously and that fast-moving events (which could not be imaged by a slow-scan confocal laser scanning microscope) can be recorded and analyzed subsequently by slow play-back and single-frame analysis. As examples of the need to conduct this kind of work, we can cite the use of the TSM in clinical opthalmology [Jester et al., 1988a–c, 1990a–c], in the vital microscopy of any tissue type where one is trying to study blood flow [Jester et al., 1990b,c], and in the analysis of high-speed cutting and fracturing events [Watson, 1990].

B. To Permit Computer Image Analysis

A further advantage of the use of TV cameras concerns the ability to transfer the image to an image-analyzing computer. The TSM then becomes the exact equivalent of any confocal laser scanning microscope, since these must necessarily have a frame-store based computer as a part of the system. Details and examples of the kind of manipulation of confocal through-focus image series which may be achieved using computer frame store will be found in other chapters of this book and in innumerable articles in the literature. The reader should be aware of the fact that all such procedures can be, and have been, undertaken with TSM images [Boyde, 1989b; Boyde and Freire, 1990; Boyde et al., 1990a,c; Wright et al., 1989]. The TSM may have advantages here in the great rapidity of acquisition of the image series (Fig. 13 and Plate 1).

C. For Low Light Level Work, Including Fluorescence

Yet another advantage of the use of video cameras concerns the ability to record images at very low light levels. The amount of light available in a TSM may be restricted because one is examining a rapidly changing field of view in reflection microscopy. It may also be inherently limited by the fluorescence intensity in, for example, immunocytochemical work [Boyde et al., 1990c].

We have explored the use of a variety of types and manufacture of low light level TV cameras. Any can be used, but practical decisions are perhaps best illustrated by our current choice of equipment. For stationary objects, we prefer a Peltier element cooled charge-coupled device (CCD) camera, in which images may be captured over time ranges from 20 msec to seconds or minutes. Stereo images acquired by oblique through-focusing can be captured directly in such a device. Thus 3D images of cellular immunofluorescence can be captured on two passes of one to a few seconds' duration with very much less total irradiation than would normally be used in confocal laser scanning microscopy [Boyde, 1989b; Boyde et al., 1990c; Wright et al., 1989].

For moving objects, the combination of an image intensifier with a sensitive TV camera will be preferred. Such devices are now assembled as compact, lightweight units which are ideal for use in the TSM context.

X. APPLICATIONS OF TSM

TSMs have been used in the study of a wide variety of problems where it is necessary to map the microtopography of surfaces of opaque structures (Fig. 13) or surfaces within translucent objects, or to study scattering or fluorescence in semitransparent objects in some depth. The depth to which sharp images will be obtained depends upon the intensity of distribution of scattering objects. The more features of interest, the less far one will be able to image into the object. Indeed, the ideal object for any confocal microscope would be one having an entirely uniform retractive index with the occasional bright fluorescent feature.

Of living tissues, the most deeply penetrating applications so far, in terms of focus depth, have been to the eye [Lemp et al., 1985; Dilly, 1988; Cavanagh et al., 1988; Jester et al., 1988a–c, 1990a–c; Petroll et al., 1990]. However, many other living tissues have been studied, including kidney, thyroid, liver, adrenals, epididymis, muscle, nerve, and connective tissues. A good example is the work of Andrews et al. (in press) who were able, using a TSM with a ×20/0.38 NA applanating objective, to view through the intact connective tissue capsule of Munich-Wistar rat kidneys and to study the individual glomerular capillary loops, glomerular arterioles, proximal and distal tubules, the macula densa, and intertubular capillaries. Using fluorescein, they were able to trace movement from the vasculature to the glomerular filtrate, etc.

The TSM has found many applications in material science, geology, and paleontology, as well as in botany, zoology, cytology, and histology. Some of its problem-solving capabilities are illustrated in the following examples.

The TSM made it possible to examine internal structural details in valuable fossils, including fossil teeth, and particularly in teeth belonging to larger fossil museum samples which would not be conveniently placed below another kind of microscope without some pre-preparation. Internal structural features of the dental enamel, which reveal the life history of the ameloblasts that secreted this tissue, can then be reconstructed. We then obtain information which may be important for interpretation of tooth structure as well as in examining taxonomic affinities and in understanding aspects of cell behavior which occurred a long time ago [Boyde, 1985a, 1987; Boyde and Martin, 1984, 1987; Boyde et al., 1983, 1986; Fortelius, 1985; Lester et al., 1986; Petran et al., 1985a]. Enamel prism boundary discontinuities (Figs. 3, 6a) are also important in understanding the functional, failure-related properties of this tissue [Boyde, 1985a] and the TSM has been used to study fracture propagation in real time in this context [Watson, 1990].

The TSM can similarly be used to examine the spaces occupied by cells in those hard tissues where cells leave a space for themselves, as, for example, osteocyte lacunae and canaliculi in bone [Fig. 5; Baddeley et al., 1987; Boyde et al., 1983, 1987a; Boyde, 1985a; Petran et al., 1985a; Howard et al., 1985], and dentine tubules (odontoblast processes) in dentine [Watson, 1989a–c; Watson and Boyde, 1984, 1985, 1986, 1987a–c, 1989] or the equivalent tubules in dental enamels which have them [Lester et al., 1986]. Acquired porosity in enamel and dentine, for example, as a consequence of dental caries and/or the resultant iatrogenic operative interference, can be similarly examined to advantage in both reflection and fluorescence modes in the TSM [Fig. 6; Benn and Watson, 1988; Boyde, 1985b; Boyde et al., 1990c,d; Jones and Boyde, 1987; Watson, 1989a–c; Watson and Boyde, 1984, 1985, 1986, 1987a–c, 1989].

As already noted, the TSM was invented to look at living nervous tissue [Egger and Petran, 1967; Egger et al., 1969; Petran et al., 1968, 1985b, 1986, 1987; Sallam-Sattar, 1974; Sallam-Sattar and Petran, 1974; Petran and hadravsky, 1974]. It is particularly valuable in examining prepared (fixed and stained) specimens in which reflection is confined to the few cells which acquire impregnation in one of the so-called Golgi staining routines [Figs. 8 to 11; Boyde, 1985a,b, 1989b]. In combination with through-focusing to produce extended-focus or -range images and stereo imaging, one can now obtain 3D maps which would have taken a great deal of operator intervention in the past [Boyde, 1985b, 1987a,b, 1989a; Freire and Boyde, 1990].

In cytology, the TSM gives an excellent high-contrast image due to an interference reflection contrast mechanism at the point of contact of cells with a substrate [Figs. 12a,c; Boyde, 1989; Paddock, 1989]. TSM can then be used to image and film the migration of cells across substrates using rather low-intensity irradiation.

Given that one has made the choice of TSM rather than confocal laser scanning microscope, it will be necessary to add a low light level TV camera to perform immunocytochemical work, but the results will be equivalent [Wright et al., 1989; Boyde et al., 1990].

ACKNOWLEDGMENTS

Work on tandem scanning microscopy in the author's laboratory has been supported by grants from the Medical Research Council and Science and Engineering Research Council. The author acknowledges the long-term collaboration of Professors M. Petran and M. Hadravsky, Plzen, Czechoslovakia, and the technical assistance of Roy Radcliffe.

REFERENCES

Andrews PM, Petroll WM, Cavanagh HD, Jester JV (1991): Tandem scanning confocal microscopy (TCSM) of normal and ischemic living kidney (submitted).

Baddeley AJ, Howard CV, Boyde A, Reid SA (1987): Three-dimensional analysis of the spatial distribution of particles using the tandem scanning reflected light microscope. Acta Stereologica 6(Suppl 2):87–100.

Benn DK, Watson TF (1988): Correlation between film position, bitewing shadows, clinical pitfalls and the histological size of approximal carious lesions. Quintessence 20:131–141, 1989.

Boyde A (1985a): The tandem scanning reflected light microscope. Part 2. Pre-MICRO 84 applications at UCL. Proc R Microsc Soc 20:130–139.

Boyde A (1985b): Stereoscopic images in confocal (tandem scanning) microscopy. Science 230:1270–1272.

Boyde A (1985c): Anatomical considerations relating to tooth preparation. In Van Herle G, Smith DC (eds): "Posterior Composite Resin Dental Restorative Materials." pp 377–403. St Paul, MN: 3M Co, ISBN-088159-601-9.

Boyde A (1987a): Applications of tandem scanning reflected light microscopy and 3-dimensional imaging. Ann NY Acad Sci 483:428–439.

Boyde A (1987b): Colour-coded stereo images from the tandem scanning reflected light microscope. J Microsc 146:137–142.

Boyde A (1989a): Combining confocal and conventional modes in tandem scanning reflected light microscopy. Scanning 11:147–152.

Boyde A (1989b): Direct recording of stereo-pairs from disc-scanning confocal light microscopes. In Pawley JB (ed): "The Handbook of Confocal Microscopy," pp 147–151. Madison, WI: IMR Press.

Boyde A, Ali NN, Jones SJ (1985): Optical and scanning electron microscopy in the single osteoclast resorption assay. Scan Elec Microsc 1985/3:1259–1271.

Boyde A, Dillon CE, Jones SJ (1990a): Measurement of osteoclastic resorption pits with a tandem scanning microscope. J Microsc 158:261–265.

Boyde A, Hadravsky M, Petran M, Jones SJ, Martin LB, Watson TF, Reid SA (1986a) TSRLM: How it works and applications. In Bailey GN (ed): Proc 44th Ann Meeting Electron Microscopy Soc Am, pp 84–87. San Francisco: San Francisco Press.

Boyde A, Hendel P, Hendel R, Maconnachie E, Jones SJ (1990b): Human cranial bone structure and the healing of cranial bone grafts: A study using backscattered electron imaging and confocal microscopy. Anat Embryol 181:235–251.

Boyde A, Jones SJ, Taylor ML, Wolfe LA, Watson TF (1990c): Fluorescence in the tandem scanning microscope. J Microsc 157:39–49.

Boyde A, Maconnachie E, Reid SA, Delling G, Mundy GR (1986b): SEM in bone pathology: Review of methods, potential and applications. Scan Elect Microsc 1986/4:1537–1554.

Boyde A, Martin L (1984): A non-destructive survey of prism packing patterns in primate enamels. In Fearnhead RW, Suga S (eds): "Tooth Enamel IV," pp 417–421. Amsterdam: Elsevier.

Boyde A, Martin L (1987): Tandem scanning reflected light microscopy of primate enamel. Scan Microsc 1:1935–1948.

Boyde A, Petran M (1990): Light budgets, light and heavy losses: One or two sided tandem scanning (real time, direct view, confocal) microscopy. J Microsc 160:335–342.

Boyde A, Petran M, Hadravsky M (1983): Tandem scanning reflected light microscopy of internal features in whole bone and tooth samples. J Microsc 132:1–7.

Boyde A, Reid SA (1986): 3-D analysis of tetracycline fluorescence in bone by tandem scanning reflected microscopy. Bone 7:148–149.

Boyde A, Watson TW (1989): Fluorescence mode in the tandem scanning microscope. Proc R Microsc Soc 24:7.

Boyde A, Wolfe LA (1986): Block face microscopy for the interfacial region between bone and implant materials. J Dent Res 66:859 (abst 223).

Boyde A, Xiao GQ, Corle T, Watson TF, Kino GS (1990d): An evaluation of the unilateral TSM for biological applications. Scanning 20:(4) (in press).

Cavanagh HD, Jester JV, Mathers W, Lemp MA (1988): In vivo confocal microscopy of the eye. Proc Int Soc Eye Res 5:132.

Davidovits P, Egger MD (1971): Scanning laser microscope for biological investigations. Appl Opt 10:1615–1619.

Dilly PN (1988): Tandem scanning reflected light microscopy of the cornea. Scanning 10:153–156.

Dilly PN, Cavanagh HD, Jester JV, Lemp MA (1987): In vivo scanning reflected light microscopy: II. Morphological demonstration of ouabain inhibition of rabbit corneal endothelial pump. Inv Ophth Vis Sci (Suppl) 28:326.

Egger MD, Petran M (1967): New reflected-light microscope for viewing unstained brain and ganglion cells. Science 157:305–307.

Egger MD, Gezari W, Davidovits P, Hadravsky M, Petran M (1969): Observation of nerve fibers in incident light. Experientia (Basel) 25:1225–1226.

Fortelius M (1985): Ungulate cheek teeth: Developmental, functional and evolutionary interrelations. Acta Zool Fennica 180:1–76.

Francon M (1961): "Progress in Microscopy," p 161. Evanston, IL: Row Peterson & Co.

Freire M, Boyde A (1990): Study of Golgi-impregnated material using the confocal tandem scanning reflected light microscope. J Microsc 158:285–290.

Howard V, Reid SA, Baddeley A, Boyde A (1985): Unbiased estimation of particle density in the tandem scanning reflected light microscope. J Microsc 138:203–212.

Jester JV, Cavanagh HD, Lemp MR (1988a): In vivo confocal imaging of the eye using tandem scanning confocal microscopy. Proc SPIE 1028:122–126.

Jester JV, Cavanagh HD, Lemp MR (1988b): In vivo confocal imaging of the eye using tandem scanning confocal microscopy (TSCM). Proc Electr Microsc Soc Am 46:56–57.

Jester JV, Cavanagh HD, Essepian J, Shields WJ, Lemp MR (1990a): Confocal microscopy of the living eye. CLAO J 16:65–73.

Jester JV, Cavanagh HD, Lemp MR (1990b): Confocal microscopic imaging of the eye and tandem scanning confocal microscope (TSCM). In Masters BR (ed): "Noninvasive Diagnostic Techniques in Ophthalmology." New York: Springer Verlag.

Jester JV, Cavanagh HD, Mathers W, Lemp MA (1988c): In vivo responses to injury using confocal microscopy. Proc Int Soc Eye Res 5:132.

Jester JV, Petroll WM, Andrews P, Cavanagh HD, Lemp MA (1990c): In vivo confocal microscopy. J Electron Microsc Tech.

Jones SJ, boyde A, Ali NN, Maconnachie E (1985): A review of bone cell and substratum interactions. Scanning 7:5–24.

Jones SJ, Boyde A (1987): Scanning microscopic observations on dental caries. Scanning Microsc 1:1991–2002.

Kino GS (1989): Efficiency in Nipkow disc microscopes. In Pawley JB (ed): "Handbook of Biological Confocal Microscopy," pp 93–97. Madison, WI: IMR Press.

Kino GS, Chou C-H, Xiao GQ (1989): Imaging theory for the scanning optical microscope. In Wilson T (ed): "Scanning Imaging." Proc SPIE 1028:104–113.

Lemp MA, Dilly PN, Boyde A (1985): Tandem-scanning (confocal) microscopy of the full-thickness cornea. Cornea 4:205–209.

Lemp MA, Dilly PN, Cavanagh HD, Jester JV (1987): In vivo scanning reflected light microscopy: 1. Morphology of the normal rabbit cornea. Inv Ophth Vis Sci (Suppl) 8:221.

Lester KS, Boyde A, Gilkeson C (1986): Marsupial and monotreme enamel. Scanning Microsc 1:401–420.

Melki T, Cavanagh HD, Jester JV, Dilly PN, Lemp MA, Foegh M (1988): Correlation of in vivo confocal tandem scanning microscopy (TSM) observations of effect of thromboxane antagonist AH23848B (Glaxo, GB) and ouabain on transmission electron microscopy and lanthanum penetration studies. Inv Opth Vis Sci 257 (Suppl).

Minsky M (1961): United States Patent No. 3,013,467. Microscopy apparatus. Filed Nov. 7, 1957, granted Dec. 19, 1961.

Paddock SW (1989): Tandem scanning reflected-light microscopy of cell-substratum adhesions and stress fibers in Swiss 3T3 cells. J Cell Sci 93:143–146.

Petran M, Hadravsky M (1966a): Zpusob a zarizeni pro omezeni rozptylu svetla v mikroskopu pro osvetleni shora. Czechoslovak Patent No. 128936, application July 5, 1966, granted Feb. 15, 1968, published Sept. 15, 1968.

Petran M, Hadravsky M (1966b): Zpusob a zarizeni pro zlepseni rozlisovaci schopnosti a kontrastu optickeho mikroskopu. Czechoslovak Patent No. 128937, application July 5, 1966, granted Feb. 15, 1968, published Sept. 15, 1968.

Petran M, Hadravsky M (1974): Employment of tandem scanning microscope in morphological research of living sensory systems. Activitas Nervosa Superior (Praha) 16(4):289.

Petran M, Boyde A, Hadravsky M (1990) Direct view confocal microscopy. In Wilson T (ed): "Confocal Microscopy," pp 245–283. London: Academic Press.

Petran M, Hadravsky M, Benes J, Boyde A (1987): In vivo microscopy using tandem scanning microscope. Ann NY Acad Sci 483:440–448.

Petran M, Hadravsky M, Benes J, Kucera R, Boyde A (1985b): The tandem scanning reflected light microscope. Part 1. The principle, and its design. Proc R Microsc Soc 20:125–129.

Petran M, Hadravsky M, Boyde A (1985a): The tandem scanning reflected light microscope. Scanning 7:97–108.

Petran M, Hadravsky M, Boyde A, Muller M (1986): Tandem scanning reflected light microscopy. In Muller M, Becker RP, Boyde A, Wolosewick J (eds): "Science of Biological Specimen Preparation," pp 85–94. SEM Inc., AMF O'Hare Il.

Petran M, Hadravsky M, Egger MD, Galambos R (1968): Tandem scanning reflected light microscope. J Opt Soc Am 58:661–664.

Petran M, Sallam-Sattar M (1974): Microscopical

observations of the living (unprepared and unstained) retina. Physiologia Bohemoslovenica 23:369.

Petroll WM, Cavanagh HD, Jester JV (1990): *In vivo* digital image acquisition in confocal microscopy. Trans R Microsc Sco 90: (in press).

Reid SA, Smith R, Boyde A (1985): Some scanning microscopies of fibrogenesis imperfecta ossium. Bone 6:275–276.

Sallam-Sattar M (1974): "New Technique for In Vivo Correlation of Morphological and Electrophysiological Events in the Central Nervous System." PhD Thesis, deposited in the Inst of Biophysics, Czechoslovak Academy of Sciences, Brno, Kralovopolska, Vol 1 text, Vol 2 illustrations.

Sallam-Sattar M, Petran M (1974): Dynamic alterations accompanying spreading depression in chick retina. Physiologia Bohemoslovenica 23:373.

Sheppard CJR, Mao XQ (1988): Confocal microscopes with slit apertures. J Modern Opt 35:1169–1185.

Sheppard CJR, Wilson T (1981): The theory of the direct-view confocal microscope. J Microsc 124: 107–117.

Watson TF (1989a): A confocal optical microscope study of the morphology of the tooth/restoration interface using Scotchbond 2 dentin adhesive. J Dent Res 68:1124–1131.

Watson TF (1989b): Tandem scanning reflected light microscopy for rapid histological evaluation of carious lesions labelled with fluorescent markers. In Elderton RJ (ed): "Evolution in Dental Care." Proc Dental Conf Bristol 1988 (in press).

Watson TF (1989c): "The Effect of Cavity Preparation and Adhesive Restorations on the Microstructure of Enamel and Dentine: A Confocal Optical Microscope Study." PhD Thesis, University of London.

Watson TF (1990): Real-time confocal microscopy of high speed dental bur/tooth cutting interactions. J Microsc 157:51–60.

Watson TF, Boyde A (1984): The tandem scanning reflected light microscope (TSRLM) in conservative dentistry. J Dent Res 62:512.

Watson TF, Boyde A (1985): Tandem scanning reflected light microscopy of fluorescent labelled composite interfaces. J Dent Res 64:664.

Watson TF, Boyde A (1986): *In vitro* dentine penetration of a GLUMA bonding agent. J Dent Res 66:835 (abst 8).

Watson TF, Boyde A (1987a): Tandem scanning reflected light microscopy: A new method for *in vitro* assessment of dental operative procedures and restorations. Clin Materials 2:33–43.

Watson TF, Boyde A (1987b): The use of fluorescent markers for studying the distribution of a dentine bonding agent between a composite restoration and tooth. Clin Materials 2:45–53.

Watson TF, Boyde A (1987c): Tandem scanning reflected light microscopy: Applications in clinical dental research. Scanning Microsc 1:1971–1981.

Watson TF, Boyde A (1989): An *in vitro* study, using a confocal optical microscope, of the marginal adaptation of vitrabond TM light cured glass ionomer. J Dent Res 68:577.

Wolfe LA (1988): "The Incorporation of Titanium and Hydroxyapatite Reinforced Polyethylene Implants into Rabbit Bone." PhD Thesis, University of London.

Wright SJ, Walker JS, Schatten H, Simerly C, McCarthy JJ, Schatten G (1989): Confocal fluorescence microscopy with the tandem scanning light microscope. J Cell Sci 94:617–624.

Xiao GQ, Kino GS (1987): A real-time confocal scanning optical microscope. In Wilson T, Balk L (eds): "Scanning Imaging Technology." Proc SPIE 809:107–113.

Xiao GQ, Kino GS, Masters BR (1990): Observations of the rabbit cornea and lens with a new real-time confocal scanning optical microscope. Scanning 12:161–166.

CHAPTER 14

Charge-Coupled Devices for Quantitative Nipkow Disk Real-Time Scanning Confocal Microscopy

Barry R. Masters and Gordon S. Kino

I. INTRODUCTION

This chapter describes the advantages of using a two-dimensional cooled charge-coupled device (CCD) camera as a quantitative electronic imager for quantitative confocal light microscopy. Typically, video cameras are used as electronic imaging devices for Nipkow disk scanning confocal microscopes. Video cameras have the advantage of providing images at video rates (30 frames per second), but they have many negative characteristics when used as confocal imaging devices. They have a low dynamic range and low sensitivity, and are electronically noisy systems. Therefore, they are not suitable for quantitative microscopy. Furthermore, video cameras exhibit geometric distortions that vary in magnitude across the field. Perhaps the most serious problem is their nonlinear response to light. These disadvantages can be eliminated through the use of a cooled CCD camera [Aikens et al., 1989; Hiraoka et al., 1987].

The real-time Nipkow disk scanning optical microscope used is that designed by Xiao and Kino [Xiao et al., 1988] who have described its construction and its optical characteristics. This microscope yields images with high resolution, image quality, and contrast. In addition to these properties, the microscope provides true color capability, real-time operation, and easy alignment [Kino and Corle, 1989]. These properties result in a real-time scanning optical microscope that has advantages over previous confocal systems. The analytical treatment of the axial resolution of the microscope for plane and point reflectors is presented, and leads to a theoretical analysis of the optimal pinhole size in the Nipkow disk of the microscope.

We have coupled a cooled CCD camera to this real-time scanning optical microscope and illustrate its performance in reflected light imaging. The test object was the living removed eye of a rabbit which provides a low-reflectance, low-contrast, thick sample of known structure. The present results are compared to previous studies using the same microscope coupled to a silicon intensifier target (SIT) video camera as the imaging device [Masters and Kino, 1990a; Xiao et al., 1990].

Finally, the characteristics of a new generation of CCD imaging devices are described [Janesick et al., 1987a,b]. These imaging devices, with their high quantum efficiencies, will further improve the quality of the images obtained in conjugation with confocal light microscopes.

Before describing the use of this microscope coupled to a cooled CCD imaging device, it is valuable to describe the quantitative details of the microscope itself.

II. THE REAL-TIME SCANNING OPTICAL MICROSCOPE

A. Microscope Design

The confocal scanning optical microscope (CSOM) has the major advantages over the standard microscope of extremely good axial resolution and optical sectioning capability;

it also has somewhat better transverse resolution, and is well adapted to quantitative measurements, as described by Wilson in Chapter 10 of this volume [Wilson and Sheppard, 1984; Xiao et al., 1988; Wilson, 1990].

Xiao and Kino have designed a derivative of the tandem scanning optical microscope (TSM) of Petran and Hadravsky, described by Boyde in Chapter 13 of this volume. Such a microscope gives a real-time image rather than, as in a single beam scanning confocal microscope, an image which takes several seconds to form. (Petran et al., 1968, 1985; Xiao et al., 1988; Kino and Corle, 1989). They have called their microscope the real-time scanning optical microscope (RSOM). Like the TSM, the RSOM, illustrated in Figure 1, makes use of a rotating Nipkow disk containing a large number of pinholes spaced approximately 10 pinhole diameters apart in an interleaved spiral configuration. The RSOM has approximately 160,000 pinholes of 20–30 μm diameter placed in the disk. As shown in Figure 1, several thousand of these pinholes are illuminated at once by an incident beam from a mercury vapor arc or a simple lamp source, and are imaged on the object. Light returns through the same pinhole by which it entered, thus simplifying

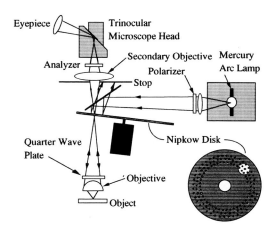

Fig. 1. The real-time scanning optical microscope. The important optical and mechanical parts are labeled, and are described in the text.

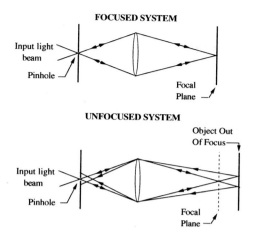

Fig. 2. Principles of confocal imaging. A focused optical system is compared to an unfocused optical system in order to illustrate optical sectioning in a confocal reflection Nipkow disk imaging system, in which light returns through the illuminating pinhole.

the alignment problem. The image is observed through transfer lenses and an eyepiece. The disk is rotated at approximately 2,000 rpm to form an image with approximately 5,000 lines at a frame rate of approximately 700 frames/sec.

The basic reason for the optical sectioning capability of this microscope, as illustrated in Figure 2, is that it uses direct illumination of the objective lens through a pinhole and brings back the light through the same pinhole to the detector. Light first passes through the pinhole to a microscope objective and forms a diffraction-limited spot on the object. Light reflected from the object passes through the objective and back through the pinhole. If the object is moved out of focus, the reflected light reaching the pinhole is defocused and the light does not pass through it. Consequently, if a detector is placed behind the pinhole, the amplitude of the signal received drops off rapidly with the defocus distance, and the image disappears rather than blurs as it does with the standard microscope.

Initially, the main problem in this one-sided Nipkow disk configuration was to eliminate the light reflected from the disk. This was done by adopting the following principles:

1. The disk is made by photolithographic techniques and consists of specularly reflecting black chrome, with a low reflectivity of a few percent, laid down on a glass disk. The highly polished chrome yields a well-directed reflected beam, which is easier to eliminate with a stop than a diffusely scattered beam.

2. The disk is tilted and a stop is placed at the position where the light reflected from the disk is focused, thus further differentiating against the reflected light. The light hitting the disk is convergent, not parallel, and comes to a focus and is thus easier to block.

3. To eliminate the remaining reflected light, the input light is polarized by a polarizer and the light that is received at the eyepiece is observed through an analyzer, with its plane of polarization rotated at right angles to that of the input light. A quarter-wave plate is placed in front of the objective lens so that the plane of polarization of light reflected by the sample is rotated by 90 degrees, and thus light from the sample, but not that reflected by the disk, passes the analyzer and can be observed with the eyepiece.

When these precautions are taken, reflected light from the disk can be decreased to a level well below that of the image reflected from the sample, even one from a weak reflector such as the cornea of the eye. The main source of unwanted reflections, as in any optical microscope, is now that from the internal components such as the objective lens or the quarter-wave plate itself; this latter effect can be eliminated by tilting the plate. Alternatively, an antiflex objective, which has a rotatable quarter-wave plate attached to the front of the objective, may be used, eliminating reflections from within the objective itself.

A Köhler illumination system is used, so that, when the disk is not present, a mercury vapor or filament lamp source is focused to a point at the back focal plane of the objective lens. With the disk present, this implies that the central axes of the diffracted beams passing through individual pinholes all pass through the center of the back focal plane of the objective lens. Thus, the illumination, and hence the definition, of the system are made as uniform as possible over the field of view.

B. Resolution and Pinhole Size

It follows from the Rayleigh-Sommerfeld scalar diffraction theory, using the Fraunhoffer approximation, that at a distance h_1 (the tube length of the objective) from a circular pinhole, the field Ψ_1 at the pupil plane of the objective at radius r_1 from the axis varies as:

$$\Psi_1 = A \Psi_0 \frac{J_1 (kr_1/h_1)}{kr_1/h_1} \tag{1}$$

where Ψ_0 is the field at the pinhole, $k = 2\pi/\lambda$, A is a constant, λ is the optical wavelength in free space, and $J_1(x)$ is a Bessel function of the first kind and first order [Goodman, 1968].

Using this formula, we may make a rough estimate of the optimum pinhole radius $a(opt)$ by choosing the radius at the half-power points (one-half of the maximum normalized power or intensity) of the beam diffracted by the pinhole to be equal to the pupil radius b of the objective. This leads to the result:

$$a(opt) = \frac{0.25 \lambda h_1}{b} \tag{2}$$

where h_1 is the spacing between the pinhole and the objective (the tube length). For a tube length $h_1 = 160$ mm, $\lambda = 540$ nm, and $b = 2$ mm; this leads to a pinhole radius $a(opt) = 10.8$ μm. It is apparent from Equa-

tion 1 that if the pinholes are made too large, the fields at the pupil plane will be highly nonuniform and may even reverse in sign across the pinhole. This condition gives rise to a highly nonuniform, undesirable focused spot at the object. When the size of the pinhole is infinitesimal, the normalized intensity of the signal reflected from a perfect mirror a distance z from the focal plane is given by the approximate formula:

$$I(z) = \left[\frac{\sin k_2 z(1 - \cos \theta_0)}{k_2 z(1 - \cos \theta_0)} \right]^2 \tag{3}$$

where $k_2 = nk$, n is the refractive medium, and the numerical aperture (NA) of the objective is equal to $n \sin \theta_0$.

We may derive from Equation 3 a very useful formula for the spacing d_z of the half-power points of the response:

$$d_z = \frac{0.45 \lambda}{n(1 - \cos \theta_0)}. \tag{4}$$

The definition of resolution depends to a large extent on what type of object is imaged and what criteria are important to the observer. For integrated circuits, one is often interested in measuring profiles of stepped surfaces. For biological applications of confocal microscopy, one is often more interested in distinguishing two neighboring point reflectors. When a confocal microscope images a point reflector, the intensity $I(z)$ of the optical signal at the detector varies with distance z from the focus as follows:

$$I(z) = \left\{ \frac{\sin \dfrac{k_2 z(1 - \cos \theta_0)}{2}}{\dfrac{k_2 z(1 - \cos \theta_0)}{2}} \right\}^4. \tag{5}$$

It should be noted that this formula is different from that for the reflection from a plane mirror and gives an axial resolution approximately 1.4 times greater than that given by Equation 4 for the reflection from a plane mirror. The intensity of the signal due to small scatterers falls off far more rapidly with increasing distance than does the reflection from a mirror. Consequently, a large number of scatterers some distance from the focus give very little glare.

It is apparent that in all confocal microscopes, the size of the pinhole is of critical importance. If the pinhole is too large, the transverse and axial resolution are impaired, as discussed by Wilson in Chapter 10. In the RSOM, if the pinholes are too small, the amount of light passing through the disk is decreased and the light budget becomes critical. The major source of light loss is the relatively small fraction of pinhole area to total illuminated area of the disk. As this fractional area α is increased, the light efficiency increases, but the rejection of glare from out-of-focus layers in the object gets worse, for a fraction α of the defocused light will pass back to the detector through the pinholes. Thus the fractional area of the pinholes relative to that of the disk is normally kept in the 1–2% range.

C. Beam Splitters and Dichroic Mirrors

A standard beam splitter provides an additional source of light loss, a factor of 0.25 since half the power is lost in transmission and half in reflection; this factor is the same as would be expected in a standard reflecting microscope but is more serious here due to the additional large loss of light through the Nipkow disk. The loss of light in the beam splitter can be eliminated by the use of polarizing beam splitters, or in the case of a fluorescence microscope, a dichroic beam splitter. In the RSOM, the polarizer provides one more source of loss (a factor of 0.5). However, because polarizing beam splitters are far more efficient than typical commercial polarizers and analyzers, a striking improvement in the light efficiency by as much as a factor of eight may be realized by using a polarizing beam splitter rather than a polarizer and analyzer along with a standard beam splitter.

For fluorescence microscopy, it is unnecessary to use a polarizer and analyzer to eliminate reflected light from the disk, since the use of a dichroic mirror and wavelength-selective filters accomplishes this purpose. The main problem when any Nipkow disk microscope is employed for fluorescence imaging is the lack of light due to the loss of light at the disk and the fact that the light source is not necessarily as intense as that of a laser-illuminated confocal microscope. It is, indeed, possible to use a laser source with the RSOM. With fluorescence imaging, speckle or interference between different pinholes is not a problem. Interference problems are normally more severe from internal reflection from the quarter-wave plate or the lenses. But since the fluorescent light is an incoherent source, these problems do not occur.

It is because of the low light efficiency of Nipkow disk confocal optical microscopes that we need an integrating electronic detector, namely a cooled CCD array camera, to record the weak images obtained.

III. CHARGE-COUPLED DEVICES FOR QUANTITATIVE CONFOCAL MICROSCOPY

In order to demonstrate the efficacy of the charge-coupled device as an imaging detector for confocal light microscopy, we coupled a cooled CCD camera to the real-time scanning optical microscope. With this combined detector–microscope system, we obtained a series of optical sections from a rabbit eye. The cornea, which is the clear tissue in the anterior region of the eye, was imaged. This semitransparent, low-contrast biological tissue is a severe test of both the microscope and the detector system, and can serve as a standard benchmark of the optical–electronic system performance.

A. Sources of Biological Material

The freshly removed rabbit eye was used as the test biological specimen in order to evaluate the CCD confocal microscope system. The rabbits were maintained and handled in accordance with the ARVO (The Association for Research in Vision and Ophthalmology, Bethesda, MD, USA) Resolution on the Use of Animals in Research. The eyes were obtained from male New Zealand white rabbits weighing 2.5–3.0 kg. The rabbits were anesthetized with an intramuscular injection of ketamine HCl (40 mg/kg) and xylazine (5 mg/kg). The eyes were freed of adhering tissue and were swiftly removed. The eyes were immediately placed in a beaker containing bicarbonate Ringer's solution with glucose (5 mM) and calcium (2 mM) at 25°C. The removed eye was then transferred to a black plastic chamber containing the same Ringer's solution and placed on the stage of the confocal microscope.

B. Confocal Microscope

The microscope used in these studies was a K2 BIO Nipkow disk confocal scanning optical microscope (Technical Instrument Company, San Francisco, CA, USA), a commercial version of the microscope described in the preceding section. This microscope incorporates a Nipkow disk pinhole size of 20–30 μm, which was attached to a Leitz Diaplan microscope equipped with a standard Leitz trinocular microscope head, designed so that 100% of the light is either directed into the eyepieces or into the vertical phototube, to which the CCD camera is attached. The microscope objective used was a Leitz $\times 25$, 0.6 NA water-immersion objective, which was focused onto the cornea by immersion in the Ringer's solution overlying the eye. Since it had a free working distance of 680 μm, it could be used to focus through the full thickness of the cornea (400 μm) without physical contact with the eye. A lens of $\times 2$ power in the body of the confocal microscope increased the effective power of the objective to $\times 50$. A 100 W mercury arc lamp was used without attenuation for broad-band illumination. All of the imaging discussed in this chapter was

made with the microscope in the confocal reflected-light mode, with a depth of field of less than 1 μm.

For real-time viewing of the specimen, the light from the microscope is directed into the eyepieces. This is useful and important for positioning the microscope stage and for focusing the microscope. All of the images shown in this chapter (Figs. 3–5) can also be observed with the naked eye using the eyepieces of the confocal microscope.

C. The Charge-Coupled Device (CCD) Detector and Associated Camera and Electronics

The charge-coupled device detector used for the images presented here was the Kodak KAF-1400 CCD chip, which consists of 1,320 $\times$ 1,035 pixels, each 6.8 μm^2. The full-well capacity of this CCD is 40,000 electrons per pixel. Other CCD imaging devices with larger pixels but lower resolution have full-well capacities some 10-fold greater, which may be important, since they permit longer integration times before saturation, which results in larger signals.

The CCD chip was installed in a cooled camera head (CH220) supplied by Photometrics Ltd. (Tucson, AZ, USA). The CCD was cooled to $-50°C$ with a three-stage thermoelectric (Peltier) cooler in conjunction with a liquid heat exchanger using a mixture of 50% ethylene glycol:50% deionized water, and was run in the clock-inverted mode to reduce dark current. The readout speed of the CCD camera was 500 kHz, and the images were digitized to 12 bits with a read noise of 17 electrons per pixel.

The image acquisition and shading correction were performed in the following manner: The exposure or integration time for each image was 4 sec. A total of six 4-sec "dark" exposures made with the camera shutter closed were averaged, and the average "dark" exposure was subtracted from each 4-sec specimen image obtained to correct for the dark-current accumulation during the integration and for readout noise.

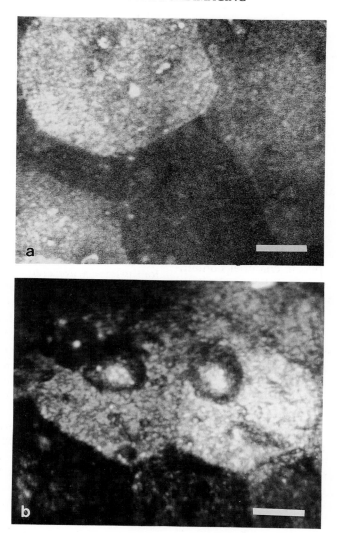

Fig. 3. a: Reflected-light confocal image of the superficial epithelial cells of a living rabbit cornea, recorded as described in the text. The polygonal margins of the cells are visible. With the focal plane of the microscope at the midpoint of the cells **(b)**, the nuclei are imaged as oval objects in the center of the cells. The different reflectances of cells in these images are caused by variations in their axial position relative to the plane of focus. Scale bar, 10 μm.

Six "flat-field" images were then captured by taking 4 sec exposures of the background light, by moving the field of view to an area with no sample, and then going out of focus to be sure that the empty field was not producing any pattern. These were then averaged. The previously averaged "dark image" was then subtracted from the averaged "flat-field" image to yield a corrected "flat-field" image, and the mean pixel value of this image was then calculated. The "dark"-subtracted specimen images were finally multiplied by the corrected "flat-field" mean, and then divided pixel by pixel by the corrected flat-field image. This flat-field operation was performed to compensate for

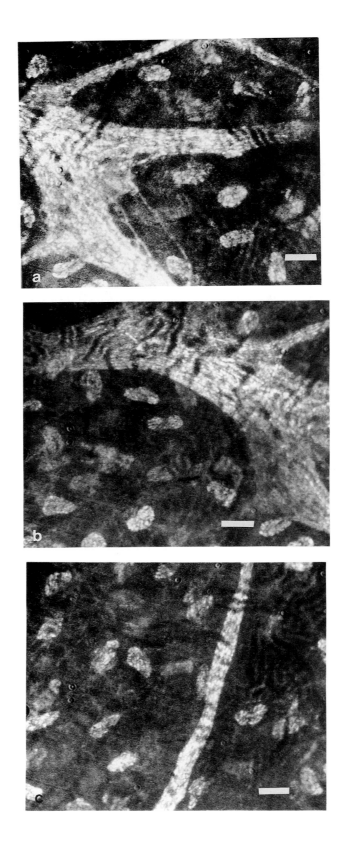

Fig. 5. Reflected-light confocal image of the endothelial cells of a living rabbit cornea. The focal plane of this confocal image is 400 μm below the superficial epithelial cells shown in Figure 3. On the left side of the figure, the focal plane is at the interface between the aqueous humor and the polygonal endothelial cells, at the level of the microvilli, which appear as granular regions on the surface. On the right side of the figure, the focal plane intercepts a region of lower reflectance within the endothelial cells, within which nuclei can be seen. The cell borders in this region are highly invaginated, resulting in the dark regions between the cells appearing thicker. Scale bar, 25 μm.

uneven illumination due to the microscope (light source inhomogeneities, dust on the Nipkow disk), as well as differences in the quantum sensitivities of each CCD pixel to light.

The Photometrics camera was controlled by a Photometrics CC200 camera controller interfaced to an Apple MAC II computer via a National Instruments NB-DMA-8 IEEE-488 interface board. The MAC II was run in a terminal emulation mode while it commanded the camera. Recently, a new camera controller from Photometrics became available which is constructed on a single board and which occupies an expansion slot within the MAC II computer.

Fig. 4. Reflected-light confocal images of nerve plexus with branching nerve fibers in the anterior stromal region of a living rabbit cornea **(a,b)** and another nerve fiber **(c)**. Fine longitudinal filaments and transverse bands are observed. The oval objects are the nuclei of the stromal keratocytes, within which there are regions of differing reflectance. Scale bar, 25 μm.

One advantage of using the MAC II computer as the system controller for the cooled, slow-scan CCD imaging device is the availability of a digital image processing software package, IMAGE, from the United States National Institutes of Health, and of several commercial packages, which offer a variety of morphological, filtering, and Fourier domain operations.

While this particular Photometrics camera digitized the images to 12 bits, and other CCD chips have outputs which can be digitized to 14 or 16 bits for a larger dynamic range, the monitor used with the MAC II computer could only display an 8-bit image. Therefore, the original 12-bit images were windowed to an 8-bit image (256 gray levels) to optimize the contrast. This was done as an interactive process based subjectively on picture quality.

Since the monitor could only display a 512 × 512 array, a selected 512 × 512 subarray region of the 1,320 × 1,035 pixel array was stored. The images were stored on a computer hard disk drive as 8-bit 512×512 TIFF format images, as well as 12-bit 512×512 Photometrics images. The images were photographed off the monitor of the MAC II computer using a 35 mm camera which contained Kodak TMAX 100 black-and-white film.

IV. CONFOCAL MICROSCOPY OF THE LIVING CORNEA

A. The Use of a Cooled CCD as an Imaging Device with a Real-Time Scanning Optical Confocal Microscope

The clear transparent tissue of the cornea in a freshly removed rabbit eye is an extremely difficult object to image in reflected light, since the specimen has a very low reflectivity and low contrast. For example, the amount of light reflected from the interface between the corneal endothelium and the aqueous humor is about 0.02% of the light incident at this interface. The observed contrast in the confocal reflection images of this living, unstained, unfixed tissue is due to the very small differences in refractive index between different regions of the sample, which result in differences in reflectivity.

Typical images of the cells of the corneal epithelium are shown in Figure 3. When the focal plane of the confocal microscope was changed by less than 0.5 μm, the image was observed to change its appearance, confirming that the axial resolution of the microscope was less than 1 μm. Since the images are due to specularly reflected light, they naturally select for objects in the focal plane which are perpendicular to the optic axis.

Figure 4 illustrates the ability of the combined microscope-imaging system to obtain high-resolution, high-contrast images of the nerve plexus and nerve fibers in the anterior region of the rabbit cornea. These images were taken about 100 μ below the images shown in Figure 3. The smallest resolvable objects are less than 1 μ in size. Note the fine structures within the neural plexus and within the nuclei of the stromal keratocytes.

The variation in pixel intensity within a given image may be due to different structural elements or to variations in reflectivity within a given structural element. Deeper into the cornea, the corneal endothelium presents a reflecting layer at its interface with the aqueous humor, shown in Figure 5.

The quality of the CCD imaging system is evident from the contrast and detail shown in these images. The 12-bit dynamic range of the original CCD images is required in order to visualize the full intensity range of the specimen. Consequently, the original images were saved to disk, and from them various 8-bit windows of the full dynamic range could be selected for display on the monitor, as illustrated in Figures 3–5.

B. Confocal Microscopic Imaging of the Cornea Based on Other Detector–Microscope Combinations

It is interesting to compare the present work with previous investigations. A tandem scanning confocal microscope developed by Petran provided early confocal images of the rabbit cornea [Petran et al., 1968, 1985; Lemp et al., 1986]. More recently, the BioRad MRC-500 confocal laser scanning microscope has been used [Masters, 1989a,b, 1990; Masters and Kino, 1990b; Masters and Paddock, 1990a,b]. Since intense illumination on the sample is necessary to image the cornea, due to its low reflectivity, this system must operate at maximum laser power, and the electronic amplifiers are set to the highest gain, providing high-contrast, high-resolution images.

Separately, the Technical Instruments K2 BIO confocal microscope described above was used with a Leitz ×25/0.6 NA water-immersion objective and an SIT video camera (DAGE 66 SIT) in place of the CCD camera to image the rabbit eye. A series of five video frames were averaged to obtain each image, and a "glare" image was digitally subtracted from the averaged image to remove some glare from the internal reflections of the microscope. The images produced with the video camera had less resolution, less dynamic range, and less contrast than the equivalent confocal images obtained with the CCD camera.

V. NEW DEVELOPMENTS IN CHARGE-COUPLED DEVICES AS IMAGERS

The development of charge-coupled devices is a continuing process [Hsieh and Hosack, 1987; Beal et al., 1987]. The coupling of a charge-coupled device to a Nipkow disk confocal microscope has several advantages for quantitative confocal light imaging, since the CCD imaging device possesses high dynamic range, photometric accuracy, high resolution, high sensitivity, and geometric stability, which are important for accurate two-dimensional imaging and three-dimensional reconstructions from volume data sets [Masters and Paddock, 1990c].

In order to select an appropriate CCD device, it is important to understand both the performance characteristics of the CCD imager and the specific imaging requirements of the specimens to be observed [Boreman, 1987; Marien and Pitz, 1987]. In this section we review some of the performance characteristics of current CCD arrays, discuss the advantages and disadvantages of these characteristics, and, finally, present some of the choices that are available. We hope this section will be a helpful guide for the selection of an appropriate CCD imaging device for confocal imaging.

The following properties are desirable in a CCD imaging device: high quantum efficiency from the ultraviolet (UV) to the infrared (IR), mechanical rigidity of the chip, low dark current, high dynamic range, high charge transfer efficiency at low signal levels, on-chip amplifiers for low noise readout, rapid readout rates, and high resolution. Charge-coupled devices are available with the above characteristics in a variety of formats, including: 512 × 512, 1,024 × 1,024, 1,320 × 1,035, and 2,048 × 2,048.

Two of these characteristics deserve special consideration: quantum efficiency and spatial resolution. For low light level confocal imaging it is important to detect as many photons as possible. It is thus advisable to obtain a CCD with the highest possible quantum efficiency in the wavelength region of interest. The quantum efficiency of most front-illuminated CCDs varies from 20% to 40% in the visible wavelength region; however, it rapidly falls to zero at 400 nm. Photometrics Ltd. has developed a proprietary phosphor (METACHROME II) which can be applied as a 0.4 to 0.6 mm-thick coating and extends the sensitivity in the blue and near-ultraviolet regions. The spatial resolution obtainable from the CCD array will depend on the number of pixels within the imaging area. Obviously, for a given imaging area, a larger number of smaller pixels will give a higher spatial resolution. However, the actual spatial resolution of the image will depend upon the magnification at which the optical image is projected onto the CCD array. For accurate digital sampling, it is only necessary to have pixel spacings at twice the spatial frequency of the highest spatial frequency which occurs in the optical image (see Castleman, Chapter 3).

While one may think it desirable to increase the number of pixels in the CCD imaging device to 2,048 × 2,048, there are two further considerations which may dissuade one. The first is cost. Large CCD arrays are disproportionately more expensive than smaller ones. The second is computer memory, since digital memory requirement and readout time both increase with the square of the increase in linear resolution. The image of a CCD with 1,024 × 1,024 pixels requires four times the memory as one with 512 × 512 pixels. However, current developments in data compression techniques may help to offset this memory requirement.

We have recently installed a new CCD imaging device on our real-time scanning optical microscope with the following design and performance characteristics. The CCD chip is a TK512 imager from Tektronix, designed for imaging at low light levels from the UV to the near-IR. It derives this characteristic by being a thinned, back-illuminated device with antireflection coatings. The device is optically flat and built on a stable, fully supported single die. The sensor is a full-frame area imager, with 512×512 pixels fabricated using a buried channel, three-level polysilicon gate process that results in high transfer efficiency and low dark current. The quantum efficiency is 93% at 700 μm and 86% at 400 μm, with a full well capacity of 830,000 electrons per pixel and a charge-transfer efficiency of greater than 99.999%. The low-noise on-chip amplifiers provide an interface to external preamplifiers with readout noise typically less than 10 electrons per pixel [Schempp, 1989; Schempp and Toker, 1989; Lesser, 1987]. This Tektronix CCD array chip was installed in the Photometrics Ltd. cooled camera previously described in this chapter. It digitizes the images to 14 bits per pixel and has a readout rate of 50 kHz. In contrast, the Kodak KAF-1400 CCD chip previously installed in the Photometrics camera digitized the images to 12 bits and had a readout rate of 500 kHz. For applications involving confocal fluorescent microscopy, the larger dynamic range available with the 14-bit digitization is an advantage. However, certain applications may require the faster readout speed available with the 12-bit digitization. CCD imaging devices such as these are ideal for low-light imaging when coupled either to a conventional or to a confocal light microscope.

VI. CONCLUSIONS

This chapter demonstrates the utility of coupling a cooled slow-scan charge-coupled imaging detector to a real-time Nipkow disk confocal scanning optical microscope. The advantages of the CCD imager include the properties of high resolution, high dynamic range, and linear photometric accuracy. The use of a CCD imaging detector can help to achieve the full optical capabilities of the confocal microscope. The quality of the confo-

cal images of the living cornea illustrate the advantages of the CCD imaging device for quantitative confocal light microscopy.

ACKNOWLEDGMENTS

This work was supported by grant EY-06958 from the NIH (to B.R.M.) and by the National Science Foundation under contract No. ECS-86-11638 (to G.S.K.). The authors thank Mr. Frank Lundy and Mr. Vlasta Cejna of the Technical Instrument Company for the use of the K2 BIO confocal microscope, and Mr. Tim Bruchman, of Photometrics Ltd., for the use of the Photometrics cooled CCD camera. The Tektronix TK512 thinned, back-illuminated CCD was donated by Tektronix for the development of medical imaging using CCD arrays. We thank Drs. S.P. Srinivas and D.M. Maurice of Stanford University for supplying the biological materials.

REFERENCES

Aikens RS, Agard DA, Sedat JW (1989): Solid state imagers for microscopy. In Wang Y-L, Taylor DL (eds): "Methods in Cell Biology," vol 29, pp 291–313, Orlando, FL: Academic Press.

Beal G, Boucharlat G, Chabbal J, Dupin JP, Fort B, Mellier Y (1987): Thompson-CSF frame-transfer charge-coupled-device imagers: Design and evaluation at very low flux level. Opt Engin 29:902–910.

Boreman GD (1987): Fourier spectrum techniques for characterization of spatial noise in imaging arrays. Opt Engin 26:985–991.

Goodman JW (1968): "Introduction to Fourier Optics." New York: McGraw-Hill.

Hiraoka Y, Sedat JW, Agard D (1987): The use of a charge-coupled device for quantitative optical microscopy of biological structures. Science 238:36–41.

Hsieh SM, Hosack HH (1987): Low light level imaging with commercial charge-coupled devices. Opt Engin 26:884–889.

Janesick JR, Elliot T, Colline S, Blouke MM, Freeman J (1987a): Scientific charge-coupled devices. Opt Engin 26:692–714.

Janesick JR, Klaasen KP, Elliot T (1987b): Charge-coupled device collection efficiency and the photon-transfer technique. Opt Engin 26:972–980.

Kino GS, Corle TR (1989): Confocal scanning optical microscopy. Physics Today 42:55–62.

Lemp MA, Dilly PN, Boyde A (1986): Tandem-scanning (confocal) microscope for optically sectioning the living cornea. Cornea 4:205–209.

Lesser M (1987): Antireflection coatings for silicon charge-coupled devices. Opt Engin 26:911–915.

Marien K-H, Pitz E (1987): Measurement of the absolute quantum efficiency of a charge-coupled device in the ultraviolet. Opt Engin 26:742–746.

Masters BR (1989a): Confocal microscopy of the eye. In Wampler J (ed): "New Methods in Microscopy and Low-Light Imaging." Proc SPIE 1161: 350–359.

Masters BR (1989b): Scanning microscope for optically sectioning the living cornea. In Wilson T (ed): "Scanning Imaging." Proc SPIE 1028: 133–143.

Masters BR (1990): Confocal microscopy of ocular tissue. In Wilson T (ed): "Confocal Microscopy," pp 305–324. London: Academic Press.

Masters BR, Kino GS (1990a): Real-time confocal scanning imaging of the eye: Instrument performance of reflectance and fluorescence imaging. Inst Phys Conf Ser 98, ch 14, pp 625–628. Institute of Physics, Bristol.

Masters BR, Kino GS (1990b): Confocal microscopy of the eye. In Masters BR (ed): "Noninvasive Diagnostic Techniques in Ophthalmology," pp 152–171. New York: Springer-Verlag.

Masters BR, Paddock SW (1990a): Confocal bioimaging the living cornea with autofluorescence and specific fluorescent probes. Smith L (ed). Proc SPIE 1205:164–178.

Masters BR, Paddock SW (1990b): In vitro confocal imaging of the rabbit cornea. J Microsc 158: 267–275.

Masters BR, Paddock SW (1990c): Three-dimensional reconstruction of the rabbit cornea by confocal scanning optical microscopy and volume rendering. Applied Optics 29:3816–3822.

Petran M, Hadravsky M, Boyde A (1985): The tandem scanning reflected light microscope. Scanning 7:97–108.

Petran M, Hadravsky M, Egger MD, Galambos R (1968): Tandem scanning reflected light microscope. J Opt Soc Am 58:661–664.

Schempp WV (1989): "How to Convert Between Photometric Units and the Number of Photons." Application note, Photometrics Ltd, Tucson, AZ.

Schempp WV, Toker E (1989): "Signal-to-Noise Ratio Calculations." Application note, Photometrics Ltd, Tucson, AZ.

Wilson T (1990): "Confocal Microscopy." London: Academic Press.

Wilson T, Sheppard CJR (1984): "Theory and Practice of Scanning Optical Microscopy." Orlando, FL: Academic Press.

Xiao GQ, Corle TR, Kino GS (1988): Real-time confocal scanning optical microscope. Appl Phys Lett 53:716–718.

Xiao GQ, Kino GS, Masters BR (1990): Observation of the rabbit cornea and lens with a new real-time confocal scanning optical microscope. Scanning 12:161–166.

Index